Human Retroviral Infections

Immunological and Therapeutic Control

INFECTIOUS AGENTS AND PATHOGENESIS

Series Editors: Mauro Bendinelli, *University of Pisa*
Herman Friedman, *University of South Florida
College of Medicine*

Recent volumes in this series:

DNA TUMOR VIRUSES
Oncogenic Mechanisms
Edited by Guiseppe Barbanti-Brodano, Mauro Bendinelli,
and Herman Friedman

ENTERIC INFECTIONS AND IMMUNITY
Edited by Lois J. Paradise, Mauro Bendinelli, and Herman Friedman

FUNGAL INFECTIONS AND IMMUNE RESPONSES
Edited by Juneann W. Murphy, Herman Friedman,
and Mauro Bendinelli

HERPESVIRUSES AND IMMUNITY
Edited by Peter G. Medveczky, Herman Friedman,
and Mauro Bendinelli

HUMAN RETROVIRAL INFECTIONS
Immunological and Therapeutic Control
Edited by Kenneth E. Ugen, Mauro Bendinelli,
and Herman Friedman

MICROORGANISMS AND AUTOIMMUNE DISEASES
Edited by Herman Friedman, Noel R. Rose, and Mauro Bendinelli

OPPORTUNISTIC INTRACELLULAR BACTERIA AND IMMUNITY
Edited by Lois J. Paradise, Herman Friedman, and Mauro Bendinelli

PSEUDOMONAS AERUGINOSA AS AN OPPORTUNISTIC PATHOGEN
Edited by Mario Campa, Mauro Bendinelli, and Herman Friedman

PULMONARY INFECTIONS AND IMMUNITY
Edited by Herman Chmel, Mauro Bendinelli, and Herman Friedman

RAPID DETECTION OF INFECTIOUS AGENTS
Edited by Steven Specter, Mauro Bendinelli, and Herman Friedman

RICKETTSIAL INFECTION AND IMMUNITY
Edited by Burt Anderson, Herman Friedman, and Mauro Bendinelli

---

Human Retroviral Infections

Immunological and Therapeutic Control

Edited by

Kenneth E. Ugen

University of South Florida College of Medicine
Tampa, Florida

Mauro Bendinelli

University of Pisa
Pisa, Italy

and

Herman Friedman

University of South Florida College of Medicine
Tampa, Florida

Kluwer Academic / Plenum Publishers
New York, Boston, Dordrecht, London, Moscow

Library of Congress Cataloging-in-Publication Data

Human retroviral infections: immunological and therapeutic control/edited by Kenneth
E. Ugen, Mauro Bendinelli, and Herman Friedman.
        p.   ; cm. — (Infectious agents and pathogenesis)
    Includes bibliographical references and index.
    ISBN 0-306-46222-2
        1. Retrovirus infections.   I. Ugen, Kenneth E.   II. Bendinelli, Mauro.   III. Friedman,
    Herman, 1931–   IV. Series.
        [DNLM: 1. HIV Infections—therapy.   2. HIV Infections—immunology. WC 503.2
    H918 2000]
    QR201.R47 H835 2000
    616.9′25—dc21

                                                                                00-037101

ISBN: 0-306-46222-2

©2000 Kluwer Academic / Plenum Publishers, New York
233 Spring Street, New York, N.Y. 10013

http://www.wkap.nl

10 9 8 7 6 5 4 3 2 1

A C.I.P record for this book is available from the Library of Congress.

Contributors

MICHAEL G. AGADJANYAN • Department of Pathology and Laboratory Medicine, University of Pennsylvania School of Medicine, Philadelphia, Pennsylvania 19104, and Institute of Viral Preparations, Russian Academy of Medical Science, Moscow, Russia 129028

J. ARP • Gene Therapy and Molecular Virology Group, The John P. Robarts Research Institute, and the Department of Microbiology and Immunology, University of Western Ontario, London, Ontario, Canada, N6A 5K8

VELPANDI AYYAVOO • Department of Infectious Diseases and Microbiology, University of Pittsburgh, Pittsburgh, Pennsylvania 15261

MOSI K. BENNETT • Department of Pathology and Laboratory Medicine, University of Pennsylvania School of Medicine, Philadelphia, Pennsylvania 19104

EWA BJÖRLING • Microbiology and Tumorbiology Center, Karolinska Institute, S-171 77 Stockholm, Sweden

JEAN D. BOYER • Stellar-Chance Laboratories, Department of Pathology and Laboratory Medicine, University of Pennsylvania, Philadelphia, Pennsylvania 19104-6100

CYNTHIA L. BRISTOW • McLendon Clinical Laboratories, Department of Pathology and Laboratory Medicine, University of North Carolina Hospitals, Chapel Hill, North Carolina 27514

GEORGE J. CIANCIOLO • Department of Pathology, Duke University Medical Center, Durham, North Carolina 27710

JOSEPH P. COTROPIA • BioClonetics, Inc., Philadelphia, Pennsylvania 19147, and Department of Medicine, University of Pennsylvania School of Medicine, Philadelphia, Pennsylvania 19104

G. A. DEKABAN • Gene Therapy and Molecular Virology Group, The John P. Robarts Research Institute, and the Department of Microbiology and Immunology, University of Western Ontario, London, Ontario, Canada, N6A 5K8

MAITE DE LA MORENA • Washington University School of Medicine, St. Louis, Missouri 63110

ROBERT W. DOMS • Department of Pathology and Laboratory Medicine, University of Pennsylvania, Philadelphia, Pennsylvania 19104

BENJAMIN J. DORANZ • Department of Pathology and Laboratory Medicine, University of Pennsylvania, Philadelphia, Pennsylvania 19104

RALPH DORNBURG • The Dorrance H. Hamilton Laboratories, Center for Human Virology, Division of Infectious Diseases, Department of Medicine, Jefferson Medical College, Thomas Jefferson University, Philadelphia, Pennsylvania 19107

PATRICIA J. EMMANUEL • Department of Pediatrics, College of Medicine, University of South Florida, Tampa, Florida 33612

G. FRANCHINI • Basic Research Laboratory, National Cancer Institute, National Institutes of Health, Bethesda, Maryland 20892

TERESA C. GENTILE • Department of Medicine, SUNY Health Sciences Center, Syracuse, New York 13210

MAUREEN M. GOODENOW • Department of Pathology, Immunology, and Laboratory Medicine, and Division of Immunology and Infectious Diseases, Department of Pediatrics, College of Medicine, Health Science Center, University of Florida, Gainesville, Florida 32610

THOMAS KIEBER-EMMONS • Department of Pathology and Laboratory Medicine, University of Pennsylvania, Philadelphia, Pennsylvania 19104-6082

SAGAR KUDCHODKAR • Stellar-Chance Laboratories, Department of Pathology and Laboratory Medicine, University of Pennsylvania, Philadelphia, Pennsylvania 19104

THOMAS P. LOUGHRAN • Program Leader, Hematological Malignancies, H. Lee Moffitt Cancer Center and Research Institute, Veterans Administration Hospital, and Departments of Medicine and Microbiology/Immunology, University of South Florida College of Medicine, Tampa, Florida 33612

JERRY R. McGHEE • Department of Microbiology, Immunobiology Vaccine Center, University of Alabama at Birmingham, Birmingham, Alabama 35294

T. NAGASHUNMUGAM • Stellar-Chance Laboratories, Department of Pathology and Laboratory Medicine, University of Pennsylvania, Philadelphia, Pennsylvania 19104

ROBERT P. NELSON, JR. • Indiana University School of Medicine, Indianapolis, Indiana 46202

ELENA E. PEREZ • Department of Pathology, Immunology, and Laboratory Medicine, and Medical Scientist Training Program, College of Medicine, Health Science Center, University of Florida, Gainesville, Florida 32610

A. PETERS • Gene Therapy and Molecular Virology Group, The John P. Robarts Research Institute, and the Department of Microbiology and Immunology, University of Western Ontario, London, Ontario, Canada, N6A 5K8

ROGER POMERANTZ • The Dorrance H. Hamilton Laboratories, Center for Human Virology, Division of Infectious Diseases, Department of Medicine, Jefferson Medical College, Thomas Jefferson University, Philadelphia, Pennsylvania 19107

AMI R. SHAH • Rollins School of Public Health, Emory University, Atlanta, Georgia 30322

JOHN W. SLEASMAN • Department of Pathology, Immunology, and Laboratory Medicine, and Division of Immunology and Infectious Diseaes, Department of Pediatrics, College of Medicine, Health Science Center, University of Florida, Gainesville, Florida 32610

HERMAN F. STAATS • Departments of Medicine and Immunology, Center for AIDS Research, Duke University Medical Center, Durham, North Carolina 27710

KENNETH E. UGEN • Department of Medical Microbiology and Immunology, University of South Florida College of Medicine, Tampa, Florida 33612

DAVID B. WEINER • Stellar-Chance Laboratories, Department of Pathology and Laboratory Medicine, University of Pennsylvania, Philadelphia, Pennsylvania 19104-6100

Preface to the Series

The mechanisms of disease production by infectious agents are presently the focus of an unprecedented flowering of studies. The field has undoubtedly received impetus from the considerable advances recently made in the understanding of the structure, biochemistry, and biology of viruses, bacteria, fungi, and other parasites. Another contributing factor is our improved knowledge of immune responses and other adaptive or constitutive mechanisms by which hosts react to infection. Furthermore, recombinant DNA technology, monoclonal antibodies, and other, newer methodologies have provided the technical tools for examining questions previously considered too complex to be successfully tackled. The most important incentive of all is probably the regenerated idea that infection might be the initiating event in many clinical entities presently classified as idiopathic or of uncertain origin.

Infectious pathogenesis research holds great promise. As more information is uncovered, it is becoming increasingly apparent that our present knowledge of the pathogenic potential of infectious agents is often limited to the most noticeable effects, which sometimes represent only the tip of the iceberg. For example, it is now well appreciated that pathologic processes caused by infectious agents may emerge clinically after an incubation of decades and may result from genetic, immunologic, and other indirect routes more than from the infecting agent in itself. Thus, there is a general expectation that continued investigation will lead to the isolation of new agents of infection, the identification of hitherto unsuspected etiologic correlations, and, eventually, more effective approaches to prevention and therapy.

Studies on the mechanisms of disease caused by infectious agents demand a breadth of understanding across many specialized areas, as well as much cooperation between clinicians and experimentalists. The series *Infectious Agents and Pathogenesis* is intended not only to document the state of the art in this fascinating and challenging field, but also to help lay bridges among diverse areas and people.

M. Bendinelli
H. Friedman

Preface

The discovery of the human T cell leukemia virus type I in the late 1970s heralded a new era in retrovirology. For the first time, it was demonstrated that a retrovirus could play a role in the development of a human disease, in this case adult T cell leukemia (ATL). Several years later, the acquired immunodeficiency syndrome (AIDS) epidemic began, and it was demonstrated that a retrovirus, originally designated the human T cell lymphotropic virus type 3, was the causal agent of this syndrome. This virus, later named the human immunodeficiency virus type 1 (HIV-1), has since been extensively studied in terms of its pathogenesis as well as its ability to elicit immune responses. In that time, a tremendous amount of information has been obtained about the virus. Although recent drug regimens have been useful in significantly lowering viral loads and perhaps maintaining an asymptomatic state among individuals infected with HIV-1, an established "cure" for AIDS eludes us. In addition, the effective drug therapies are very expensive, and are not available to infected people in the third world, where greater than 90% of new infections occur. Furthermore, the development of viral resistance against the drug therapies is an additional concern.

Despite extensive study, no effective vaccine has been developed. One of the problems in developing an effective vaccine against HIV-1 is the ability of the virus, particularly in the immunogenic envelop glycoprotein, to undergo amino acid hypervariability. Therefore, vaccines generated against one envelop glycoprotein are ineffective against other viruses with more hypervariable envelopes.

Although the rate of new HIV-1 infections in the United States has slowed and appears to have reached a plateau, nearly 16,000 people per day worldwide become newly infected with this devastating virus. Ominously, it is predicted that at the turn of the century, over 40 million people across the globe will be infected. Clearly, the development of new effective drug therapies, as well as immune prophylactic and therapeutic regimens against

HIV and AIDS is warranted. The chapters in this volume, written by experts in the field of human retroviral pathogenesis, vaccine development, and the clinical treatment of AIDS, summarize the current status of work in these areas as well as future directions for research and therapeutic development.

The first chapter, by Dr. Bristow, deals with the important descriptions of the two principal viremias of HIV-1, sexual (heterologous) transmission and autologous transmission of viruses between cells within the same host. This information is relevant for the development of methods to limit the spread of HIV-1 infection. Drs. Gentile and Loughran then describe evidence for the role of the human T cell leukemia viruses (HTLV) in the etiology and pathogenesis of disorders other than ATL. This is an interesting and important area because of evidence suggesting that the HTLVs may have a role in a number of autoimmune diseases. Next, Dr. Cianciolo discusses the evidence for the immunomodulatory and immunosuppressive activity for a number of HIV-1 proteins including gp120 and gp41 as well as the regulatory proteins Tat and Nef. Dr. Kieber-Emmons describes the importance of carbohydrate moieties and glycosylation patterns in HIV-1 and their role in masking potentially immunogenic and therapeutically important epitopes. Drs. Bennett and Agadjanyan summarize some of the immunologic and molecular aspects of HTLV-I and HTLV-II infection, emphasizing the role of adhesion molecules in the binding of these two viruses to infectible target cells.

The next set of chapters deals with issues concerning the development of immunologic interventions against HTLV, HIV-2, and HIV-1, including vaccine development and passive immunotherapies. Dr. Franchini and colleagues describe the current status of efforts to develop a vaccine against HTLV-I, a retrovirus that infects 10–20 million people worldwide. Dr. Björling discusses immune responses against HIV-2 and how these responses are important for the development of vaccine strategies against this retrovirus, which primarily infects individuals in West Africa. Drs. Staats and McGhee deal with the important need to develop vaccine strategies against HIV-1 which will protect against mucosal infection. Since HIV-1 is transmitted primarily by mucosal routes, this is a very timely and important issue in the area of HIV-1 vaccinology. In terms of novel vaccine strategies against HIV-1, Dr. Boyer and colleagues describe the nucleic acid (DNA) vaccination technology. This approach, which appears to mimic live, attenuated viral vaccines without major safety concerns, elicits both humoral and cytotoxic T cell responses against HIV-1, and is currently in several human clinical trials. Drs. Cotropia and Ugen describe the current status of passive immunotherapy and immunoprophylaxis against HIV-1, including the use of human polyclonal HIV immunoglobulins and specific neutralizing human monoclonal antibodies. The set of chapters dealing with immunologic interventions ends with the description by Dr. Ayyavoo and colleagues of efforts to target the accessory genes of HIV-1 for vaccine development and immunotherapy.

The final set of chapters deals with developments in therapeutic interventions against HIV-1 infection and AIDS using drugs or gene therapy. Drs. Doranz and Doms discuss the role of chemokine receptors in mediating entry of HIV-1 into cells as well as the design of new antiviral therapeutics that may prevent HIV from using these receptors. Dr. Goodenow and colleagues then summarize the activities of the therapeutically important protease inhibitors and, in particular, describe the role of genetic variability of the HIV-1 protease in modulating the response to these drugs in pediatric patients. Drs. Dornburg and Pomerantz follow with a chapter that describes current ideas and implementations of gene therapeutic regimens against HIV-1, as well as future possibilities of such an approach. The volume concludes with a chapter by Dr. Nelson and colleagues on the current status of the clinical use of antiretroviral therapy against pediatric HIV, the relative effectiveness of this therapy, and the need for the development of new therapies.

It is anticipated by the editors as well as by the authors of the individual chapters that this volume will provide a useful summary of the current status of immunologic and therapeutic interventions against human retroviral infections, most notably HIV-1. It is hoped that the need is made apparent for the development of efficacious vaccines against both HIV-1 and HTLV-I, as well as for the development of novel drug therapies and regimens such as gene therapy against HIV-1. The editors thank Ilona Friedman for excellent editorial assistance in coordinating and assisting in the preparation of the manuscripts for this volume.

Kenneth E. Ugen
Mauro Bendinelli
Herman Friedman

Contents

6. Vaccine Approaches for Human T-Cell Lymphotropic Virus Type I 109

G. A. DEKABAN, A. PETERS, J. ARP, and G. FRANCHINI

7. Immune Responses against HIV-2 143

EWA BJÖRLING

**10. Passive Immunotherapy against HIV-1: Current Status and
Potential** 217

JOSEPH P. COTROPIA and KENNETH E. UGEN

**11. Human Immunodeficiency Virus Type 1 Accessory Genes:
Targets for Therapy** 239

SAGAR KUDCHODKAR, T. NAGASHUNMUGAM,
and VELPANDI AYYAVOO

12. A New Generation of Antiviral Therapeutics Designed to Prevent the Use of Chemokine Receptors for Entry by HIV-1   269

BENJAMIN J. DORANZ and ROBERT W. DOMS

13. Protease Inhibitors and HIV-1 Genetic Variability in Infected Children 287

MAUREEN M. GOODENOW, ELENA E. PEREZ, and JOHN W. SLEASMAN

14. **Gene Therapy and HIV-1 Infection: Experimental Approaches, Shortcomings, and Possible Solutions** 307

RALPH DORNBURG and ROGER POMERANTZ

15. **Pediatric HIV: Antiretroviral Therapy** 325

ROBERT P. NELSON, JR., PATRICIA J. EMMANUEL,
and MAITE DE LA MORENA

1

The Two Principal Viremias of HIV

A Comparison of Viral and Host Characteristics

CYNTHIA L. BRISTOW

1. INTRODUCTION

Viruses are biological entities that do not inherently possess life since significant reproduction is dependent on internalization by a susceptible host cell. Ultimate parasitism by certain viral species allows continued replication of both the host cell and the virus, and this requires mutual equanimity. At this point in history, human immunodeficiency virus (HIV) has acquired an evolutionary state of very efficient, but not ultimate, parasitism. In fact, it is the insufficiency of equanimity between host cells and virus that produces the chronic inflammation which culminates in consummation of the mononuclear phagocyte system.

According to the 1993 classification system of HIV disease defined by the Centers for Disease Control, HIV infection can be described in terms of three categories. Category A is defined as a mononucleosis-like syndrome associated with seroconversion for HIV antibody. Category A is accompanied by acute viremia occurring 2–4 weeks after exposure and which is no longer detected systemically after the first 3 months. The initial viremia, represented by passage of the virus between heterologous cells (sexual

CYNTHIA L. BRISTOW • McLendon Clinical Laboratories, Department of Pathology and Laboratory Medicine, University of North Carolina Hospitals, Chapel Hill, North Carolina 27514.

Human Retroviral Infections, edited by Kenneth E. Ugen *et al.*
Kluwer Academic / Plenum Publishers, New York, 2000.

1

transmission), is recognized and contained, but not cleared, by the mononuclear phagocyte system. A postacute, asymptomatic period as long as 10 years is characterized by diminutive systemic viral replication and is followed by Category B, the onset of constitutional symptoms which concurrently manifests systemically with a second chronic viremia. The second principal viremia, represented by passage of the virus between autologous cells (autologous transmission), is not contained, and it is evident that both the virus and infected cell populations have dramatically changed since the initial viremia. Finally, Category C is defined as the onset of multiple opportunistic infections and acquired immunodeficiency syndrome (AIDS) and results in death after 2–3 years.

The differences which exist during heterologous and autologous transmission are readily demonstrated by culturing susceptible cells in heterologous versus autologous serum (Table I). A primary patient isolate (third passage) was incubated with peripheral blood mononuclear cells cultured in autologous serum, heterologous human serum, or fetal calf serum. The 50% tissue culture infectious dose ($TCID_{50}$) under these conditions decreased as the incompatibility between cells and serum increased.

Investigation by virologists worldwide has greatly increased our understanding of the structural and functional characteristics of HIV. On the other hand, limited information is available on the precise host cell characteristics which exist during the moment of susceptibility. For example, host cells and their receptors have been examined with little attention to the effect of the microenvironment on secondary receptor interactions. Significantly, neither the host cells which have the capacity to become susceptible nor the environments where they reside are uniform, and this would suggest diversity during cellular uptake of HIV with a complexity now only scarcely imagined. The diversity of HIV infectious units, target cells, and conditions under which they make contact are inspected in this chapter with the intent of expanding our conceptualization of molecular events involved during the initiation of infection.

TABLE I

Differences between Autologous and Heterologous Transmission[a]

Subject	Autologous serum	Heterologous human serum	Fetal calf serum
1	$10^{3.7}$	ND	$10^{4.4}$
2	$10^{4.4}$	ND	$10^{5.1}$
3	ND	$10^{3.7}$	$10^{4.4}$
4	ND	$10^{3.7}$	$10^{4.4}$

[a]Viral titer determined by p24 capture following *in vitro* infectivity using a primary HIV patient isolate with four unrelated subjects cultured in serum from different sources. ND, Not done.

2. SEXUAL TRANSMISSION

The architecture of a virion is designed to allow its safe entry into a host cell with minimal disturbance of host cell physiology, and this requires concerted contact in the milieu where the cell resides. In the majority of cases, the parameters of HIV infectivity have traditionally been measured using transformed target cells (cell lines) or peripheral blood target cells cultured in the presence of fetal calf serum, as opposed to human serum or mucosal secretions which are known to impact infectivity *in vitro* as well as *in vivo*. Neither the infectious unit, the initially infected cell population, nor the conditions in the environment during sexual transmission of HIV are definitively understood.

2.1. The Infectious Unit

There is little evidence as to whether sexual transmission occurs via the cell-free versus cell-associated form of HIV. That seminal plasma is toxic to cell-free HIV[1] suggests that as a result of sexual transmission, the target cell of the anorectal junction or genital tract is likely to receive the virus associated with either spermatozoa,[2] donor, or recipient phagocytic cells. It can be speculated that sexual transmission may involve phagocytosis or intercellular fusion of cell-associated virus; however, the identity of molecular mechanisms involved during sexual transmission have proven technically prohibitive.

2.2. The Target Cell

The transmissibility of cell-associated virus to non-CD4$^+$ mucosal target cells, e.g., epithelial cells,[3] has been demonstrated; however, the histoincompatibility of the virus or virus-associated donor cell which is transmitted through sexual contact suggests that the selectively dominant recipient target cell is most likely a mucosal antigen-processing cell, i.e., macrophage-like cell (Mo/Mφ). Evidence has suggested that the Langerhans type of tissue Mφ can be the initial recipient target cell in sexual transmission,[4,5] and similar phagocytic cell types are implicated in distinct tissues.

The names of specialized phagocytic cells derived from bone marrow promonocytes vary according to their location: Langerhans cells in the skin; macrophages (Mφ) in the bone marrow, lymphoid tissue, and serous cavities; monocytes (Mo) in the blood; histiocytes in connective tissue; microglia in the nervous system; alveolar macrophages in the lungs; and Kupffer cells in the liver.[6] The *antigen-processing* cell population including the peripheral monocyte gives rise to the tissue macrophage referred to as the histiocyte. The *antigen-presenting* cell population, referred to as the dendritic cell, is generally not phagocytic. Antigen-presenting cells can be divided into three subsets: follicular dendritic reticulum cells, interdigitating reticulum cells, and Langerhans cells. Circulating monocytes, tissue histiocytes, and macrophages normally do not proliferate except in conditions of inflammation.

Langerhans cells originate in bone marrow and reside in the suprabasal and spinous layers of epithelium, where Mφ, lymphocytes, and other immune cells are located in the subepithelial lamina propria. These cells represent 2–5% of the epidermis and are two- to three-fold more frequent in nonkeratinized than keratinized epithelium.[7] It has been shown that depletion of Langerhans cells from cutaneous surfaces allows access of antigen to the systemic immune apparatus as a tolerogenic, rather than an immunogenic signal.[8] After processing antigen, Langerhans cells migrate to and reside in the proximal lymph nodes where antigen is presented to T lymphocytes.[9] During migration to the lymph nodes, Langerhans cells have been reported to undergo phenotypic changes correlating with differentiation from predominantly antigen-processing to predominantly antigen-presenting interdigitating dendritic cells.[10] Differentiation has been reported to be enhanced by products of T lymphocytes and Mφ.[11]

In common with a subset of Mo/Mφ, Langerhans cells can possess surface CD4, major histocompatibility complex (MHC), Fc receptors, and C3 receptors.[7,12] The involvement of each of these receptors in facilitating HIV infectivity has been implicated.[13–16] It has been reported that trypsinization of Langerhans cells during isolation from tissue results in enhanced binding of HIV envelope proteins in a non-CD4-dependent manner,[17] suggesting the possibility that the initiating infection might not necessarily be a CD4-dependent event. As would be predicted, loss of Langerhans cells has been reported to occur concomitantly with severity of oral hairy leukoplakia and candidiasis as well as HIV-associated and non-HIV-associated periodontal disease,[7] supporting the participation of these cells in protecting against the attendant mucosal pathology of HIV infection.

It was reported that certain HIV-1 clades preferentially infected Langerhans cells *in vitro*,[5] and this suggests that either more than one subtype of tissue Mφ might be involved in transmission in different settings or the *in vitro* tissue culture conditions were selective.

2.3. The Environment

Mucosal secretions are a cardinal determinant in systemic immune responsiveness following mucosal pathogenic insult. The overall purpose of mucosal lymphoid tissue is to localize antigen and prevent systemic involvement. In fact, systemic unresponsiveness is a common consequence of mucosal antigen exposure. Evidence suggests the mucosa of an HIV-infected individual has atypical immune activity. Synthesis of immunoglobulin G (IgG), but neither IgA nor IgM, having specificity for HIV-1 has been reported in the genital tracts of male and female seropositive individuals.[18,19] The loss of mucosal secretory IgA correlates with clinical presentation, suggesting the loss of protective immunity in the mucosa.[20] Increased IgG in the mucosa suggests a breakdown in the mucosal barrier and concomitant inflammation, and this correlates with the clinical spectrum of HIV infection.

The state of the mucosa of the recipient partner is also not likely to be quiescent. It has been found in many epidemiologic studies that risk of sexual transmission of HIV increases with number of partners, and may increase with number of encounters with the same partner.[21,22] The incidence of transmission after exposure with a single partner has been linked to increased viral inoculum with greater likelihood during the second principal viremia of advanced AIDS in the infected partner. Even though excessive viral inoculum increases the probability of a "hit" in an otherwise quiescent mucosal environment, there is a conspicuous lack of evidence supporting infection following a single exposure.[21]

Three situations for optimum sexual transmission have been proposed: (1) in repeated exposure with many partners, a state of inducible immunity may preexist in the recipient mucosa (lymphoblasts, activated Mφ, neutrophils, cytokines, immunoglobulin); (2) in repeated exposure with a single partner, excessive viral inoculum in the presence of the concomitant inflammatory components of seminal plasma (immunoglobulin, complement activation, cytokines, lymphoblasts, activated Mφ, neutrophils) may perturb the microenvironment of the target cell population; or (3) the viral genome, having interacted with or integrated with the recipient tissue during a prior exposure, may be taken up by a permissive cell in the presence of activating cofactors (neutrophils, cytokines, coinfecting pathogens, endotoxin). Each of these proposed conditions suggests that the state of the mucosal target cell population during the conclusive infectious event is probably altered from its normal physiological state.

Besides the perturbing effects of repetitive pathogenic insult to the mucosal environment, there is the significant, yet poorly understood contribution of the mucosal secretions during HIV encounter. Transmission of Simian immunodeficiency virus (SIV) by the genital route in the rhesus macaque has been demonstrated in the cell-free, but never in a cell-associated manner, and the importance of seminal plasma has been postulated.[23] Immunosuppressive mechanisms of saliva, seminal plasma, and other mucosal secretions have not been carefully compared, independently or in combination, even though there is ample evidence for such an effect. Seminal plasma has immunosuppressive character not only as a mechanism to facilitate fertilization by these gametes in an immunologically unfriendly environment, but also to protect spermatozoa against autoimmunity.[24]

Predominant in seminal plasma are proteinases and proteinase inhibitors, some of which have been implicated in HIV internalization.[25–29] One proteinase in canine seminal plasma accounts for over 90% of the total protein.[30] A primary immunosuppressive component of seminal plasma has been identified as a prostasome. Prostasomes are multilamellar vesicles which, in addition to the potentially immunosuppressive effect of lipid components, manifest surface-associated immunoreactive components including IgG, IgA, purine nucleotides, and enzymes. Prostasomes are rapidly bound and internalized by neutrophils and Mϕ, but not lymphocytes, and internalization results in decreased superoxide anion production and phagocytosis of latex particles.[31] The potential for prostasomes or other noncellular vesicles to participate in transmission of HIV has not been investigated.

Saliva has been shown to be immunoregulatory. It has been found that oral tolerance results from ingestion of antigen by inducing T cell, but not B cell, suppression in humans and has been suggested to be a mechanism for avoiding immunoreactivity to certain foods.[32] Prevention of atopic reactions to food has also been proposed as a role for the immunosuppressive effect of human milk and colostrum.[33]

Not surprisingly, unique functional and therefore structural attributes of each tissue and organ dictate that the generic qualities of immunity be fashioned to fit the special needs of each tissue and organ.[34] The regionally specialized gut-associated (GALT), bronchus-associated (BALT), nasal-associated (NALT), or mucosal-associated lymphoreticular tissue (MALT) migrate to the draining lymph nodes, thoracic lymph duct, bloodstream, and finally the lamina propria representative of their origin, where they mature and produce IgA-secreting plasma cells. It has been reported that the mucosal immune system is compartmentalized in a manner resulting in preferential distribution of lymphocytes to mucosal regions adjacent to the site of immunization.[35] This suggests the possibility that HIV could prime the recipient and HIV-specific plasma cells might establish in the local environment prior to infection. Recipient-derived or donor-derived IgG or IgA might facilitate or even be requisite during the initial infectious event. Evidence has been found for the involvement of host factors during disease progression following cell-free, intravenous SIV infection of macaques, and these factors appeared to influence infection outcome equivalently *in vivo* and *in vitro.*[36]

In addition to the migration of infected cells through the lymphatics, the initial viremia is manifested in blood, which results in global distribution of infected cells, primarily T lymphocytes.[37] It has been suggested that the primary role of GALT is twofold: to prime an immune response (antigen-processing and antigen-presenting) and to amplify a response by dissemination of antigen-specific lymphocytes (antigen-presenting).[38] By analogy, MALT might be involved in priming as well as in amplification of a response already primed at another site. Once infection has been established, the diffuse lymphatic tissue of the mucosal surfaces may be seeded with macrophages and lymphocytes capable of releasing viral particles from the basal and/or apical surfaces, initiating a chronic, carrier state of infection. The manifestations of pathology at mucosal surfaces (oral, upper respiratory, gastrointestinal) may result from the release of inflammatory viral proteins and host mediators to the epithelial cell surface and interstitium of the mucosa.

In summary, neither the infectious unit, target cell, nor the environment that exists during sexual transmission has been clearly identified. At best, it can be speculated that cell-associated HIV is transmitted to a mucosal phagocytic cell in an environment containing products characteristic of inflammation. The significance of previously identified HIV cofactors (β-chemokine receptors) and receptors (CD4 and proteinases) during sexual transmission has not been explored.

3. AUTOLOGOUS TRANSMISSION

Once the initial viremia has abated, the infection becomes difficult to detect and symptoms are minimal. The predominant infected cell in tissue is the Mφ and the predominant infected cell in blood is the CD4$^+$ T lymphocyte.[37] The *in vivo* CD4$^+$ T lymphocyte population is maintained at a functionally intact, but relatively diminished level, which persists until an inevitable, unascertained precipitous event heralds the final dysregulation of immune competence. This threshold is defined by a capacitance which may be affected by both lymphocytes and Mo/Mφ directly or indirectly; however, the properties determining capacitance are precisely the mechanisms of transmission between heterologous cells and between autologous cells.

3.1. The Infectious Unit

The fate of the infected Mo/Mφ is not clear. Ultrastructural analysis of HIV-infected Mo/Mφ demonstrates little or no cytopathic effects, and little or no expression of viral protein or release of virions from the cytoplasmic membrane.[39,40] Instead, HIV accumulates within intracellular vacuoles derived from the Golgi complex, assembles, and buds from the vacuolar membranes.[41] Exocytosis into the extracellular milieu appears suppressed, and infected cells can persist in tissue for months containing infectious particles within the intracytoplasmic vacuoles.[37] When these cells become stimulated, cell-free virions are released which have been demonstrated to have dual tropism, infecting monocytes and T lymphocytes equivalently.[37]

Activation of Mo/Mφ was defined by Adams and Johnson[42] as the acquisition of competence to complete a complex function such as the destruction of microbes or tumor cells. After entering tissues, Mφ generally remain inactive until stimulated by one or a combination of inductive signals. Receptor ligation results in one of several inductive or suppressive cascades of signal transduction which ultimately impact on specific genes. Transcription of HIV DNA has been found to be dependent on the specific characteristics of the region of the cellular genome in which the proviral DNA has integrated,[43] and this reflects the specificity of the activation cascade initiated by receptor ligation.

The fate of the infected T cell is ultimately cell death through mechanisms including apoptosis, cytolysis, and syncytia formation,[44] and this suggests that transmission of HIV to T lymphocytes in circulation can be either cell-free or cell-associated. Evidence has suggested unidirectional transfer of HIV from the Mo/Mφ to the T lymphocyte[37]; however, phagocytosis of infected T cell debris by neutrophils, Mφ, or other phagocytic cells could influence autologous transmission.

Significantly, as the syndrome develops, the virus adopts host cell receptors and ligands,[14,45] and several of these have been shown to enhance infectivity of Mo/Mφ and T and B lymphocytes.[46,47] Whereas the viral and cellular membranes and coupled receptors during autologous transmission are derived from autologous cells, these are derived from heterologous donor cells during sexual transmission. Therefore, the infectious unit can be cell-associated or cell-free during autologous transmission, and the capacity for immune recognition may be masked by autologous membrane components.

3.2. The Target Cell

Receptor ligation by cells in the mononuclear phagocyte system is tightly regulated to minimize the potential for autoreactivity. Ligation of lymphocyte receptors in an inappropriate order induces apoptosis. To hijack a cell requires a threshold of viral compatibility that will permit replication to occur prior to death of the cell. For this reason, the physiological role of a cell-surface molecule is preempted by its ligation with a virion. A molecule that acts as a proteinase physiologically might act as a viral receptor, and a molecule that acts as a receptor physiologically might act to colocalize the virus without directly acting as a viral receptor. The temporal and topographic parameters of receptor ligation allow multiple signaling cascades to be produced permitting the resulting multiplicity of discrete cell functions. This implies that the ligation of a signaling receptor by a virion requires uncoupling of the activating signal at some subsequent step.

The saturable and specific binding of HIV to CD4 and rate-dependent infectivity of CD4[+] T lymphocytes[13,48] *in vitro* support the involvement of the CD4 receptor during infection of T lymphocytes with cell-free virus *in vivo*. A lack of cellular activation in response to CD4 ligation has been demonstrated.[49,50] Evidence for unique CD4 involvement during internalization by Mo/Mφ is not quite so convincing.[51–53] Evidence suggests that HIV permissiveness can be conferred to fibroblasts,[15] CD4[−]CD8[+] T lymphocytes,[54] and epithelial cells and skeletal muscle cells[55] in a CD4-independent manner through receptors for IgG or C3. A family of chemokines has been demonstrated to influence infectivity both *in vitro*[56] and *in vivo*.[57] Although strong evidence suggests the influence of chemokine receptors during HIV internalization, a direct interaction between these receptors and HIV has not been found. The involvement of additional host-specific molecules has been proposed,[29,58,59] and it has been empirically demonstrated that chemokine receptors lack the capacity to definitively confer susceptibility.[60]

3.3. The Environment

In situ infection of Mo/Mφ could occur in skin, bone marrow, the serous cavities, blood, connective tissue, nervous system, lungs, liver, and lymphoid tissue. By Southern blot hybridization, large quantities of HIV DNA are found in the lymph node and brain, but not lysates of lung, liver, or spleen.[61] Application of more sensitive molecular genetic techniques in comparative analysis of HIV burden in Mo isolated from peripheral blood and Mφ isolated from bronchoalveolar lavage fluid during the asymptomatic period has revealed that the proviral burden is substantial and equivalent, but that levels of expression are minimal.[62] Evidence suggests that alveolar Mφ from HIV-infected patients rarely contain the HIV-1 p24 protein or HIV-1 RNA, but do exhibit proviral DNA.[63] Alveolar Mφ from these infected individuals were frequently found to produce infectious virions when induced by a combination of cytokines. Blood monocytes that traffic to the respiratory tract in response to infection or inflammation were significantly less susceptible to HIV infection than the resident alveolar Mφ. The quiescence of HIV in infected Mφ can occur on at least two levels: transcription of integrated proviral DNA and release of virus sequestered within vacuoles.

The virions released by the infected T lymphocytes have not been found to have monocytic tropism *in vitro*,[37] and this suggests that any systemic uptake of virus by Mo/Mφ may involve a different process, e.g., phagocytosis. Infected progenitor cells can be detected in bone marrow in a subset (14%) of seropositive individuals,[64] and these cells contribute to the infected cell population in circulation in these individuals. A possible site of infectivity of Mo/Mφ is during transmigration from circulation into tissue through the lamina propria of endothelium containing resident HIV-infected Mo/Mφ.

In skin biopsies of HIV-infected individuals, Langerhans cells are reported to demonstrate altered morphology, and viral particles have been shown to bud into the intercellular space between Langerhans cells and adjacent epithelial cells.[65] This suggests that infection of Langerhans cells during the second viremia may not induce their migration to regional lymph nodes, but that released viral products might subsequently infect cells located in the lamina propria with focal bystander cytopathology of the epithelial cells.

Evidence now suggests the predominant site in autologous transmission may be lymphoid tissue.[66] Lymph node biopsies in the SIV model suggest that virus can be detected by the seventh day postinoculation of cell-free virus, and is exclusively cell-associated.[67,68] Binding of HIV to follicular dendritic cells has been determined to be C3-dependent, and not CD4-dependent.[69] The precipitous event which heralds the second principal viremia may be the depletion of interactive immune components not directly related to the CD4$^+$ cell population. Instead, these components may be related to the immune containment of HIV, e.g., lymph node architecture[66] or serum components such as inhibitors of complement activation and cytokines.

Langerhans cells and interdigitating cells are found in the cortex of the lymph node (the T cell zone), whereas the dendritic cells are located in the germinal centers (the B cell zone). The germinal centers of lymph nodes are primarily composed of memory B lymphocytes, but T lymphocytes and Mφ can be demonstrated. Antigen localization by interdigitating and dendritic cells is a mechanism for concentrating antigen to regions with heavy lymphocyte traffic and allows clonal selection and expansion. In a primary response, the germinal center is a highly active site for blast transformation, cytokine release, and Mφ recruitment. In HIV infection, localization of HIV to the germinal centers of lymph nodes would optimize conditions for transmission of cell-free HIV from Mφ to lymphocytes. Mφ within germinal centers contain a distinctive type of phagolysosome, the tingible body, which includes nuclear material of phagocytized lymphocytes[70] providing a mechanism for countertransmission of cell-associated HIV from T lymphocytes to Mφ. Further, clonal selection, expansion, and dissemination of immunoreactive cells would vary depending on variation in HIV antigens as well as on variation in the residual functionally active resident cells.

In summary, the infectious unit in autologous transmission may be cell-free or cell-associated. The primary target cells during the asymptomatic period appear to be Mφ and lymphocytes located in the germinal cells of lymph nodes where inflammatory immune activity could permit transmission of cell-free HIV in a CD4-dependent manner or cell-associated HIV in a CD4-independent manner. During the second principal viremia, the CD4$^+$ T lymphocyte population has been significantly depleted suggesting the final primary target cells are the Mo/Mφ.

4. CONCLUSIONS

The infectious units, target cells, and environments during sexual transmission and autologous transmission are fundamentally different. The infectious unit during sexual transmission is probably cell-associated and histoincompatible, whereas the infectious unit during autologous transmission may be cell-free or cell-associated and host-adapted. The target cell during sexual transmission is probably a mucosal phagocytic cell. The target cells during autologous transmission are probably Mo/Mφ and T lymphocytes in lymphoid tissue and finally in circulation. The mucosal environment during sexual transmission is probably inflammatory and possibly facilitated by inflammatory mediators. The lymph node environment during autologous transmission is probably inflammatory and possibly facilitated by cell death. Chronic nonspecific and HIV-specific inflammation, at the expense of protective immunity for commonly encountered pathogens, results in depletion of immune resources, and finally immunodeficiency.

Understanding viral-internalization steps is paramount to intervention in the spread of HIV. Our previous attempts to define the molecular mechanisms involved in viral internalization have used target cells and infectious units derived from heterologous peripheral blood cells and cultured in the presence of calf serum, conditions which we now recognize as inadequately representing either sexual or autologous transmission. HIV drug susceptibility has traditionally been determined in a similar manner, sometimes leading to artefactual interpretations of efficacy, e.g., recombinant soluble CD4,[71–73] hypergammaglobulins,[74] and vitamin C.[75] Our knowledge regarding the target cells, infectious units, and physiological conditions during transmission is evolving, and thus we are poised to discover the precise molecular mechanisms that define the conundrum of chronic HIV infection and AIDS.

REFERENCES

1. Anderson, D., Tucker, L., and Gallion, T., 1992, Optimization of culture conditions for detection of infectious HIV-1 in semen, *Int. Conf. AIDS* **8**:19–24.

2. Bagasra, O., Farzadegan, H., Seshamma, T., Oakes, J. W., Saah, A., and Pomerantz, R. J., 1994, Detection of HIV-1 proviral DNA in sperm from HIV-1 infected men, *AIDS* **8:**1669–1674.

3. Milman, G., and Sharma, O., 1994, Mechanisms of HIV/SIV mucosal transmission, *AIDS Res. Hum. Retrovir.* **10:**1305–1312.

4. Hussain, L. A., and Lehner, T., 1995, Comparative investigation of Langerhans' cells and potential receptors for HIV in oral, genitourinary and rectal epithelia, *Immunology* **85:**475–484.

5. Soto-Ramirez, L. E., Renjifo, B., McLane, M. F., Marlink, R., O'Hara, C., Sutthent, R., Wasi, C., Vithayasai, P., Vithayasai, V., Apichartpiyakul, C., Auewarakul, P., Cruz, V. P., Chui, D.-S., Osathanondh, R., Mayer, K., Lee, T.-H., and Essex, M., 1996, HIV-1 Langerhans' cell tropism associated with heterosexual transmission of HIV, *Science* **271:**1291–1292.

6. Egeler, R. M., 1993, Langerhans cell histiocytosis and other histiocytic disorders, Thesis, University of Amsterdam, Amsterdam, Netherlands.

7. Lombardi, T., Hauser, C., and Budtz-Jorgensen, E., 1993, Langerhans cells: Structure, function and role in oral pathological conditions, *J. Oral Pathol. Med.* **22:**193–202.

8. Toews, G. B., Bergstresser, P. R., and Streilein, J. W., 1980, Epidermal Langerhans cell density determines whether contact hypersensitivity or unresponsiveness follows skin painting with DNFB, *J. Immunol.* **124:**445–451.

9. Schmitt, D., and Dezutter-Dambuyant, C., 1994, Epidermal and mucosal dendritic cells and HIV1 infection, *Path. Res. Pract.* **190:**955–959.

10. Van Wilsem, E. J., Van Hoogstraten, I. M., Breve, J., Scheper, R. J., and Kraal, G., 1994, Dendritic cells of the oral mucosa and the induction of oral tolerance. A local affair, *Immunology* **83:**128–132.

11. Cella, M., Scheidegger, D., Palmer-Lehmann, K., and Lane, P., 1996, Ligation of CD40 on dendritic cells triggers production of high levels of interleukin-12 and enhances T cell stimulatory capacity: T–T help via APC activation, *J. Exp. Med.* **184:**747–752.

12. Pimentel-Muinos, F. X., Lopez-Geurrero, J. A., Fresno, M., and Alonso, M. A., 1992, CD4 gene transcription is transiently repressed during differentiation of myeloid cells to macrophage-like cells, *Eur. J. Biochem.* **207:**321–325.

13. Maddon, P. J., Dalgleish, A. G., McDougal, J. S., Clapham, P. R., Weiss, R. A., and Axel, R., 1986, The T4 gene encodes the AIDS virus receptor and is expressed in the immune system and the brain, *Cell* **47:**333–348.

14. Arthur, L. O., Bess, Jr., J. W., Sowder II, R. C., Benveniste, R. E., Mann, D. L., Chermann, J.-C., and Henderson, L. E., 1992, Cellular proteins bound to immunodeficiency viruses: Implications for pathogenesis and vaccines, *Science* **258:**1935–1938.

15. McKeating, J. A., Griffiths, P. D., and Weiss, R. A., 1990, HIV susceptibility conferred to human fibroblasts by cytomegalovirus-induced Fc receptor, *Nature* **343:**659–661.

16. Delibrias, C. C., Kazatchkine, M. D., and Fischer, E., 1993, Evidence for the role of CR1 (CD35), in addition to CR2 (CD21), in facilitating infection of human T cells with opsonized HIV. *Scand J. Immunol.* **38:**183–189.

17. Dezutter-Dambuyant, C., Schmitt, D. A., Dusserre, N., Hanau, D., Kolbe, H. V. J., Kieny, M. P., Gazzolo, L., Mace, K., Pasquali, J.-L., Olivier, R., and Schmitt, D., 1991, Trypsin-resistant gp120 receptors are upregulated on short-term cultured human epidermal Langerhans cells, *Res. Virol.* **142:**129–138.

18. Wolff, H., Martinez, A., Politch, J., Saltzman, S., Mayer, K., and Anderson, D. J., 1989, Detection and quantification of HIV-1 antibodies in semen, *Int. Conf. AIDS* **5:**181.

19. Belec, L., Tevi-Benissan, C., Lu, X. S., Prazuck, T., and Pillot, J., 1995, Local synthesis of IgG antibodies to HIV within the female and male genital tracts during asymptomatic and pre-AIDS stages of HIV infection, *AIDS Res. Hum. Retroviruses* **11:**719–729.

20. Kotler, D. P., Scholes, J. V., and Tierney, A. R., 1987, Intestinal plasma cell alternations in acquired immunodeficiency syndrome, *Dig. Dis. Sci.* **32:**129–138.

21. Padian, N. S., Shiboski, S. C., and Jewell, N. P., 1990, The effect of number of exposures on the risk of heterosexual HIV transmission, *J. Infect. Dis.* **161:**883–887.

22. Levin, L. I., Peterman, T. A., Renzullo, P. O., Lasley-Bibbs, V., Shu, X. O., Brundage, J. F., and McNeil, J. G., 1995, HIV-1 seroconversion and risk behaviors among young men in the US army. The Seroconversion Risk Factor Study Group, *Am. J. Public Health* **85:**1500–1506.

23. Miller, C. J., 1994, Animal models of viral sexually transmitted disease, *Am. J. Reprod. Immunol.* **31:**52–63.

24. Wolff, H., 1995, The biologic significance of white blood cells in semen, *Fertil. Steril.* **63:**1143–1157.

25. Hattori, T., Koito, A., Takatsuki, K., Kido, H., and Katunuma, N., 1989, Involvement of tryptase-related cellular protease(s) in human immunodeficiency virus type, *FEBS Lett.* **248:**48–52.

26. Clements, G. J., Price-Jones, M. J., Stephens, P. E., Sutton, C., Schulz, T. F., Clapham, P. R., McKeating, J. A., McClure, M. O., Thomson, S., Marsh, M., Kay, J., Weiss, R. A., and Moore, J. P., 1991, The V3 loops of the HIV-1 and HIV-2 surface glycoproteins contain proteolytic cleavage, *AIDS Res. Hum. Retroviruses* **7:**3–16.

27. Harvima, I. T., Harvima, R. J., Nilsson, G., Ivanoff, L., and Schwartz, L. B., 1993, Separation and partial characterization of proteinases with substrate specificity for basic amino acids from human MOLT-4 T lymphocytes: Identification of those inhibited by variable-loop V3 peptides of HIV-1 (human immunodeficiency virus-1) envelope glycoprotein, *Biochem. J.* **292:**711–718.

28. Avril, L.-E., di Martino-Ferrer, M., Pignede, G., Seman, M., and Gauthier, F., 1994, Identification of the U-937 membrane-associated proteinase interacting with the V3 loop of HIV-1 gp120 as cathepsin G, *FEBS Lett.* **345:**81–86.

29. Bristow, C. L., Fiscus, S. A., Flood, P. M., and Arnold, R. R., 1995, Inhibition of HIV1 by modification of a host membrane protease, *Int. Immunol.* **7:**239–249.

30. Isaacs, W. B., and Coffey, D. S., 1984, The predominant protein of canine seminal plasma is an enzyme, *J. Biol. Chem.* **259:**11520–11526.

31. Skibinski, G., Kelly, R. W., Harkiss, D., and James, K., 1992, Immunosuppression by human seminal plasma-extracellular organelles (prostasomes) modulate activity of phagocytic cells, *Am. J. Reprod. Immunol.* **28:**97–103.

32. Husby, S., Mestecky, J., Moldoveanu, Z., Holland, S., and Elson, C. O., 1994, Oral tolerance in humans, *J. Immunol.* **152:**4663–4670.

33. Crago, S. S., and Mestecky, J., 1983, Immunoinhibitory elements in human colostrum, *Surv. Immunol. Res.* **2:**164–169.

34. Streilein, J. W., Stein-Streilein, J., and Head, J., 1986, Regional specialization in antigen presentation, in: *The Reticuloendothelial System* (S. M. Phillips and M. R. Escobar, eds.), Plenum Press, New York, pp. 37–93.

35. Moldoveanu, Z., Russell, M. W., Wu, H. Y., Huang, W.-Q., Compans, R. W., and Mestecky, J., 1995, Compartmentalization within the common mucosal immune system, in: *Advances in Mucosal Immunology* (J. Mestecky, ed.), Plenum Press, New York, pp. 97–101.

36. Lifson, J. D., Nowak, M. A., Goldstein, S., Rossio, J. L., Kinter, A., Vasquez, G., Wiltrout, T. A., Brown, C., Schneider, D., Wahl, L., Lloyd, A. L., Williams, J., Elkins, W. R., Fauci, A. S., and Hirsch, V. M., 1997, The extent of early viral replication is a critical determinant of the natural history of simian immunodeficiency virus infection, *J. Virol.* **71:**9508–9514.

37. Meltzer, M. S., Skillman, D. R., Gomatos, P. J., Kalter, D. C., and Gendelman, H. E., 1990, Role of mononuclear phagocytes in the pathogenesis of human immunodeficiency virus infection, *Annu. Rev. Immunol.* **8:**169–194.

38. McDermott, M. R., Befus, A. D., and Bienenstock, J., 1983, The structural basis of immunity in the respiratory tract, in: *International Review of Experimental Pathology* (G. W. Richter and M. A. Epstein, eds.), Academic Press, New York, pp. 47–112.
39. Gendelman, H. E., Orenstein, J. M., Martin, M. A., Ferrua, C., Mitra, R. T., Phipps, T., Wahl, L. A., Lane, H. C., Fauci, A. S., Burke, D. S., Skillman, D., and Meltzer, M. S., 1988, Efficient isolation and propagation of human immunodeficiency virus on recombinant colony stimulating factor 1-treated monocytes, *J. Exp. Med.* **167:**1428–1434.
40. Orenstein, J. M., Meltzer, M. S., Phipps, T., and Gendelman, H. E., 1988, Cytoplasmic assembly and accumulation of human immunodeficiency virus types 1 and 2 in recombinant human colony stimulating factor-1 treated human monocytes: An ultrastructural study, *J. Virol.* **62:**2578–2585.
41. Meltzer, M. S., Hoover, D. L., Finbloom, D. S., Turpin, J. A., Kalter, C., Friedman, R. M., Moyer, M. P., Nara, P., and Gendelman, H. E., 1993, Mononuclear phagocytes in the pathogenesis of human immunodeficiency virus disease, in: *Mononuclear Phagocytes in Cell Biology* (G. Lopez-Berestein and J. Klostergaard, eds.), CRC Press, Boca Raton, Florida, pp. 147–175.
42. Adams, D. O., and Johnson, S. P., 1992, Molecular bases of macrophage activation: Regulation of class II MHC genes in tissue macrophages, in: *Mononuclear Phagocytes. Biology of Monocytes and Macrophages* (R. vanFurth, ed.), Kluwer, Dordrecht, The Netherlands, pp. 425–436.
43. Winslow, B. J., Pomerantz, R. J., Bagasra, O., and Trono, D., 1993, HIV-1 latency due to the site of proviral integration, *Virology* **196:**849–854.
44. Chun, T. W., Finzi, D., Margolick, J., Chadwick, K., Schwartz, D., and Siliciano, R. F., 1996, *In vivo* fate of HIV-1-infected T cells: Quantitative analysis of the transition to stable latency, *Nature Med.* **1:**1284–1290.
45. Montefiori, D. C., Cornell, R. J., Zhou, J. Y., Zhou, J. T., Hirsch, V. M., and Johnson, P. R., 1994, Complement control proteins, CD46, CD55, and CD59, as common surface constituents of human and simian immunodeficiency viruses and possible targets for vaccine protection, *Virology* **205:**82–92.
46. Tremblay, M., Meloche, S., Sekaly, R.-P., and Wainberg, M. A., 1990, Complement receptor 2 mediates enhancement of human immunodeficiency virus 1 infection in Epstein–Barr virus-carrying B cells, *J. Exp. Med.* **171:**1791–1796.
47. Rizzuto, C. D., and Sodroski, J. G., 1997, Contribution of virion ICAM-I to human immunodeficiency virus infectivity and sensitivity to neutralization, *J. Virol.* **71:**4847–4851.
48. Brighty, D. W., Rosenberg, M., Chen, I. S. Y., and Ivey-Hoyle, M., 1991, Envelope proteins from clinical isolates of human immunodeficiency virus type 1 that are refractory to neutralization by soluble CD4 possess high affinity for the CD4 receptor, *Proc. Natl. Acad. Sci. USA* **88:**7802–7805.
49. Horak, I. D., Popovic, M., Horak, E. M., Lucas, P. J., Gress, R. E., June, C. H., and Bolen, J. B., 1990, No T-cell protein kinase signalling or calcium mobilization after CD4 association with HIV-1 or HIV-1 gp120, *Nature* **348:**557–560.
50. Cefai, D., Debre, P., Kaczorek, M., Idziorek, T., Autran, B., and Bismuth, G., 1990, Human immunodeficiency virus-1 glycoproteins gp120 and gp160 specifically inhibit the CD3/T cell-antigen receptor phosphoinositide transduction pathway, *J. Clin. Invest.* **86:**2117–2124.
51. Collman, R., Godfrey, B., Cutilli, J., Rhodes, A., Hassan, N. F., Sweet, R., Douglas, S. D., Friedman, H., Nathanson, N., and Gonzalez-Scarano, F., 1990, Macrophage-tropic strains of human immunodeficiency virus type 1 utilize the CD4 receptor, *J. Virol.* **64:**4468–4476.
52. Stamatatos, L., Werner, A., and Cheng-Meyer, C., 1994, Differential regulation of cellular tropism and sensitivity to soluble CD4 neutralization by the envelope gp120 of human immunodeficiency virus type 1, *J. Virol* **68:**4973–4979.

53. Sonza, S., Maerz, A., Uren, S., Violo, A., Hunter, S., Boyle, W., and Crowe, S., 1995, Susceptibility of human monocytes to HIV type 1 infection *in vitro* is not dependent on their level of CD4 expression, *AIDS Res. Hum. Retroviruses* **11:**769–776.

54. de Maria, A., Pantaleo, G., Schnittman, S. M., Greenhouse, J. J., Baseler, M., Orenstein, J. M., and Fauci, A. S., 1991, Infection of CD8[+] T lymphocytes with HIV: Requirement for interaction with infected CD4[+] cells and induction of infectious virus from chronically infected CD8[+] cells, *J. Immunol.* **146:**2220–2226.

55. Lusso, P., Lori, F., and Gallo, R. C., 1990, CD4-independent infection by human immunodeficiency virus type 1 after phenotypic mixing with human T-cell leukemia viruses, *J. Virol.* **64:**6341–6344.

56. Feng, Y., Broder, C. C., Kennedy, P. E., and Berger, E. A., 1996, HIV-1 entry cofactor: Functional cDNA cloning of a seven-transmembrane, G protein-coupled receptor, *Science* **272:** 872–877.

57. Cocchi, F., DeVico, A. L., Garzino-Demo, A., Arya, S. K., Gallo, R. C., and Lusso, P., 1995, Identification of RANTES, MIP-1α, and MIP-1β as the major HIV suppressive factors produced by CD8[+] T cells, *Science* **270:**1811–1815.

58. Bristow, C. L., 1996, HIV fusion, *Science* **273:**1642–1643.

59. Mackewicz, C. E., Barker, E., and Levy, J. A., 1996, Role of β-chemokines in suppressing HIV replication, *Science* **274:**1393–1394.

60. Moriuchi, H., Moriuchi, M., and Fauci, A. S., 1997, Nuclear factor-kB potently up-regulates the promoter activity of RANTES, a chemokine that blocks HIV infection, *J. Immunol.* **158:**3483–3491.

61. Shaw, G. M., Harper, M. E., Hahn, B. H., Epstein, L. G., Gajdusek, D. C., Price, R. W., Navia, B. A., Petito, C. K., O'Hara, C. J., Groopman, J. E., *et al.*, 1985, HTLV-III infection in brains of children and adults with AIDS encephalopathy, *Science* **227:**177–182.

62. Cohen, O. J., Pantaleo, G., Walker, R., Ognibene, F. P., Shelhamer, J. H., and Fauci, A. S., 1993, Comparative analysis of HIV burden in mononuclear cells (MC) from peripheral blood (PB) versus bronchoalveolar lavage (BAL) of the same individuals, *Int. Conf. AIDS* **9:**62.

63. Lebargy, F., Branellec, A., Deforges, L., Bignon, J., and Bernaudin, J. F., 1994, HIV-1 in human alveolar macrophages from infected patients is latent *in vivo* but replicates after *in vitro* stimulation, *Am. J. Respir. Cell Mol. Biol.* **10:**72–78.

64. Stanley, S. K., Kessler, S. W., Justement, J. S., Schnittman, S. M., Greenshouse, J. J., Brown, C. C., Musongela, L., Musey, K., Kapita, B., and Fauci, A. S., 1992, CD34[+] bone marrow cells are infected with HIV in a subset of seropositive individuals, *J. Immunol.* **149:**689–697.

65. Rappersberger, K., Gartner, S., Schenk, P., Stingl, G., Groh, V., Tschachler, E., Mann, D. L., Wolff, K., Konrad, K., and Popovic, M., 1988, Langerhans' cells are an actual site of HIV-1 replication, *Intervirology* **29:**185–194.

66. Fauci, A. S., 1993, Multifactorial nature of human immunodeficiency virus disease: Implications for therapy, *Science* **262:**1011–1018.

67. Reimann, K. A., Tenner-Racz, K., Racz, P., Montefiori, D. C., Yasutomi, Y., Lin, W., Ransil, B. J., and Letvin, N. L., 1994, Immunopathogenic events in acute infection of rhesus monkeys with simian immunodeficiency virus of macaques, *J. Virol.* **68:**2362–2370.

68. Chakrabarti, L., Isola, P., Cumont, M.-C., Claessens-Maire, M. A., Hurtrel, M., Montagnier, L., and Hurtrel, B., 1994, Early stages of simian immunodeficiency virus infection in lymph nodes. Evidence for high viral load and successive populations of target cells, *Am. J. Pathol.* **144:**1226–1237.

69. Joling, P., Bakker, L. J., Van Strijp, J. A. G., Meerloo, T., de Graaf, L., Dekker, M. E., Goudsmit, J., Verhoef, J., and Schuurman, H.-J., 1993, Binding of human immunodeficiency virus type-1 to follicular dendritic cells *in vitro* is complement dependent, *J. Immunol.* **150:**1065–1073.

70. Klaus, G. G. B., and Humphrey, J. H., 1993, The fate of antigens, in: *Clinical Aspects of Immunology*, 5th ed. (P. J. Lachmann, K. Peters, F. S. Rosen, and M. J. Walport, eds.), Blackwell, Boston, pp. 107–126.
71. Hussey, R. E., Richardson, N. E., Kowalski, M., Brown, N. R., Chang, H.-C., Siliciano, R. F., Dorfman, T., Walker, B., Sodroski, J., and Reinherz, E. L., 1988, A soluble CD4 protein selectively inhibits HIV replication and syncytium formation, *Nature* **331:**78–81.
72. Deen, K. C., McDougal, S., Inacker, R., Folena-Wasserman, G., Arthos, J., Rosenberg, J., Maddon, P. J., Axel, R., and Sweet, R. W., 1988, A soluble form of CD4 (T4) protein inhibits AIDS virus infection, *Nature* **331:**82–84.
73. Fisher, R. A., Bertonis, J. M., Werner, M., Johnson, V. A., Costopoulos, D. S., Liu, T., and Tizard, R., 1988, HIV infection is blocked *in vitro* by recombinant soluble CD4, *Nature* **331:**76–78.
74. Cummins, L. M., Weinhold, K. J., Matthews, T. J., Langlois, A. J., Perno, C. F., Condie, R. M., and Allain, J.-P., 1991, Preparation and characterization of an intravenous solution of IgG from human immunodeficiency virus-seropositive donor, *Blood* **77:**1111–1117.
75. Harakeh, S., Jariwalla, R. J., and Paulling, L., 1990, Suppression of human immunodeficiency virus replication by ascorbate in chronically and acutely infected cells, *Proc. Natl. Acad. Sci. USA* **87:**7245–7249.

2

Potential Role of Human T-Cell Leukemia/Lymphoma Viruses (HTLV) in Diseases Other Than Acute T-Cell Leukemia/Lymphoma (ATL)

TERESA C. GENTILE and THOMAS P. LOUGHRAN

1. INTRODUCTION

Isolated in 1978 from patients with aggressive T cell malignancies with skin involvement,[1] human T-cell leukemia/lymphoma virus type I (HTLV-I) was the first retrovirus identified in humans. Association with adult T cell leukemia/lymphoma (ATLL) and a degenerative neurologic disease, tropical spastic paresis (TSP) or HTLV-I-associated myelopathy (HAM), was more recently described. Although most infected persons are asymptomatic, there is considerable evidence to definitively link HTLV-I with these two diseases.[2] In addition, multiple case reports associate HTLV-I with a variety of other diseases including hematologic, neurologic, dermatologic, and autoimmune/inflammatory disorders; however, direct evidence of a

TERESA C. GENTILE • Department of Medicine, SUNY Health Sciences Center, Syracuse, New York 13210. THOMAS P. LOUGHRAN • Program Leader, Hematological Malignancies, H. Lee Moffitt Cancer Center and Research Institute, Veterans Administration Hospital, and Departments of Medicine and Microbiology/Immunology, University of South Florida College of Medicine, Tampa, Florida 33612.

Human Retroviral Infections, edited by Kenneth E. Ugen *et al.*
Kluwer Academic / Plenum Publishers, New York, 2000.

TABLE I
Possible Disease Associations of HTLV-I

Disease	References	Disease	References
Peripheral neuropathy	21	Antithyroid antibodies	58, 59, 61
Myasthenia gravis	24, 25	Graves' disease	62
Multiple sclerosis	26, 27	Hashimoto's thyroiditis	63
Cutaneous T cell lymphoma	32–35	Infective dermatitis	65, 66
Arthropathy	48–50	Hypercalcemia	70, 71
Sjogren's syndrome	51–57		

causal relationship is lacking. Table I summarizes clinical entities reported in the literature possibly associated with HTLV-I.

In 1982, human T-cell leukemia/lymphoma virus type II (HTLV-II) was isolated from a patient thought to have a T cell variant of hairy cell leukemia.[3] Although another case of HTLV-II infection was subsequently identified in a patient who had the more common B cell hairy cell leukemia,[4] an additional 21 cases of hairy cell leukemia had no serologic evidence of HTLV-II infection.[5] Subsequently, infection with HTLV-II was noted to be common among intravenous drug users and endemic in certain North American tribes. Since initially described, HTLV-II has also been linked to a neurologic disorder similar to HTLV-I-associated HAM/TSP, several hematologic diseases, and dermatologic and autoimmune/inflammatory conditions. There is no definitive evidence, however, of a direct etiologic role for HTLV-II in any of these diseases. A review of the literature describing possible disease associations of HTLV-II is summarized in Table II.

2. NEUROLOGIC DISEASE

In 1985–1986, patients with TSP in Martinique and a similar myelopathy in southern Japan were independently noted to have antibodies against HTLV-I.[6,7] Comparative studies subsequently determined these diseases to be identical and the collective term HAM/TSP was adopted to describe this syndrome of progressive paresis. The estimated cumulative lifetime risk of developing HAM/TSP in persons infected with HTLV-I is reported to be 0.25%.[8] Clinical manifestations include lower extremity spasticity and weakness, urinary bladder dysfunction, constipation, impotence in males, sensorineural deafness, and other sensory disturbances produced by involvement of the pyramidal tracts and posterior columns.[7,9] The mean age of

TABLE II
Possible Disease Associations of HTLV-II

Disease	References	Disease	References
HAM/TSP-like syndrome	12–17	Antithyroid antibodies	60
Peripheral neuropathy	22	Graves/Hashimoto's	64
Hairy cell leukemia	3	thyroiditis	
LGL leukemia	38, 39	Acquired ichthyosis	67
Atypical CLL	41	Infiltrative dermatitis	68
Prolymphocytic leukemia	43	Chronic fatigue syndrome	69
Thrombotic microangiopathy	44, 45	Hypercalcemia	70, 71

onset is 40–50 years. Women develop HAM/TSP approximately twice as often as males, perhaps reflecting the excess of female seropositives, which may be due to the more efficient sexual transmission from male to female.[10,11] Laboratory evaluation demonstrates HTLV-I antibodies in the serum as well as the cerebrospinal fluid (CSF) of these patients. In addition, the CSF may also contain increased protein and lymphocyte pleocytosis.[7]

A similar syndrome of chronic progressive myelopathy characterized by spastic gait and bladder dysfunction has been reported in patients infected with HTLV-II.[12–17] Although the first cases were reported in patients dually infected with human immunodeficiency virus type I (HIV-I),[18,19] several subsequent case reports[12–16] and a recently published series of four patients by Lehky *et al.*[17] document sole infection with HTLV-II. Clinical manifestations include a progressive paraparesis with bladder and bowel dysfunction and variable sensory involvement. Patients with HTLV-II-associated myelopathy appear to have a more rapid progression of disease[17] compared with the time course usually observed with HAM/TSP.[20] Laboratory findings include HTLV-II seropositivity in both serum and CSF as well as a mild CSF pleocytosis.[17] The reports documenting an association between HTLV-II infection and a progressive myelopathy are increasing; however, compared with HTLV-I infected individuals, the incidence of CNS disease in HTLV-II infected persons appears to be less.[8] Further study is necessary to establish a definite etiologic role for HTLV-II in this syndrome of chronic progressive myelopathy.

Although HAM/TSP and a similar myopathy in HTLV-II-infected individuals have been most extensively studied, the spectrum of neurologic diseases associated with HTLV-I and HTLV-II may in fact be quite diverse as suggested by an increasing number of reports describing peripheral neuropathy,[21,22] myasthenia gravis,[23–25] and multiple sclerosis[26,27] in HTLV-I- and HTLV-II-infected individuals. Douen *et al.*[21] recently reported a case study of patients developing peripheral neuropathy and myositis in the absence of spastic paresis. HTLV-I seropositivity was sufficiently high in these patients to suggest that patients with neuropathies of unknown origin from areas endemic for HTLV-I should be screened for the retrovirus. All patients in this report were HIV-negative. Peripheral neuropathy has also been reported in patients coinfected with HTLV-II and HIV. Zehender *et al.*[22] compared a group of HIV-1-infected individuals with predominantly sensory neuropathy to a control group of HIV-1-infected individuals without neuropathy and found a significantly higher prevalence of HTLV-II antibodies in the patients with neuropathy. In addition, polymerase chain reaction (PCR) analysis of femoral nerve tissue from one patient demonstrated HTLV-II proviral sequences, suggesting a possible role for HTLV-II in the pathogenesis of peripheral neuropathy.

There have been two case reports of patients seropositive for HTLV-I developing myasthenia gravis. Fukui *et al.*[24] described a patient with HAM and myasthenia gravis, and Ijichi *et al.*[25] reported a carrier of HTLV-I with myasthenia gravis who subsequently developed acute transverse myelitis. These authors suggest an association between HTLV-I and myasthenia gravis and postulate that modulation of cellular immunity by HTLV-I may be involved in the pathogenesis of this disorder. While this is an interesting hypothesis, additional data need to be presented before a causal relationship can be established. To our knowledge, no association between myasthenia gravis and HTLV-II has been reported.

The clinical syndrome typical of HAM/TSP has many similarities to the progressive spinal form of multiple sclerosis (MS), including inflammation, demyelination, probable cell-mediated pathogenesis, and cerebral white matter pathology. On magnetic resonance imaging (MRI), periventricular lesions and isolated white matter lesions are seen in both TSP and MS.[28] These observations suggest a clinical and radiologic overlap between these two diseases which has led investigators to hypothesize a common etiologic agent. A few authors have attempted to identify an association between HTLV-I and MS with inconclusive results.[26,27] Not enough data exist to draw definitive conclusions concerning other potential neurologic sequelae resulting from infection with HTLV-I/II.

3. HEMATOLOGIC DISEASE

The pathogenic role of HTLV-I in ATL has been firmly established. Based on studies from Japan, the lifetime risk of developing this disease among carriers of HTLV-I is estimated to be 2–4%.[29] A spectrum of clinical and pathologic features have been described with this disease.[30,31] Clinical manifestations of acute ATL include lymphadenopathy, hepatospleno-megaly, and leukemic infiltration of skin. Hypercalcemia, lytic bone lesions, and elevated levels of serum bilirubin and lactic dehydrogenase are common. Peripheral blood leukocytosis with abnormal multilobulated lymphocytes is seen on peripheral blood smear. Median survival from time of diagnosis is 11 months. A chronic, less symptomatic form of ATL is called smoldering ATL. Diffuse lymphadenopathy with fewer circulating leukemic cells is typical of this form. It usually follows a more indolent course, unless it progresses to the acute form. Finally, a preleukemic state has been described in asymptomatic patients with oligoclonally integrated HTLV-I. About half will eventually develop smoldering or acute ATL while the other half will remain in the carrier state.

In addition, infection with HTLV-I and/or HTLV-II has been reported in patients with a variety of other lymphoproliferative disorders; however, definitive evidence of a causal relationship is lacking. An association between cutaneous T-cell lymphoma (CTCL) and HTLV-I and HTLV-II has long been suspected. CTCL involving malignant CD4[+] cells (mycosis fungoides and Sezary syndrome) has clinical and histopathologic features similar to those of ATL; however, only a minor fraction of patients with this disease have antibodies to HTLV-I. Deleted variants of HTLV-I have been described in patients with CTCL.[32–34] Some investigators have reported HTLV-like particles in the peripheral blood mononuclear cells (PBMC) of such patients, which has led to a continued effort at providing evidence to link HTLV-I to this disease.[35] CTCL of the CD8[+] phenotype have also been described.[36,37] There has been no association of CTCL of CD8[+] lymphocytes with HTLV-II. Although multiple observations have suggested a possible role for HTLV-II in this disease, it remains speculation.

As mentioned previously, the first isolate of HTLV-II was from a patient with hairy cell leukemia.[3] Further studies, however, failed to confirm HTLV-II as the cause of this disease. Similarly, CD3[+] large granular lymphocyte (LGL) leukemia in association with HTLV infection has also been reported.[38,39] In one report,[38] serum from only one patient was confirmed positive for HTLV-II. Frequent reactivity of patient sera to HTLV I/II *gag* protein p24 and to *env* protein p21e was noted, however, and led these authors to suggest that a deleted or variant form of HTLV I/II may be associated with LGL leukemia. Heneine *et al.*[40] screened 51 patients with LGL leukemia and also found only 1 patient who was confirmed positive for HTLV-II using conventional criteria. These observations suggest that prototypical HTLV infection is uncommon in LGL leukemia, but do not exclude the possibility of infection with a retrovirus related to HTLV I/II.

Infection with HTLV-I/II has also been reported in isolated cases of atypical chronic lymphocytic leukemia,[41] prolymphocytic leukemia,[42] pancytopenia,[43] thrombotic microangiopathy,[44,45] and pure red cell aplasia.[46] In summary, HTLV-I has been definitively linked to ATL, while its role in other lymphoproliferative disorders is not well established. HTLV-II has not been definitively linked to any hematologic disease. Although it has been reported in association with a variety of hematologic disorders, its role in the pathogenesis of these disorders has not been clearly established.

4. AUTOIMMUNE/INFLAMMATORY

The etiology of rheumatologic diseases such as rheumatoid arthritis, polymyositis, Sjogren's syndrome, thyroiditis, and systemic lupus erythromatosis has yet to be established, but appears to be multifactorial. Retroviruses have been implicated to play a role in some of these disorders.[47] HTLV-I-associated arthropathy has been described in infected individuals.[48] Evidence for direct HTLV-I involvement is supported by detection of HTLV-I antibodies in the synovial fluids of the affected joints and demonstration by PCR of HTLV-I proviral DNA in synovial tissue and synovial fluid lymphocytes.[49] Guerin *et al.*[50] retrospectively reviewed 17 cases of polyarthropathy in HTLV-I-infected patients in Martinique. All patients demonstrated peripheral, bilateral, symmetric polyarthritis, with the hands and knees being the most commonly involved sites. Approximately one half of these patients demonstrated other serologic abnormalities including positive rheumatoid factor and anti-nuclear antibodies. These observations suggest that HTLV-I may disrupt the immune system and lead to the development of a variety of rheumatologic disorders. There are no reported cases of HTLV-II-associated arthropathy.

Sjogren's syndrome is a disorder causing xerostomia or keratoconjunctivitis sicca or both. Although an autoimmune etiology has been suggested, the pathogenesis of Sjogren's syndrome remains unclear. An association between HTLV-I and Sjogren's syndrome was first suggested by Vernant *et al.*[51] when it was noted that patients with HAM/TSP commonly had clinically or histologically confirmed Sjogren's syndrome. Eguchi *et al.*[52] later showed a high seroprevalence of HTLV-I among patients with Sjogren's syndrome in an endemic area, and salivary IgA antibodies to HTLV-I are commonly detected among HTLV-I-seropositive patients with Sjogren's syndrome.[53,54] In addition, expression of Tax protein, a transactivator encoded by HTLV-I, in transgenic mice produces an exocrinopathy affecting lacrimal and salivary glands similar to Sjogren's syndrome in humans.[55] Recently, A-type retroviral particles have been observed in cocultured cell lines as well as in epithelial cells of salivary glands from patients with Sjogren's syndrome by transmission electron microscopy.[56,57] Specific immunohistologic staining for gene products of HTLV-I (p19 and p28 proteins) has been demonstrated in the epithelium of the salivary ducts of one patient.[58] Taken together, these observations provide substantial indirect evidence implicating a retrovirus in the pathogenesis of Sjogren's syndrome. Whether this is prototypical HTLV-I or another retrovirus is yet to be determined.

A number of investigators have reported serologic abnormalities of the thyroid in association with HTLV-I and HTLV-II. Mine *et al.*[59] examined a group of blood donors for antithyroid antibodies and anti-HTLV-I antibodies and found a high frequency of HTLV-I seropositivity in blood donors with antithyroid antibodies. Other investigators have looked at patients with autoimmune thyroid disease and found a high prevalence of HTLV-I carriers[60] and HTLV-II-related genes in DNA from peripheral blood leukocytes[61] in these patients. Akamine *et al.*[62] looked for antithyroid and antithyroglobulin antibodies in the sera of patients with ATL, carriers of HTLV-I, and healthy controls. They found a statistically significant difference between these groups, with positivity for thyroid autoantibodies of 40.4% in patients with ATL, 30% in HTLV-I carriers, and 13.7% in healthy controls. Finally, HTLV-I has been described in association with specific autoimmune thyroid disorders. Kawai *et al.*[63] described three carriers of HTLV-I who also developed Graves' disease. All had uveitis and one had chronic arthropathy. This same group also detected virus envelope protein and messenger RNA for HTLV-I in many of the follicular epithelial cells of the thyroid tissue from a patient with Hashimoto's thyroiditis.[64] Yokai *et al.*[65] detected HTLV-II proviral elements in 52% of patients with Hashimoto's thyroiditis and 12% of patients with Graves' disease, compared with 1% of healthy individuals. Infection with HTLV-II (by classic criteria), however, could not be confirmed, as none of the patients were HTLV-II-seropositive. More information is needed to determine the significance of these observations.

5. DERMATOLOGIC DISEASE

HTLV-I and HTLV-II infection have both been reported in association with dermatologic diseases distinct from cutaneous T-cell lymphomas. HTLV-I infection has been associated with infective dermatitis in Jamaican children.[66] Further studies in these children have demonstrated abnormalities of immune function and some developed other HTLV-I-associated disorders. In a review of childhood dermatitis in the tropics, LaGrenade *et al.*[67] suggested that infective dermatitis is a marker for infection with HTLV-I.

Kaplan *et al.*[68] examined the relationship between acquired ichthyosis and concomitant HIV-1 and HTLV-II infection in intravenous drug users. They noted occurrence of the dermatitis only after profound T-cell helper depletion and noted a higher incidence with dual infection with HIV-1 and HTLV-II (22.2%) compared with patients singly infected with HIV-1 (6.8%). This observation led to the suggestion that acquired ichthyosis may be a marker of concomitant infection of HIV-1 and HTLV-II in intravenous drug users. Another disorder consisting of an unusual infiltrative dermatitis with dermatolymphatic lymphadenopathy and eosinophilia in IV drug users dually infected with HIV-1 and HTLV-II was described by the same group of investigators.[69] Since this syndrome had not previously been seen in patients with HIV-1 infection, it was theorized that HTLV-II infection may play a role.

6. OTHER REPORTED DISEASE ASSOCIATIONS

A small number of patients with chronic fatigue syndrome were evaluated for evidence of HTLV-II infection.[70] Serologic and molecular evidence of HTLV-II infection was found in one half of patients with chronic fatigue syndrome. Since an association of chronic fatigue syndrome with many other infectious agents has been suggested, the exact etiology and pathogenesis remain unclear.

HTLV-I and HTLV-II Tax proteins have been shown to transactivate viral and cellular gene expression. Ejima *et al.*[71,72] demonstrated that the parathyroid hormone-related protein promoter was responsive to HTLV-I/II Tax and may contribute to the pathogenesis of hypercalcemia in ATL.

REFERENCES

1. Poiesz, B. J., Ruscetti, F. W., Gazdar, A. F., Bunn, P. A., Minna, J. D., and Gallo, R. C., 1980, Detection and isolation of type C retrovirus particles from fresh and cultured lymphocytes of a patient with cutaneous T-cell lymphoma, *Proc. Natl. Acad. Sci. USA* **77:**7415–7419.
2. Manns, A., and Blattner, W. A., 1991, The epidemiology of human T cell lymphotrophic virus type I and type II: Etiologic role in human disease, *Transfusion* **31:**67–75.
3. Kalyanaraman, V. S., Sarngadharan, M. G., and Robert-Guroff, M., 1982, A new subtype of human T cell leukemia virus (HTLV II) associated with a T cell variant of hairy cell leukemia, *Science* **218:**571–573.
4. Rosenblat, J. D., Golde, D. W., and Wachsman, W., 1986, A second isolate of HTLV II associated with atypical hairy cell leukemia, *N. Engl. J. Med.* **315:**372–377.
5. Rosenblatt, J. D., Gasson, J. C., and Glaspy, J., 1987, Relationship between human T cell leukemia virus II and atypical hairy cell leukemia: A serologic study of hairy cell leukemia patients, *Leukemia* **1:**397–401.
6. Gessian, A., Barin, F., Vernant, J. C., Gout, O., Maurs, L., and Calender, A. de The G., 1985, Antibodies to human T-lymphotropic virus type-I in patients with tropical spastic paraparesis, *Lancet* **2:**407–410.
7. Osame, M., Usuku, K., Izumo, S., Ijichi, N., Amitani, H., Igata, A., Matsumoto, M., and Tara, M., 1986, HTLV-I associated myelopathy, a new clinical entity, *Lancet* **1:**1031–1032.

8. Kaplan, J. E., Osame, M., Kubota, H., Igata, A., Nishitani, H., Maeda, Y., Khabbaz, R. F., and Janssen, R. S., 1990, The risk of development of HTLV-I-associated myelopathy/tropic spastic paraparesis among persons infected with HTLV-I, *J. Acquired Immune Defic. Syndr.* **3:**1096–1101.
9. Roman, G., 1987, Retrovirus-associated myelopathies, *Arch. Neurol.* **44:**659–663.
10. Manns, A., and Blattner, W. A., 1991, The epidemiology of the human T-cell lymphotropic virus type I and type II: Etiologic role in human diseases, *Transfusion* **31:**67–75.
11. Murphy, E. L., Wilks, R., Morgan, O. S., Hanchard, B., Cranston, B., Figueroa, J. P., Gibbs, W. N., Murphy, J., and Blattner, W. A., 1989, Modeling the risk of adult T-cell leukemia/lymphoma (ATLL) in persons infected with human T-lymphotropic virus type-I, *Int. J. Cancer* **43:**250–253.
12. Hjelle, B., Appenzeller, O., Mills, R., Alexander, S., Torrez-Martinez, N., Jahnke, R., and Ross, G., 1992, Chronic neurodegenerative disease associated with HTLV-II infection, *Lancet* **339:**645–646.
13. Harrington, W. J., Jr., Sheremata, W., Hjelle, B., Dube, D. K., Bradshaw, P., Foung, S. K., Snodgrass, S., Toedter, G., Cabral, L., and Poiesz, B., 1993, Spastic ataxia associated with human T-cell lymphotropic virus type II infection, *Ann. Neurol.* **33:**411–414.
14. Sheremata, W. A., Harrington, W. J., Bradshaw, P. A., Foung, S. K., Raffanti, S. P., Berger, J. R., Snodgrass, S., Resnick, L., and Poiesz, B. J., 1993, Association of "(tropical) ataxic neuropathy" with HTLV-II, *Virus Res.* **29:**71–77.
15. Jacobson, S., Lehky, T., Nishimura, M, Robinson, S., McFarlin, D. E., and Dhib-Jalbut, S., 1993, Isolation of HTLV-II from a patient with chronic progressive neurological disease clinically indistinguishable from HTLV I. Associated myelopathy/trophic spastic paraparesis, *Ann. Neurol.* **33:**392–396.
16. Murphy, E. L., Engstrom, J. W., Miller, K., Sacher, R. A., Busch, M. P., and Hollingsworth, C. G., 1993, HTLV-II associated myelopathy in 43-year-old woman, *Lancet* **341:**757–758.
17. Lehky, T. J., Flerlage, N., Katz, D., Houff, S., Hall, W. S., Ishii, K., Monekn, C., Dhib-Jalbut, S., McFarland, H. F., and Jacobsen, S., 1996, Human T-cell lymphotrophic virus type II-associated myelopathy: Clinical and immunologic profiles, *Ann. Neurol.* **40:**714–723.
18. Berger, J. R., Svenningsson, A., Raffanti, S., and Resnick, L., 1991, Tropical spastic paraparesis-like illness occurring in a patient dually infected with HIV-1 and HTLV-II, *Neurology* **41:**85–87.
19. Rosenblatt, J. D., Tomkins, P., Rosenthal, M., Kacena, A., Chan, G., Valerama, R., Harrington, W., Jr., Saxton, E., Diagne, A., and Zhao, J. Q., 1992, Progressive spastic myelopathy in a patient coinfected with HIV-1 and HTLV-II: Autoantibodies to the human homologue of rig in blood and cerebrospinal fluid, *AIDS* **6:**1151–1158.
20. Gessain, A., and Gout, O., 1992, Chronic myelopathy associated with human T-lymphotropic virus type-1 (HTLV-1), *Ann. Intern. Med.* **117:**933–946.
21. Douen, A. G., Pringle, C. E., and Guberman, A., 1997, Human T-cell lymphotropic virus type 1 myositis, peripheral neuropathy, and cerebral white matter lesions in the absence of spastic paraparesis, *Arch. Neurol.* **54:**896–900.
22. Zehender, G., DeMaddalena, C., Osio, M., Cavalli, B., Parravicini, C., Moroni, M., and Galli, M., 1995, High prevalence of human T-cell lymphotropic virus type II infection in patients affected by immunodeficiency virus type 1-associated predominantly sensory polyneuropathy, *J. Infect. Dis.* **172:**1595–1598.
23. Verma, A., and Berger, J., 1995, Myasthenia gravis associated with dural infection of HIV and HTLV-I, *Muscle Nerve* **18:**1355–1356.
24. Fukui, I., Sugita, K., Ichikawa, H., Negishi, A., Kasai, H., and Tsukagoshi, H., 1994, Human T lymphotropic virus type I associated myelopathy and myasthenia gravis: A possible association? *Eur. Neurol.* **34:**158–161.

25. Ijichi, T., Adachi, Y., Nishio, A., Kanaitsuka, T., Ohtomo, T., and Nakamura, M., 1995, Myasthenia gravis, acute transverse myelitis, and HTLV-I, *J. Neurol. Sci.* **133**:194–196.

26. Garcia, F., Castillo, L. C., Larreategui, M., Roberts, B., Cedeno, V., Heneine, W., Blattner, W., Kaplan, J. E., and Levine, P. H., 1995, Relation between human T-lymphotropic virus type I and neurologic diseases in Panama: 1985–1990, *J. Acquired Immune Defic. Syndr. Hum. Retrovirol.* **10**:192–197.

27. Lisby, G., 1993, Search for an HTLV-I like retrovirus in patients with MS by enzymatic DNA amplification, *Acta Neurol. Scand.* **88**:385–387.

28. Bhagavati, S., Ehrlich, G., Kula, R., Kowk, S., Sninsky, J., Udani, V., and Poiesz, B. J., 1988, Detection of human T-cell lymphoma/leukemia virus Type-I DNA and antigen in spinal fluid and blood of patients with chronic progressive myelopathy, *N. Engl. J. Med.* **318**:1141–1147.

29. Kaplan, J. E., and Khabbaz, R. F., 1993, The epidemiology of human T-lymphotropic virus types I and II, *Rev. Med. Virol.* **3**:137–148.

30. Bunn, P. A., Jr., Schechter, G. P., Jaffe, E., Blayney, D., Young, R. C., Matthews, M. J., Blattner, W., Broder, S., Robert-Guroff, M., and Gallo, R. C., 1983, Clinical course of retrovirus-associated adult T-cell lymphoma in the United States, *N. Engl. J. Med.* **309**:257–264.

31. Kawano, F., Yamaguchi, K., Nishimura, H., Tsuda, H., and Takatsuki, K., 1985, Variation in the clinical course of adult T-cell leukemia, *Cancer* **55**:851–856.

32. Hall, W. W., Liu, C. R., Schneewind, O., Takahashi, H., Kaplan, M. H., Roupte, G., and Vahlne, A., 1991, Deleted HTLV-I provirus in blood and cutaneous lesions of patients with mycosis fungoides, *Science* **253**:317–320.

33. Zucker-Franklin, D., Coutavas, E. E., Rush, M. G., and Zuozias, D. C., 1991, Detection of human T-lymphotropic virus-like particles in cultures of peripheral blood lymphocytes from patients with mycosis fungoids, *Proc. Natl. Acad. Sci. USA* **88**:7630–7634.

34. Manca, N., Piacentini, E., Gelmi, M., Calzavara, P., Manganoni, M. A., Glukhov, A., Gargiulo, F., DeFrancesco, M., Pirali, F., DePanfilis, G., and Turano, A., 1994, Persistence of human T-cell lymphotropic virus type I (HTLV-I) sequences in peripheral blood mononuclear cells from patients with mycosis fungoides, *J. Exp. Med.* **180**:1973–1978.

35. Zucker-Franklin, D., and Pancake, B. A., 1994, The role of human T-cell lymphotropic viruses (HTLV-I and II) in cutaneous T-cell lymphomas, *Semin. Dermatol.* **13**:160–165.

36. Fujiwara, Y., Abe, Y., Kuyama, M., Arata, Y., Yoshino, T., Akagi, T., and Miyoshi, K., 1990, CD8$^+$ cutaneous T-cell lymphoma pagetoid epidermotropism and angiocentric and angiodestructive infiltration, *Arch. Dermatol.* **126**:801.

37. Agnarsson, B. A., Vonderheid, E. C., and Kadin, M. E., 1990, Cutaneous T cell lymphoma suppressor/cytotoxic (CD8) phenotype: Identification of rapidly progressive and chronic subtypes, *J. Am. Acad. Dermatol.* **22**:569.

38. Loughran, T. P., Jr., Sherman, M. P., Ruscetti, F. W., Frey, S., Coyle, T., Montagna, R. A., Jones, B., Starkebaum, G., and Poiesz, B. J., 1994, Prototypical HTLV-I/II infection is rare in LGL leukemia, *Leuk. Res.* **18**:423–429.

39. Loughran, T. P., Jr., Coyle, T., Sherman, M. P., Starkebaum, G., Ehrlich, G. D., Ruscetti, F. W., and Poiesz, B. J., 1992, Detection of human T-cell leukemia/lymphoma virus, type II, in a patient with large granular lymphocyte leukemia, *Blood* **80**:1116–1119.

40. Heneine, W., Chan, W., Lust, J., Sinha, S. D., Zaki, S. R., Khabbaz, R. F., and Kaplan, J. E., 1994, HTLV-II infection rare in patients with large granular lymphocyte leukemia, *J. Acquired Immune Defic. Syndr.* **7**:736–737.

41. Sohn, C. C., Blayne, D. W., Missett, J. L., Mathe, G., Flandrin, G., Moran, E. M., Jensen, F. C., Winberg, C. D., and Rappaport, H., 1986, Leukopenic chronic T-cell leukemia mimicking hairy-cell leukemia: Association with human retroviruses, *Blood* **67**:949–956.

42. Cervantes, J., Hussain, S., Jensen, F., and Schwartz, J. M., 1986, T-prolymphocytic leukemia associated with human T-cell lymphotropic virus II, *Clin. Res.* **34:**454A [Abstract].
43. Kalyanaraman, V. S., Narayanan, P., Feorino, P., Ramsey, R. B., Palmer, E. L., Chorba, T., McDougal, S., Getchell, J. P., Holloway, B., and Harrison, A. K., 1985, Isolation and characterization of a human T-cell leukemia virus type II from a hemophilia-A patient with pancytopenia, *EMBO J.* **4:**1455.
44. Ucar, A., Fernandez, H. F., Byrnes, J. J., Lian, E. C., and Harrington, W. J., Jr., 1994, Thrombotic microangiopathy and retroviral infections: A 13-year experience, *Am. J. Hematol.* **45:**304–309.
45. Dixon, A. C., Kwock, D. W., Nakamura, J. M., Yanagihara, E. T., Saiki, S. M., Bodner, A. J., and Alexander, S. S., 1989, Thrombotic thrombocytopenic purpura and human T-lymphotrophic virus, type I (HTLV-I), *Ann. Intern. Med.* **110:**93.
46. Levitt, L. J., Reyes, G. R., Moonka, D. K., Bensch, K., Miller, R. A., and Engleman, E. G., 1988, Human T cell leukemia virus-1-associated T-suppressor cell inhibition of erythropoiesis in a patient with pure red cell aplasia and chronic Tγ-lymphoproliferative disease, *J. Clin. Invest.* **81:**538.
47. Krieg, A. M., and Steinberg, A. D., 1990, Retroviruses and autoimmunity, *J. Autoimmun.* **3:**137–166.
48. Nishioka, K., Marnyanna, I., Sato, K., Kitajima, I., Nakajima, Y., and Osane, M., 1989, Chronic inflammatory arthropathy associated with HTLV I, *Lancet* **1:**441.
49. Kitajima, I., Yamamoto, K., Sato, K., Nakajima, Y., Nakajima, T., Maruyami, I., Osame, M., and Nishioka, K., 1991, Detection of human TCLVI proviral DNA and its gene expression in synovial cells in chronic inflammatory arthropathy, *J. Clin. Invest.* **88:**1315–1322.
50. Guerin, B., Arfi, S., Numeric, P., Jean-Baptiste, G., LeParc, J. M., Smadja, D., and Grollier-Bois, L., 1995, Polyarthritis in HTLV-1-infected patients. A review of 17 cases, *Rev. Rhum. Ed. Fr.* **62:**21–28.
51. Vernant, J. C., Buisson, G., Magdelein, J., DeThore, J., Jouannelle, A., Neisson-Vernant, C., and Monplaisir, N., 1988, T-lymphocyte alveolitis tropical spastic-paresis and Sjogren's syndrome, *Lancet* **1:**177.
52. Eguchi, K., Matsuoka, N., Ida, H., Nakashima, M., Sakai, M., Sakito, S., Kawakami, A., Terada, K., Shimada, H., and Kawabe, Y., 1992, Primary Sjogren's syndrome with antibodies to HTLV-1: Clinical and laboratory features, *Ann. Rheum. Dis.* **51:**769–776.
53. Terada, K., Katamine, S., Eguchi, K., Moriuchi, R., Kita, M., Shimada, H., Yamashita, I., Iwata, K., Tsuji, Y., Nagataki, S., and Miyamoto, T., 1994, Prevalence of serum and salivary antibodies to HTLV-1 in Sjogren's syndrome, *Lancet* **344:**1115.
54. Nakamura, H., Eguchi, K., Nakamura, T., Mizokami, A., Shirabe, S., Kawakami, A., Matsuoka, N., Migita, K., Kawabe, Y., and Nagataki, S., 1997, High prevalence of Sjogren's syndrome in patients with HTLV-1 associated myelopathy, *Ann. Rheum. Dis.* **56:**167–172.
55. Green, J. E., Hinrichs, S. H., Vogel, J., and Jay, G., 1989, Exocrinopathy resembling Sjogren's syndrome in HTLV-1 tax transgenic mice, *Nature* **341:**72–74.
56. Garry, R. F., Fermin, C. D., Hart, D. J., Alexander, S. S., Donehower, L. A., and Luo-Zhang, H., 1990, Detection of a human intracisternal A-type retroviral particle antigenically related to HIV, *Science* **250:**1127–1129.
57. Yamano, S., Renard, J. N., Mizuno, F., Narita, Y., Uchida, Y., Higashiyama, H., Sakurai, H., and Saito, I., 1997, Retrovirus in salivary glands from patients with Sjogren's syndrome, *J. Clin. Pathol.* **50:**223–230.
58. Matsumoto, Y., Muramatsu, M. O., and Sato, K., 1993, Mixed connective tissue disease and Sjogren's syndrome, accompanied by HTLV-1 infection, *J. Intern. Med.* **32:**261–264.
59. Mine, H., Kawai, H., Yokoi, K., Akaike, M., and Saito, S., 1996, High frequencies of human

T-lymphotropic virus type I (HTLV-I) infection and presence of HTLV-II proviral DNA in blood donors with anti-thyroid antibodies, *J. Mol. Med.* **74**:471–477.

60. Mizokami, T., Okamura, K., Ikenoue, H., Sato, K., Kuroda, T., Maeda, Y., and Fujishima, M., 1994, A high prevalence of human T-lymphotropic virus type I carriers in patients with antithyroid antibodies, *Thyroid* **4**:415–419.

61. Yokoi, K., Kawai, H., Akaike, M., Mine, H., and Saito, S., 1995, Presence of human T-lymphotropic virus type II-related genes in DNA of peripheral leukocytes from patients with autoimmune thyroid diseases, *J. Med. Virol.* **45**:392–398.

62. Akamine, H., Takasu, N., Komiya, I., Ishikawa, K., Shinjyo, T., Nakachi, K., and Masuda, M., 1996, Association of HTLV-1 with autoimmune thyroiditis in patients with adult T-cell leukemia (ATL) and in HTLV-1 carriers, *Clin. Endocrinol.* **45**:461–466.

63. Kawai, H., Yokoi, K., Akaike, M., Kunishige, M., Abe, M., Tanouchi, Y., Mine, H., Mimura, Y., and Saito, S., 1995, Graves' disease in HTLV-1 carriers, *J. Mol. Med.* **73**:85–88.

64. Kawai, H., Mitsui, T., Yokoi, K., Akaike, M., Hirose, K., Hizawa, K., and Saito, S., 1996, Evidence of HTLV-1 in thyroid tissue in an HTLV-I carrier with Hashimoto's thyroiditis, *J. Mol. Med.* **74**:275–278.

65. Yokoi, K., Kawai, H., Akaike, M., Mine, H., and Saito, S., 1995, Presence of human T-lymphotropic virus type II-related genes in DNA of peripheral leukocytes from patients with autoimmune thyroid disease, *J. Med. Virol.* **45**:382–398.

66. LaGrenade, L., 1994, HTLV-I, infective dermatitis, and tropical spastic paraparesis, *Mol. Neurobiol.* **8**:147–153.

67. LaGrenade, L., Schwartz, R. A., and Janniger, C. K., 1996, Childhood dermatitis in the tropics: With special emphasis on infective dermatitis, a marker for infection with human T-cell leukemia virus-I, *Cutis* **58**:115–118.

68. Kaplan, M. H., Sadick, N. S., McNutt, N. S., Talmor, M., Coronesi, M., and Hall, W. W., 1993, Acquired ichthyosis in concomitant HIV-I and HTLV-II infection: A new association with intravenous drug abuse, *J. Am. Acad. Dermatol.* **29**:701–708.

69. Kaplan, M. H., Hall, W. W., Susin, M., Pahwa, S., Salahuddin, S. Z., Heilman, C., Fetten, J., Coronesi, M., Farber, B. F., and Smith, S., 1991, Syndrome of severe skin disease, eosinophilia, and dermatopathic lymphadenopathy in patients with HTLV-II complicating human immunodeficiency virus infection, *Am. J. Med.* **91**:300–309.

70. DeFreitas, E., Hilliard, B., Cheney, P. R., Bell, D. S., Kiggundu, E., Sankey, D., Wroblewska, Z., Palladino, M., Woodward, J. P., and Koprowski, H., 1991, Retroviral sequences related to human T-lymphotropic virus type II in patients with chronic fatigue immune dysfunction syndrome, *Proc. Natl. Acad. Sci. USA* **88**:2922–2926.

71. Ejima, E., Rosenblatt, J. D., Ou, J., and Prager, D., 1995, Parathyroid hormone-related protein gene expression and human T-cell leukemia virus-I infection, *Miner. Electrolyte Metab.* **21**:143–147.

72. Ejima, E., Rosenblatt, J. D., Massari, M., Quan, E., Stephens, D., Rosen, C. A., and Prager, D., 1993, Cell-type-specific transactivation of the parathyroid hormone-related protein gene promoter by the human T-cell leukemia virus type I (HTLV-I) tax and HTLV-II tax proteins, *Blood* **81**:1017–1024.

3

Viral-Related Proteins in Immune Dysfunction Associated with AIDS

GEORGE J. CIANCIOLO

The mechanisms of pathogenesis associated with infection with human immunodeficiency virus (HIV), initially discovered in 1983,[1,2] and the development of acquired immune deficiency syndrome (AIDS) are complex and are undiscovered or poorly understood. Perhaps for no disease in history has there been such a focused research effort to understand exactly how a known etiological agent effects its pathogenesis.

Much progress has been made since the identification of HIV as the human retrovirus responsible for AIDS. However, many of the earlier concepts regarding the mechanisms of immune dysfunction, initially widely accepted, have come under closer scrutiny as we learn more about this virus and its lentivirus relative, simian immunodeficiency virus (SIV). Some of those concepts included the idea that HIV infects only CD4[+] lymphocytes and that CD4 was the only receptor for the virus. Both of those early concepts have now been clearly disproved. Another assumption now recognized as erroneous is that only a very small percentage of lymphoid cells are infected in HIV-infected individuals and that at any one time there is very little virus in the body. We now know that HIV infects a wide variety of cells and is often sequestered in sites such as the lymph nodes, and that viral replication is extensive, with a turnover rate of 10^9–10^{10} viral particles per day. The demonstration that certain chemokines can suppress viral infec-

GEORGE J. CIANCIOLO • Department of Pathology, Duke University Medical Center, Durham, North Carolina 27710.

Human Retroviral Infections, edited by Kenneth E. Ugen *et al.*
Kluwer Academic / Plenum Publishers, New York, 2000.

tivity and the identification of chemokine receptors as coreceptors for HIV has resulted in reexamination of our early concepts regarding viral binding and entry to infected cells. Furthermore, while early studies focused on the cytopathogenic potential of HIV infection of T lymphocytes, much effort is now being directed toward understanding the role of the host response in disease pathogenesis and, in particular, in the role of cytokine dysregulation in the immune dysfunction so characteristically associated with HIV infection.

Infection with different viruses has been thought to produce immunologic dysfunctions, such as immunodeficiency or autoimmune disorders, in the hosts. It has been generally assumed that such dysfunctions were related to the direct effects, such as infection, of whole virions on their target cells. This chapter seeks to examine other potential mechanisms of HIV-mediated immune dysfunction, ones that have received much less attention than many of the much more traditional mechanisms associated with viral pathogenesis. It will focus on the potential roles of soluble proteins, encoded by the HIV genome and shed from virions or infected cells, in the initiation or propagation of immune dysfunction associated with HIV infection and AIDS. These proteins may act directly on target immune cells to suppress their function or they may act indirectly by inducing soluble suppressive factors from the hosts' own cells.

The research herein reviewed is essentially in its infancy in that most of the studies describe the immunomodulatory effect on human immune cells of HIV proteins added to *in vitro* cell cultures. In the past, such studies were sometimes cursorily dismissed because the concentrations used *in vitro* appeared to be physiologically irrelevant to what one might expect to occur *in vivo*. However, with the recent evidence suggesting that the number of viral particles produced per day can be extraordinarily high and that virus may be concentrated in certain tissues, what exactly constitutes a physiologically relevant concentration remains undefined. As yet, however, only occasionally does one find published descriptions from patient studies of anecdotal evidence supporting the hypothetical mechanisms proposed from these *in vitro* investigations. Perhaps the proposition of these often intriguing, yet unproven hypotheses will prompt additional investigators to initiate new, confirmatory studies. Such studies might examine the fluids and tissues of patients not only for the presence of potentially immunoregulatory concentrations of these proposed mediators, but for the presence or absence of immune correlates which might serve to prove or disprove the hypothetical role proposed.

A review of the published HIV research literature would undoubtedly identify at least several reports demonstrating *in vitro* effects on immune cell function by any of the HIV structural, regulatory, or accessory proteins. This chapter, however, will focus only on those four proteins for which the most data seem to have been accumulated: the regulatory protein Tat, the structural proteins gp120 and gp41, and the regulatory protein Nef.

1. IMMUNOLOGIC DYSFUNCTION ASSOCIATED WITH HIV INFECTION

An important point to be considered in this discussion of the potential role of HIV-encoded soluble proteins in immune dysfunction is that it was recognized early in the AIDS epidemic that the destruction and loss of $CD4^+$ cells does not explain all of the immunopathogenic effects of HIV infection.[3] For example, the loss of T helper lymphocyte function in asymptomatic individuals, as evidenced by the ability of patients' lymphocytes to respond to soluble antigen *in vitro*, occurs even before any appreciable decline in CD4 counts is seen.[4–6] However, in most cases, the responses of patients' lymphocytes to mitogens remain normal.

2. REPORTED EFFECTS OF HIV-1 TAT ON IMMUNE CELL FUNCTION

Tat, a potent viral transcriptional *trans*-activator protein which mediates its effects by the Tat-response (TAR) element, is a 16-kDa, 86-amino acid protein with several distinct domains: residues 22–37 comprise a cysteine-rich domain which can bind cadmium and zinc, residues 37–48 constitute the core domain, and residues 48–57 contain basic residues required for nuclear and nucleolar targeting and RNA binding. Tat is a potent enhancer of viral replication and considered important since replication does not proceed in its absence. Its actions do not appear to be restricted to HIV replication, but also involve pleiotropic effects on the immune, vascular, and central nervous systems of HIV-infected individuals. Tat is known to be secreted extracellularly by infected cells and thus is readily available to interact with uninfected bystander cells.

As summarized in Table I, soluble HIV Tat protein has been reported to have a wide range of potentially deleterious effects on human immune cells which could play a role in the immune dysfunction associated with AIDS. One of the first studies reporting the ability of Tat to inhibit *in vitro* immune functions of human cells was that of Viscidi *et al.* in 1989.[7] They demonstrated that recombinant HIV-1 Tat (amino acids 1–72), expressed in *Escherichia coli,* inhibited the proliferation of human peripheral blood mononuclear cells (PBMCs) in response to stimulation by tetanus toxoid (66–

TABLE I
Potential Immunomodulating Activities Reported for HIV-1 Tat Protein

Activity	References
In human PBMCs or lymphocytes	
Increases Ig production	9, 10
Increases IL-6 release	9, 10
Upregulates Bcl-2 expression	19
Inhibits antigen-induced proliferation	7, 13, 14
Inhibits mitogen-induced proliferation	13
Inhibits anti-CD3-induced proliferation of CD4$^+$ and CD8$^+$ T cells	12, 14, 16
Induces apoptosis in uninfected T cells	18, 20, 21
In human monocytes or endothelial cells	
Induces chemotaxis and inflammatory mediator release by monocytes	22, 23
Increases chemotactic activity or monocytes	22
Increases IL-6 in endothelial cells	11
In human bone marrow cultures	
Increases TGF-β release	8
Inhibits CFU-E, CFU-M, CFU-GM	8
Enhances CFU-GM (HIV-2 Tat)	17

97%) or *Candida* antigen (75–91%). The IC_{50} for inhibition of tetanus toxoid-stimulated responses was approximately 0.5 μg/ml (ca. 50 nM). The effects of Tat did not appear to be due to toxicity since the proliferative responses of PBMCs in response to mitogens such as phytohemagglutinin (PHA), concanavalin A (ConA), or pokeweed mitogen (PWM) were unaffected. A synthetic peptide, representing amino acids 1–58, also inhibited tetanus toxoid-stimulated proliferation, although this inhibition required concentrations ten times higher than required with Tat 1–72. A second bacterially expressed Tat, representing amino acids 1–86, was as active as Tat 1–72 at inhibiting proliferative responses. The specificity of the induction of immunosuppression by Tat was demonstrated with antibody to Tat; however, the authors could not rule out that another molecule was the actual effector.

Several years later Zauli *et al.*[8] reported that recombinant Tat protein, when added to enriched normal bone marrow (BM) macrophages, induced the production of a factor that inhibited the *in vitro* growth of CD34$^+$ cells in liquid cultures and the growth of colony-forming units (CFU)-erythroid, CFU-megakaryocytic, and CFU-granulocyte/macrophage in semisolid agar. The suppressive factor was identified by use of neutralizing antibodies such as those to transforming growth factor β1 (TGF-β1) and subsequent experiments demonstrated that recombinant Tat increased the levels of both active and latent forms of TGF-β1 from BM macrophage cultures. The ability of Tat to upregulate the expression and release of TGF-β1 was confirmed by this same group[9] using human peripheral blood monocytes and they also demonstrated upregulation of IL-6, which they postulated was at least partially responsible for the upregulation of TGF-β1. Rautonen *et al.*[10] also reported increased expression of interleukin-6 (IL-6) from exogenous recombinant Tat-stimulated uninfected PBMCs. Furthermore, they reported increased production of immunoglobulin (Ig) from their cultures. The optimal concentration of Tat was 100 ng/ml (10 nM), but they saw effects at concentrations as low as 1 ng/ml (100 pM). Anti-IL-6 antibodies and IL-6 antisense oligonucleotides could block some of the Tat-induced IgG and IgA synthesis, but not the IgM synthesis, suggesting that at least some of the effects on increased Ig synthesis were secondary to the increased synthesis of IL-6. Increased synthesis of IL-6 by Tat stimulation had also been reported the previous year by Hofman *et al.*[11] in studies on human endothelial cells (EC). In addition to increased release of IL-6 from EC, they also reported increased expression of E-selectin on EC exposed to Tat.

In 1992, contrasting data to those published by Viscidi *et al.*[7] were reported by Meyaard *et al.*.[12] They found that proliferation of purified T cells to anti-CD3 monoclonal antibodies was inhibited by up to 70% by 5 μg/ml of Tat protein. Surprisingly, however, they did not observe suppression when accessory cells were present in the cultures and they could not demonstrate any inhibition of responses to recall antigen. These results led the authors to conclude that Tat did not have a significant role in the immunosuppression associated with HIV infection. However, in 1993 a study by Benjouad *et al.*[13] seemed to add to the confusion by both confirming and conflicting with the Viscidi *et al.* data. The authors reported that synthetic Tat (amino acids 2–86) inhibited not only *in vitro* antigen-induced peripheral blood lymphocyte proliferation, but mitogen-induced proliferation as well. The IC_{50} for inhibition of antigen-induced proliferation was 0.9 μM, which was substantially higher than that reported by Viscidi *et al.*, and the IC_{50} for mitogen-induced proliferation was 8 μM. They suggested that the inhibition of proliferation was due to lymphocyte cytotoxicity by Tat and that such cytotoxicity was associated with the basic region of amino acids 49–57.

Another study published that same year suggested another mechanism for Tat-induced anergy. Subramanyam *et al.*[14] reported that blockage of DP IV (CD26) by Tat partially inactivated antigen and anti-CD3 stimulated lymphocyte proliferation. As in the initial studies by Viscidi *et al.*, they found no effect on mitogen-stimulated proliferative responses. The antigen-specific inhibition they observed could be overcome by addition of exogenous IL-2 or by costimulation of PBMCs via CD28. The same group of investigators subsequently reported studies characterizing the binding of Tat to CD26.[15] The affinity of Tat for DP IV varied from 20 pM to 11 nM as the NaCl concentration was varied from 0 to 140 mM.

Although Benjouad *et al.*[13] had suggested that the inhibitory activity of

Tat might be related to its cytotoxic effects on target cells, a 1995 report by Chirmule *et al.*[16] demonstrated inhibition by Tat peptides of anti-CD3-stimulated proliferative responses of both purified $CD4^+$ and $CD8^+$ T cells. They reported that there was no effect on cell viability, as measured by trypan blue dye exclusion. The Tat concentrations tested ranged from 100 to 300 ng/ml. Another somewhat conflicting report was that of Calenda and Chermann.[17] Their 1995 study demonstrated that the HIV-2 Tat gene product enhanced growth of CFU-GM in agar, a result which contrasted with the earlier report of Zauli *et al.*[8] which showed inhibition of CFU-GM by HIV-1 Tat in $CD34^+$ cell cultures. Whether these apparently contradictory effects reflect true differences in the HIV-1 and HIV-2 Tat proteins remains to be determined.

Another potential mechanism for the immunosuppressive effects on T cells was introduced in 1995 with one of the first reports for Tat-mediated T cell apoptosis. In this study Li *et al.*[18] showed that Tat protein could induce apoptotic cell death in both a T cell line and in cultured PBMCs from un-infected donors. This apoptotic effect, observed initially in serum-deprived cells, was reversed to some extent by the inclusion of serum in the cultures, suggesting a protective effect for growth factors. In PBMC cultures, Tat induced apoptosis in T cells, but not in monocytes, and $CD4^+$ T cells did not appear to be more sensitive than $CD8^+$ T cells. Exogenous Tat added to PBMCs was found to markedly enhance total cyclin-dependent kinase (Cdk), activity and treatment of Tat-transfected Jurkat T cell lines with antisense oligonucleotides corresponding to the highly conserved regions of human cyclins A, B, and E blocked Tat-associated apoptosis. These results led the authors to propose that T cells in HIV-1-infected individuals are stimulated by Tat in lymphoid tissue, prematurely activating Cdks and preventing the cells from returning to a quiescent state. Later, when the cells are stimulated by antigen they undergo apoptosis and are depleted.

In contrast to the results of Li *et al.*,[18] who reported that Tat-induced apoptosis had no effect on Bcl-2 levels, Zauli *et al.*[19] reported that picomolar concentrations of native or recombinant Tat upregulated Bcl-2 in both Jurkat T cell lines and in primary PBMCs. They also demonstrated inhibition of apoptosis in serum-deprived Jurkat cells in which Tat had been transfected, and correlated this inhibition with Tat expression.

Katsikis *et al.*[20] reported that Tat protein had no effect on spontaneous apoptosis, but did enhance activation-induced apoptosis of both $CD4^+$ and $CD8^+$ T cells. Tat did not enhance Fas-induced apoptosis of either $CD4^+$ or $CD8^+$ T cells. Also in 1997, McCloskey *et al.*[21] published data which supported dual roles for Tat as both an inducer of, and a protector against, apoptosis. They demonstrated that while exogenous Tat induced apoptosis in uninfected T cells, T cell clones stably expressing Tat protein were protected from activation-induced apoptosis. These pleiotropic effects for Tat seem to support both of the previously described conflicting results of Li *et al.*[18] and Zauli *et al.*[19] and suggest that much remains to be learned regarding the actual role of Tat in HIV-1-associated T cell apoptosis.

Although the majority of studies have focused on the effects of Tat on T cell activation or death, several studies have suggested possible roles in regulating other cells of the immune system. For instance, in a 1996 study, Lafrenie *et al.*[22] reported that treatment with Tat at a concentration of 10 ng/ml enhanced both chemokinesis (migration in the absence of a stimulus) and chemotaxis (directed migration) to the chemotactic peptide FMLP. Tat itself was chemotactic for both treated and untreated monocytes. Pretreatment of monocytes with a similar concentration of Tat for 24 h increased their ability to invade reconstituted extracellular membrane (Matrigel)-coated filters by fivefold. The authors postulated that Tat might play a role in the recruitment of monocytes into extravascular tissues and thus contribute to the destruction of such tissues in patients with AIDS. The following year these results were confirmed by Mitola *et al.*[23] in studies in which they demonstrated that Tat induced monocyte chemotaxis at sub nanomolar concentrations and that such chemotaxis could be inhibited by cell preincubation with vascular endothelial growth factor-A (VEGF-A). Furthermore, the soluble form of VEGFR-1 could block Tat-induced monocyte chemotaxis, and specific binding of radiolabeled Tat to monocyte surface membranes was blocked by an excess of either unlabeled Tat or VEGF-A.

3. REPORTED EFFECTS OF HIV-1 gp120 ON IMMUNE CELL FUNCTION

The viral envelope protein of HIV-1 is synthesized as a 160-kDa precursor protein, gp160, which is eventually cleaved to gp120 surface (SU) and gp41 transmembrane (TM) proteins. The gp120 is subdivided into five variable (V) loops; four of these loops are bounded by disulfide-linked cysteine residues. The regions between the variable domains are designated constant (C) domains. The V3 loop is a major determinant of cell tropism, and a region of gp120, including the V2 loop and the C4 domain, is involved in CD4 interactions.

Like HIV-1 Tat, gp120 is shed from virus or virus-infected cells and is found in the plasma of HIV-infected individuals, making it a reasonable candidate to examine as a potential soluble suppressor of immune functions. Some of the immunomodulatory activities ascribed to gp120 are listed in Table II. One of the earliest studies examining gp120 was that of Shalaby

TABLE II

Potential Immunomodulating Activities Reported for HIV-1 gp120/gp160 Protein

Activity	References
In human PBMCs or lymphocytes	
Inhibits antigen-stimulated proliferation	24, 27–30, 36
Inhibits mitogen-stimulated proliferation	26
Inhibits CD3-stimulated proliferation	27, 29–31
Inhibits apoptosis	34–36
Increases Fas antigen expression	37
Reduces expression of protooncogene Bcl-2 in CD4$^+$ cells	38
Induces IL-10, IFN-α, γ, TNF-α, IL-6, IL-1α, β	44, 45, 47, 48
Inhibits PHA-induced IFN-γ and IL-2 secretion	46
Increases PHA-induced IL-4 secretion	46
Induces anergy in T helper lymphocytes stimulated with IL-2, IL-4, IL-6, anti-CD2, anti-CD3, or PMA	32
Inhibits upregulation of CD40L	43
In human monocytes	
Stimulates release of TNF-α, IL-1β, IL-6, and GM-CSF	55
Impairs chemotaxis	39
Induces release of IL-1β, PGE$_2$	33
Induces IL-10	50
Reduces accessory cell function	40, 42
Inhibits upregulation of B7-1	43
In human bone marrow cultures	
Inhibits CFU-GM	51–53, 55
Enhances growth of myeloid hematopoietic progenitors	54
Increases TGF-β	56

et al.,[24] in which they demonstrated that recombinant gp120 (rgp120), at concentrations of 1–20 μg/ml, inhibited tetanus toxoid-stimulated proliferation of human PBMCs. At 5 μg/ml, rgp120 also inhibited by 70% the number of immunoglobulin-secreting cells in PWM-stimulated PBMC cultures. Previous work by Sandstrom *et al.*[25] had suggested the possible existence of a soluble suppressor factor based on the observation that HIV-infected H9 cells could suppress the responses of uninfected PBMCs to Con A. However, studies published the same year as those of Shalaby *et al.*[24] by Mann *et al.*[26] seemed to both support and conflict with the previously described studies. They found inhibition of PHA-induced blastogenesis, but no effect on Con A-, PWM-, or alloantigen-induced proliferation. The following year, however, Chirmule *et al.*[27] demonstrated that HIV-1 gp120 could inhibit both antigen-specific and anti-CD3-stimulated proliferation of normal human lymphocytes. In that same year Krowka *et al.*[28] reported that recombinant gp120, at concentrations ≥10 μg/ml, could inhibit the proliferative responses of peripheral blood lymphocytes to UV-inactivated cytomegalovirus (CMV) and that this immunosuppression could be abrogated by recombinant IL-2.

Over the next 10 years additional studies[29–32] confirmed these initial reports of direct immunosuppressive effects of gp120, although the concentrations required for activity remained relatively high. Of interest, a study by Di Rienzo *et al.*[32] demonstrated that rgp120 could induce anergy in human T helper lymphocytes stimulated to proliferate with a variety of stimuli (IL-2, IL-4, IL-6, anti-CD2, anti-CD3 and PMA [phorbol 12-myristate 13-acetate]), but lymphocytes from chimpanzees, which are susceptible to HIV-1 infection, but do not easily develop immunodeficiency, were resistant to rgp120-associated anergy.

Much interest has focused on potential mechanisms by which gp120 exerts its effects on T helper cells. One report[33] described the production of prostaglandin E_2 and IL-1 from normal human monocytes exposed to low concentrations of gp120 purified from HIV-1. Since PGE_2 is known to suppress a variety of immune functions and the authors found a 12-fold increase in PGE_2 with gp120 concentrations of 200–400 ng/ml, these studies offered a plausible mechanism for gp120-mediated suppression. However, they were unable to demonstrate an effect with recombinant gp120 fragments, thus offering no explanation for the effects seen by others using rgp120.

The 1991 studies by Terai *et al.*[34] suggested that the direct cytopathologic effect of HIV-1 in T cells might be due to apoptosis and implicated gp120 in this mechanism. They reported that acute HIV-1 infection of MT2 lymphoblasts and activated PBMCs induces apoptosis and that addition after infection of anti-gp120 neutralizing antibody permitted sustained high levels of infection, but blocked apoptosis and cell death. The following year Banda *et al.*[35] described studies in which cross-linking of bound gp120 on human CD4$^+$ cells, followed by antigen signaling through the T cell receptor resulted in activation-dependent cell death, or apoptosis. Impressively, they were able to prime T cells for cell death with picomolar concentrations of gp120 and they offered these results as a potential mechanism for T cell depletion in AIDS. However, Liegler and Stites[31] found no evidence for significant cell death by apoptosis in gp120 treated and TCR-stimulated PBMCs and they reported that, as in earlier studies,[28] suppression of proliferation by gp120 could be reversed by addition of IL-2, thus suggesting that suppressed proliferation was not an apoptotic mechanism. Although based on the use of human T-cell clones, the studies of Amendola *et al.*[36] reported that preincubation of clones specific for influenza virus hemagglutinin with gp120 induced a significant inhibition of their antigen-stimulated proliferation which paralleled the induction of apoptosis. In these studies, however, antigen stimulation alone triggered apoptosis in a significant number of cells and gp120 merely potentiated the antigen effect. One potential mechanism for gp120-associated apoptosis was offered by the studies of Oyaizu *et al.*.[37] They showed that cross-linking of CD4 molecules (CD4XL) with HIV-1 envelope protein gp160 resulted in increased Fas expression as well as Fas mRNA in normal PBMCs and that upregulated Fas closely correlated with apoptotic cell death. The cytokines IFN and tumor necrosis factor-α (TNF-α) were implicated in CD4XL-mediated effects since antibodies to both cytokines blocked both Fas upregulation and apoptosis. This same group later showed[38] that CD4XL by either anti-CD4 monoclonal antibody or HIV gp120 reduces the expression of the protooncogene Bcl-2 in CD4$^+$ T cells, but not in CD8$^+$ T cells, concurrent with the induction of apoptosis in CD4$^+$ T cells. Addition of IL-2 to the cell cultures rescued CD4$^+$ T cells from CD4XL-induced Bcl-2 downmodulation and apoptosis, a result consistent with earlier reports[28,31] that gp120-mediated suppression of proliferative responses of PBMCs to CMV or TCR stimulation could be reversed by IL-2.

Mechanisms other than apoptosis have been suggested to account for gp120-mediated immunosuppression. For instance, there are several reports of gp120 effects on monocytes which are important for the responses of PBMCs. Wahl *et al.*[39] reported that gp120-treated monocytes were impaired in their ability to respond to chemotactic ligands due to receptor downregulation and that treated cells underwent differentiation, as evidenced by HLA-DR expression. Durrbaum-Landmann *et al.*[40] treated cultured monocytes with HIV-1 rgp120 and found that rgp120 significantly reduced the accessory function of monocytes to stimulate autologous lymphocytes with anti-CD3. In addition, they found that Fc receptor-mediated chemiluminescence was reduced as was the expression of CD4 and Fc receptor I/II, while CD14 expression and major histocompatibility complex classes I and II were unchanged. As suggested earlier, at least some of the anergy attributed to gp120 may be dependent on monocyte/macrophage-derived TNF-α. Kaneko *et al.*[41] reported that gp120 inhibition of early T cell activation and mitogen-mediated IL-2 production was blocked in the presence of antibody to TNF-α. However, inhibition of the mixed lymphocyte reaction (MLR) in CD4$^+$ T cells by gp120 was observed even in the absence of macrophage-derived TNF-α, suggesting that both TNF-α-dependent and TNF-α-independent events may play a role in T cell anergy associated with HIV. Zembala *et al.*[42] used a recombinant gp120 fragment (rp120cd; amino acids 410–511), encompassing the CD4-binding region, which had been previously shown to induce TNF-α production in monocytes, to demonstrate inhibition of antigen (PPD) presentation by monocytes to autologous T lymphocytes. Another fragment of gp120 not containing the CD4-binding region was inactive. Anti-TNF-receptor antibody blocked the depression of antigen presentation by rp120cd, suggesting a role for TNF and its receptor in impaired antigen presentation mediated by gp120.

A possible role for gp120-mediated effects on costimulatory molecules was suggested by the studies of Chirmule et al.[43] Interactions between CD28/B7-1 and CD40 ligand (CD40L)/CD40 are essential for anti-CD3 monoclonal antibody (mAb)-induced T cell proliferation as evidenced by the upregulation of B7-1 and CD40L and the ability of mAbs to B7-1 and CD40L to block anti-CD3-stimulated proliferation. Pretreatment of CD4$^+$ T cells with gp120 before CD3 ligation with anti-CD3 mAb inhibited upregulation of CD40 ligand (CD40L) on T cells and B7-1 on antigen-presenting cells (APC). Addition of anti-CD28 mAb overcame the inhibitory effect of gp120 on anti-CD3 mAb-induced T cell proliferation, supporting a role for gp120 in dysregulation of costimulatory molecules on both T cells and APC.

There are a number of reports of cytokine dysregulation by gp120. Capobianchi et al.[44] reported that recombinant gp120 could induce IFN-α from human PBMCs and that this induction could be blocked by anti-CD4 antibody or soluble CD4. In a later study, Ferbas et al.[45] reported that the major cell type responsible for the production of IFN-α in response to stimulation with HIV was the dendritic cell, as evidenced by their phenotype, large size, and veiled and ruffled morphology. Purified dendritic cells produced as much as 60-fold more IFN-α compared with HLA-DR$^+$ CD14$^+$ monocytes and IFN-α was not produced by CD3$^+$ T cells or CD56$^+$ natural killer cells. The induction of IFN-α by HIV-1 could be blocked by anti-CD4 mAb or anti-gp120 antiserum, suggesting that the gp120/CD4 interaction was required.

The potential for modulation of Th1 and Th2 cytokine profiles by HIV-1 gp160 was reported in a 1994 study by Hu et al.[46] The authors pretreated cells (unfractionated PBMCs, CD4$^+$ T cell lines, PBMCs depleted of CD8$^+$ cells) with HIV-1 gp160 and demonstrated significant reduction of PHA-induced secretion of interferon-γ (IFN-γ) and IL-2, but augmentation of IL-4 production. This effect was not observed when the PBMCs were depleted of either CD4$^+$ or CD2$^+$ cells or when the gp160 was pretreated with soluble CD4–Ig chimeric molecules, suggesting that gp120–CD4 interaction was required. Additional studies examining the balance between Th1 and Th2 cytokines have focused on IL-10. Ameglio et al.[47] reported that recombinant HIV-1 gp120 was a potent inducer in normal human PBMCs of IL-10. In addition, they also reported induction of IFN-α, IFN-γ, TNF-α, IL-6, IL-1β, and IL-1α, but not IL-2 or IL-4. Another study from the same institution also reported increased expression of IL-10. In those studies Borghi et al.[48] showed that treatment of either 1-day monocytes or 7-day monocyte-derived macrophages with recombinant gp120 induced IL-10 mRNA expression and caused a marked increase in IL-10 secretion. This effect of gp120 was abrogated by mAb to gp120. In a study[49] published a year earlier, these investigators had examined serum IL-10 levels in HIV-positive patients and correlated them with Centers for Disease Control (CDC) stages. Using a competition enzyme-linked immunosorbent assay (ELISA) for IL-10, they found that serum IL-10 levels were significantly higher in HIV-positive patients compared to HIV-negative controls. The IL-10 levels progressively increased in the subsequent CDC stages, without further changes from stage III to stage IV. Patients evaluated twice in CDC stage II, with a time interval of at least 1 year, showed significant IL-10 increases and the increases were even more pronounced when the patients progressed from CDC stage II to CDC stage III. In addition, a significant negative correlation was established between the patients' IL-10 levels and their CD4/CD8 ratios, suggesting that IL-10 might be involved in some of the immunological abnormalities associated with AIDS. More recently, Taoufik *et al.*[50] reported that in human monocyte cultures, HIV gp120 induces a significant IL-10 synthesis and that this inhibits mRNA for IL-12 subunits and the subsequent expression of IL-12 p40 and p70 proteins in response to stimulation with *Staphylococcus aureus* strain cowan I.

Still another gp120-mediated mechanism of immunosuppression is the effect of gp120 on hematopoietic stem cells. In 1991 and 1992 Zauli *et al.*[51–53] reported that treatment of human CD34$^+$ cell cultures with either HIV-1 or HIV-1 gp120 or HIV-1 gp160 caused a progressive and significant decrease in viability and a reduced percentage of committed progenitors. Both gp120 and gp160, at concentrations from 0.01 to 10 μg/ml, decreased CD34$^+$ cell viability as measured by trypan blue dye exclusion and tritiated thymidine incorporation in liquid cultures supplemented with human recombinant IL-3. In the absence of IL-3, no inhibition was seen at even the highest concentrations of gp120 or gp160. In virus-treated cells there were no signs of active virus replication and latent infection was ruled out by polymerase chain reaction (PCR). The cytotoxic activity of either HIV or gp120 could be abrogated by neutralizing antibody against gp120. Both gp120 and gp160 inhibited the *in vitro* growth of CFU-GM in a dose-dependent fashion. In contrast to the results of Zauli *et al.*,[51–53] Sugiura *et al.*[54] reported that culture of cord blood mononuclear cells with gp160 resulted in enhancement of the *in vitro* growth of myeloid hematopoietic progenitors. Using cultures of adherent cells, purified T cells, or CD34$^+$ progenitors, the authors found that gp160 had no direct effect on highly purified hematopoietic progenitors, but acted indirectly via induction of a colony-stimulating factor(s) from T cells. The enhancing activity of gp160 could be blocked by soluble CD4 or polyclonal antisera to gp120.

Maciejewski *et al.*[55] also observed suppressed hematopoiesis by gp120, but their studies suggested a role for TNF-α. They found that hematopoietic colony formation by CFU-E or CFU-GM was inhibited by both active and heat-inactivated HIV-1 virus as well as by HIV-1 gp120. Inhibition required the presence of macrophages and was not observed in cultures of highly enriched $CD34^+$ cells. The addition of anti-TNF-α neutralizing antibodies to marrow cultures abrogated the inhibition by either gp120 or virus. Neutralizing antibodies to IL-4, IFN-α, or TGF-β had no effect on inhibition of colony formation. The authors demonstrated production of TNF-α from blood monocytes and marrow mononuclear cells exposed to gp120 and suggested that viral suppression of hematopoiesis did not require direct infection of progenitor cells, but might be mediated indirectly by TNF-α-induced by virus or viral envelope protein. Zauli *et al.*[56] reported data suggesting the involvement of TGF-β1, rather than TNF-α, in mediating gp120 effects on hematopoietic precursors. Highly purified $CD34^+$ progenitor cells from the peripheral blood of 20 normal donors had impaired survival and clonogenic capacity after exposure to HIV-1 or cross-linked gp120. Cell cycle analysis suggested that HIV or gp120 cells were undergoing apoptosis. Blocking experiments with anti-TGF-β1 neutralizing serum suggested that HIV- or gp120-mediated suppression was almost entirely due to upregulation of endogenous TGF-β1. Increased levels of bioactive TGF-β1 could be detected in the supernatants of HIV-1- or HIV-1 gp120-treated $CD34^+$ cells, and anti-TGF-β1 neutralizing antiserum caused a significant increase in plating efficiency of $CD34^+$ cells from the peripheral blood of HIV-1-seropositive patients.

4. REPORTED EFFECTS OF HIV-1 gp41 ON IMMUNE CELL FUNCTION

The gp41 TM protein of HIV-1 includes several domains: an N-terminal ectodomain which is outside the lipid bilayer, a membrane-spanning hydrophobic domain, and a C-terminal domain which is inside the lipid bilayer. A hydrophobic N-terminal sequence within the ectodomain has been implicated in the fusion of the virus with cell membranes after initial binding to CD4. Also within the ectodomain is a leucine zipper domain and a helical domain which have been reported to be involved in virus–cell fusion.[57,58] Within the ectodomain of gp41 is a region, amino acid residues 583–599, which was identified[59] to be partially homologous to the immunosuppressive domain identified by Cianciolo *et al.*[60,61] within the p15E TM of type C and D retroviruses. As will be discussed below, many of the studies of gp41-mediated immunosuppression have focussed on this putative immunosuppressive region. Table III lists many of the activities described for gp41.

One of the first reports of immunosuppressive activity associated with gp41 was a 1988 study by Cianciolo *et al.*[62] using a synthetic peptide (CS-3) corresponding to the putative immunosuppressive region of HIV-1 gp41. Using human PBMCs, they demonstrated up to 81% inhibition of anti-CD3-

TABLE III
Potential Immunomodulating Activities
Reported for HIV-1 gp41 Protein

Activity	References
In human PBMCs or lymphocytes	
Inhibits mitogen and lymphokine-dependent proliferation	66, 68, 75
Inhibits antigen-dependent proliferation	62, 68, 75
Inhibits anti-CD3-stimulated proliferation	62, 65
Inhibits B lymphocyte proliferation	67
Inhibits lymphokine-activated killer activity	69
Inhibits production of IL-2, IFN-α, γ	70, 74
Enhances production of IL-1β, TNF-α	70
In human monocytes or monocytic cell lines	
Induces IL-6 or IL-10	76, 77

stimulated T cell proliferation, while they observed no effect on anti-Ig-stimulated B cell proliferation. These initial observations of T cell inhibition were later confirmed by a number of laboratories,[63–68] Ruegg et al.,[63] in addition to demonstrating inhibition of human lymphocyte proliferation with the synthetic peptide corresponding to the immunosuppressive region of gp41, also demonstrated inhibition of murine lymphocyte proliferation, which strengthened the significance of the similarities between the TM proteins of the murine and human viruses. This same group later reported that the gp41 immunosuppressive peptide could inhibit protein kinase C (PKC) and anti-CD3-stimulated Ca^{2+} influx and PKC-mediated phosphorylation of the CD3 gamma chain in Jurkat T cells.[64] In another report, they demonstrated that human lymphoproliferation induced by the T cell activation molecules CD3, CD2, or CD28 were all inhibited by this gp41 peptide.[65]

In another study using peptides corresponding to this same immunosuppressive sequence, Denner et al.[66] confirmed both the inhibition of mitogen-induced and lymphokine-dependent T lymphocyte proliferation and interspecies reactivity for this sequence. They also reported that N-terminal octamers represented the minimal immunosuppressive domain and that, in addition to its immunosuppressive activity, the peptide could inhibit the cytopathic effect of HIV-1 on human MT4 cells, suggesting interference with viral replication. This latter result would seem to agree with the data of Wild et al.,[57] who reported inhibition of viral infectivity with a gp41-encoded peptide (DP107) which partially overlapped the putative immunosuppressive region. Denner et al.[67] later reported that the GP41 immunosuppressive peptide inhibited B lymphocyte stimulation by the B cell mitogen lipopolysaccharide and by antibodies against cell surface immunoglobulins, a result which contrasts with the earlier report by Cianciolo et al.[62] in which they observed no effects on B cells. Wang et al.[68] also demonstrated inhibition of human lymphocyte proliferative responses to mitogens and recall antigens using a peptide which included the putative immunosuppressive domain, but which also included an additional C-terminal 10 amino acids encompassing a well-documented B cell epitope (amino acids 598–609). They reported inhibition of both CD4+ and CD8+ T cells, indicating that CD4 binding was not involved. Importantly, they reported that sera from HIV-infected patients, although reactive with the peptide, could not block the suppressive activity of the peptide, suggesting that the immunosuppressive domain is distinct from the immunogenic domain and is not itself immunogenic.

Although many studies have focused on the putative immunosuppressive domain of gp41, which is partially homologous to type C retroviral TM p15E, other investigators have explored additional regions of the gp41 molecule using synthetic peptides. Cauda et al.[69] found that lymphokine-activated killer (LAK) cell activity was reduced in HIV-1-infected patients and that HIV-1 gp41 synthetic peptides corresponding to amino acids 735–752 and amino acids 846–860 were able to significantly inhibit LAK activity. They found that HIV-1-positive sera and supernatant fluids from cultured PBMCs from HIV-1-infected patients had similar activity, suggesting a possible role for a gp41-encoded protein. The following year this same group reported[70] that the gp41 peptides studied earlier were also able to suppress the production of IFN-α, IFN-γ, and IL-2 from normal human PBMCs while enhancing the production of IL-1 and TNF-α.

Although data demonstrating in vitro immunosuppression with gp41 peptides were intriguing, data generated with intact gp41 were slow to be reported. Qureshi et al.[71] had identified potential cell-surface-binding proteins for HIV-1 gp41 using the CS-3 immunosuppressive peptide of gp41. Ebenbichler et al.[72] confirmed and extended the results of Qureshi et al.[71] using recombinant soluble gp41 (rsgp41) bound to Sepharose to affinity-purify cell-membrane-binding proteins. They characterized three different proteins of 44, 98, and 106 kDa which bound to rsgp41 and reported that their expression decreased from a T-lymphoid cell line to a monoblastoid cell line to a cell line representing mature monocytes. Chen et al.[73] used rsgp41 and flow cytometry to identify which cells could bind gp41. They demonstrated rsgp41 binding to normal human PBMCs, preferentially to B lymphocytes and monocytes, and independent of gp120-binding sites on CD4 molecules. The binding was dose dependent and they found that rsgp41 bound more strongly to B cells (47%) and monocytes (44%) than to CD4+ T cells (10%) or CD8+ T cells (12%).

In 1993 Oh et al.[74] reported that sera of AIDS patients contained factor(s), 30–50 kDa in size, that could exert significant inhibition on normal T lymphocyte and natural killer functions. These factor(s) inhibited the early stages of lymphocyte activation and, in addition to inhibiting production of IL-2, inhibited the transcription of IL-2 mRNA. The authors determined by Western blotting that their immunosuppressive factor(s) shared epitopes with HIV-1 gp41. Although indirect, this was one of the first suggestions that native gp41 could exert immunosuppressive activity. Two years later Chen et al.[75] reported that rsgp41 could inhibit ConA-, PHA-, and tetanus toxoid-stimulated proliferation of normal human peripheral blood lymphocytes, although the concentration required for 50% inhibition, 8 μM, was quite high. They found no effect of rsgp41 on PWM-stimulated lymphocyte proliferation, although they had earlier[73] reported that B cells bound rsgp41 to a greater extent than any other cell type.

A potential effector mechanism for immunosuppressive activity by gp41 was first suggested in 1995 by the studies of Takeshita et al.[76] Using the human monocytic cell line THP-1, they found that recombinant gp41, but not gp120 or p24, induced significant IL-6 expression in treated cells. Their studies also indicated that IL-10 was produced, with delayed kinetics, following the induction of IL-6, although the induction of IL-10 required that IFN-γ be present. When the investigators added recombinant IL-10 to gp41-treated THP-1 cells, they inhibited the expression of IL-6, and when an IL-10 neutralizing antibody was added, IL-6 production was enhanced. These studies suggested that IL-10 induced by gp41 might play a role in cytokine dysregulation associated with HIV-1 infection.

More recently, Koutsonikolis et al.[77] reported that recombinant HIV-1 gp41 induces the expression of IL-10 mRNA in human PBMCs within 3 h of exposure. Importantly, the induction of IL-10 release in the supernatants of recombinant gp41-treated PBMCs was observed at concentrations ≤10 nM, suggesting that this might be a physiologically relevant phenomenon. The authors showed that recombinant gp41 induced IL-10 only in monocytes and not in T lymphocytes and confirmed earlier reports[47,48,50] of gp120 induction of IL-10 in monocytes. They also found that gp120 induction of IL-10 was specific for monocytes and that there was no effect on T lymphocyte IL-10 production.

5. REPORTED EFFECTS OF HIV-1 NEF ON IMMUNE CELL FUNCTION

The *nef* gene is found only in the primate lentiviruses HIV and SIV and codes for a protein of approximately 25–32 kDa and 206–265 amino acids. It is N-terminally myristoylated and generally found in the cytoplasm and membrane fractions of infected cells. Studies with rhesus monkeys infected with clones of SIV demonstrated that an intact *nef* gene was essential for maintenance of high virus loads and the development of disease.[78] Additionally, investigators have described[79] an epidemiological cohort of HIV-1-infected patients with nonprogressive disease in which *nef* sequences were not intact, which supports a role for Nef protein in disease progression. However, the potential pathogenic role of Nef may be dependent on the immune status of the host, since pathogenicity was still observed when Nef-attenuated SIV was used to infect neonatal macaques.[80] A number of different biological activities have been described for Nef as well as a number of different mechanisms by which Nef might effect those activities. The majority of mechanisms described have focused on the interaction of Nef with intracellular regulatory proteins and have relied on either viral infection or *nef* gene transfection. The studies reviewed below and summarized in Table IV examined the potential extracellular role of Nef in immune dysregulation.

Early interest in an extracellular role for Nef was most likely restrained because of the known physical characteristics of the molecule and its common detection in infected cells in the cytoplasm or associated with cell membranes. However, antibodies to Nef have been routinely detected in HIV-infected patients beginning at the earliest stages of infection. Thus it is reasonable to conclude that the immune systems of patients are exposed to Nef proteins or peptide fragments early in infection; these might be shed either from virions or infected cells, expressed on the surface of infected cells, or released from dying cells.

Investigators had begun to look for antibodies to Nef and a potential

TABLE IV
Potential Immunomodulating Activities
Reported for HIV-1 Nef Protein

Activity	References
In human PBMCs or lymphocytes	
Inhibits proliferation in response to PHA	86
Induces Ig-secreting cells	93
Induces mRNA and protein secretion for IL-6	93
Induces V beta-specific expansion	96
Induces proliferation	94, 95, 97
Induces cytostasis or cytotoxicity in CD4$^+$ cells	106, 107
Induces production of IL-2 and IFN-γ	94
Inhibits production of IL-2 and IFN-γ	108
Induces mRNA and protein for IL-10	105
In human bone marrow cultures	
Inhibits proliferation of CFU-GM	110

correlation with disease as early as the late 1980s. Amiesen *et al.*,[81] using purified recombinant Nef and Nef synthetic peptides, detected antibodies in the sera of HIV-1-seronegative, viral antigen-negative, and virus culture-negative individuals at risk for HIV infection. They concluded that antibodies to Nef could precede seroconversion. Reiss *et al.*[82,83] examined antibody responses against Nef in a longitudinally studied cohort of 194 initially asymptomatic HIV-1-seropositive individuals and 72 individuals who sero-converted for antibodies to *gag/env* proteins. Antibodies to Nef were found in approximately 80% of the patients and antibodies persisted in about 85% of those patients. Although antibodies were detected early in infection, unlike the results of Amiesen *et al.*,[81] they were rarely seen prior to anti-bodies to *gag/env* proteins. Although more cases of AIDS occurred in the Nef antibody-negative group (24%) than in the Nef-antibody-positive group (16%), the difference was not statistically significant. Wieland *et al.*[84] exam-ined the sera from 136 HIV-1-infected patients from different stages of disease. Although they found antibodies in a high percentage of patients, there was no obvious correlation with stage of disease or disease progres-sion. In contrast, Rezza *et al.*[85] found that patients who were persistently positive for antibody to Nef were statistically ($p = .03$) less likely to develop AIDS than those who were transiently or persistently negative.

Studies examining a potential role for extracellular Nef began to ap-pear in the early 1990s. Fujii *et al.*[86] reported studies utilizing Nef protein expressed in bacteria as a fusion protein with a phage gene product (Nef–gene 10). When PBMCs from healthy donors were incubated with IL-2 and Nef–gene 10, but not gene 10 alone, there was a decline in the CD4/CD8 ratio and in the response to PHA. The same result occurred with nylon wool-purified T cells. Although Nef–gene 10 inhibited the proliferation of CD4$^+$ cells, it did not kill them and the inhibition of proliferation by Nef–gene 10 could be blocked by anti-Nef antibodies. These same investigators also reported[87] that Nef was partly expressed on both acutely and persis-tently infected human T cell lines. They prepared monoclonal antibodies to recombinant Nef expressed in baculovirus as a truncated 22-kDa protein containing the middle to C-terminus of the native protein. The monoclonal antibodies reacted with both the 25- and 27-kDa forms of Nef and by flow cytometry reacted with the membranes of both fixed and unfixed human Molt-4 T cells which had been infected with HIV-1. They suggested that the cell-surface form of Nef might play a role in depletion of CD4$^+$ cells. In order to determine what portion of Nef might be expressed on the cell surface the investigators developed a series of monoclonal antibodies with identified reactivities to specific regions of the Nef molecule. Using these antibodies, they reported[88] that intensity of cell surface staining on infected PBMCs was stronger with antibodies reactive to amino acid residues 158–206 and 192–206 than with antibody reactive with residues 148–157. They reported no staining with an antibody reactive to residues 1–33. Furthermore, they reported that binding of soluble Nef to the surface of uninfected $CD4^+$ cells was blocked by antibodies to the C-terminal region (amino acids 158–206 and 192–206) or by C-terminal synthetic peptides. Furthermore, they reported inhibition of syncytium formation between HIV-1-infected and uninfected cells by the same antibodies or peptides, suggesting that a cell surface domain of Nef could be important in pathogenic interactions between uninfected and infected cells.

It had been widely assumed that myristoylation of the N-terminus of Nef assured its association with membranes or insoluble fractions, although unmyristoylated Nef could be found in cell cytoplasm. Reports in 1994[89] and 1995[90] from two different laboratories offered an explanation as to how Nef fragments, perhaps biologically active, might find their way to the cell surface or the extracellular environment. Both laboratories reported that HIV-1 Nef was cleaved into two fragments, of approximately 8 and 19 kDa, by the HIV-1 protease. The cleavage site was identified by both laboratories as being located between Trp57 and Leu58. Both recombinant Nef and Nef obtained from extracts of infected cells were subject to this specific cleavage. A subsequent study[91] examined the physicochemical properties of Nef in virions and in infected cells. Quantitative analysis of radiolabeled HIV-1-infected cells and virions demonstrated that Nef was incorporated on the order of 10% of reverse transcriptase incorporation, which corresponded to 5–10 Nef molecules per virion. In infected cells Nef was detected as a full-length, 27-kDa protein, while in virions, approximately 50% of the Nef was present as a 18-kDa protein. This virion-associated 18-kDa protein co-migrated with a 18-kDa protein generated by cleavage of purified Nef by HIV-1 protease. Cleavage of Nef in virion preparations was completely abolished by an HIV-1 proteinase inhibitor. These results have been confirmed[92] by another laboratory, which found that the 18-kDa species of Nef was the major form in mature virions and that its generation was indeed dependent on HIV-1 protease. These studies also revealed that initial association of Nef with HIV-1 virions was dependent on its N-terminal myristoylation.

There have been several reports of Nef stimulating human peripheral blood cells. Chirmule et al.[93] reported that recombinant Nef was able to induce immunoglobulin-secreting cells in PBMC cultures of HIV-1-seronegative donors. Pretreatment of Nef with a polyclonal anti-Nef antibody blocked its B cell stimulatory activity, as did monoclonal antibodies to LFA-1, ICAM-1, HLA-DR, and B7, suggesting that cell-to-cell contact was important in Nef-stimulated B cell activation. mRNA for IL-6 was upregulated, primarily in monocytes, in Nef-treated PBMCs. Studies[94–96] by another group have focused on the ability of Nef to function as a CD4 T cell superantigen. They reported that Nef-induced proliferation of human PBMCs was blocked by polyclonal antisera raised against Nef synthetic peptides and that responses were T cell-specific and required antigen presenting cells. Nef-stimulated cells produced the T helper 1 cytokines IL-2 and IFN-γ and activation did not require processing of Nef since paraformaldehyde-inactivated APC were still active. Their studies indicated that HIV could replicate in PBMCs activated by Nef and that this infection and replication could be blocked by polyclonal antisera to Nef. Using a simple amplification technique involving immobilized anti-Vβ antibodies, these investigators demonstrated expansion of Vβ T cells in cultures of Nef-treated PBMCs. They reported that expansion of Vβ18 occurred in all of the donors tested, while Vβ5.3 expansion occurred in approximately 50% of the donors. The necessity for amplification suggested that Nef might be functioning as a weak, viral-encoded superantigen.

The hypothesis that Nef might enhance infection and replication by stimulating T cell replication is indirectly supported by the studies of Novembre et al.[97] In these studies in macaques, they used a variant of SIV, termed SIVsmmPBj, which induces an acute disease resulting in death in 5–14 days after infection. A molecular clone, PGj6.6 delta *nef*, was generated in which there was a deletion in the Nef-coding region. In general, PBj6.6 delta *nef* had markedly reduced replication abilities compared to the parental virus PBj6.6 when tested on pigtail or rhesus macaque PBMCs, although there was little difference in stimulated pigtail macaque PBMCs. Furthermore, PBj6.6 delta *nef* was unable to stimulate the *in vitro* proliferation of PBMCs, a feature characteristic for SIVsmmPBj.

In contrast to the earlier report by Torres et al.[94] that Nef could induce the *in vitro* production of the cytokines IL-2 and INF-γ, Beneviste et al.[98] found that deletion of *nef* gene from SIV seemed to enhance the *in vivo* production of these same cytokines. They used semiquantitative RT-PCR to monitor cytokine expression in unmanipulated PBMCs during the acute phase of infection of cynomolgus macaques. In monkeys infected with the pathogenic SIVmac251, they found that IL-2, IL-4, and IFN-γ mRNAs were either weakly detectable or undetectable. In contrast, animals infected with a *nef*-truncated SIVmac251 exhibited overexpression of these cytokines. Both groups of monkeys exhibited similar rises in IL-10 mRNA expression coincident with the peak of viral replication. These results will be discussed further below.

In 1996 Collette et al.[99] reported that the Nef protein of primate lentiviruses shared sequence similarities with the immunosuppressive domain[60,61] of the p15E TM of type C retroviruses of murine, feline, bovine, avian and human origin. This was an important observation because this immunosuppressive domain had been found to be highly conserved among virtually all of the retroviruses and had been postulated to play a role in the pathogenesis of these viruses. And yet, in the lentiviruses this region had only a limited degree of conservation in the gp41 TM. Studies with synthetic peptides corresponding to the immunosuppressive domain (CKS-17) of p15E had suggested a potential role for cytokine dysregulation as a mechanism for its biological effects (reviewed in ref. 100). Among the *in vitro* activities identified for CKS-17 on human cells are inhibition of induction of IL-2, IL-1, IFN-γ, and TNF-α and enhancement of the induction of IL-10. The potential of the CKS-17 immunosuppressive domain to effect a Th1 $\rightarrow$ Th2 type of cytokine shift, coupled with the now-identified similarity of a portion of Nef to CKS-17, was provocative since several investigators had already suggested that such a shift might be a critical step in HIV pathogenesis.[101–103] Furthermore, Clerici *et al.*[104] had reported that IL-10-specific mRNA was upregulated (although marginally) and increased levels of IL-10 were produced from PBMCs of HIV-infected individuals compared with uninfected individuals. In those studies they also reported that those patients whose T helper cell function was more severely suppressed produced higher levels of IL-10 and that defective *in vitro* antigen-specific T helper functions could be reversed by addition of anti-IL-10 antibodies. Thus, the recently identified structural similarities between Nef and the CKS-17 region of p15E, strong evidence suggesting a role for Nef in the pathogenesis of HIV-1 infections, and evidence suggesting a Th1 $\rightarrow$ Th2 type of cytokine shift potentially effected by IL-10 have led to studies investigating the potential effects of extracellular Nef on cytokine regulation and lymphocyte proliferation.

Brigino *et al.*[105] recently studied the ability of recombinant Nef to effect cytokine induction in human monocytic THP-1 cells or freshly isolated PBMCs. In those studies they found that extracellular Nef produced in either yeast (*Saccharomyces cerevisiae*) or bacteria (*Escherichia coli*) induced, in a dose-dependent manner, the release of IL-10 into the supernatants of Nef-treated human PBMCs. They could detect IL-10 with as little as 10 ng/ml (370 pM) recombinant Nef and although the *E. coli*-derived material also induced IL-10, it was not quite as effective as the yeast-produced material (E. Brigino, personal communication), which might be a function of the N-terminal myristoylation on the yeast protein or the ability of yeast-derived material to retain a more natural conformation. In addition to protein expression, Brigino *et al.*[105] also demonstrated by RT-PCR the induction of mRNA for IL-10, but not for IL-2, IL-4, IL-5, IL-12 (p40), IL-13, or IFN-γ, in PBMCs treated for 3 h with *E. coli*-derived Nef. By Northern blot analysis they estimated a threefold increase in mRNA for IL-10 at 24 h after Nef stimulation.

A report the previous year by Fujii *et al.*[106] was significant in that they reported, apparently for the first time, the detection of soluble Nef in the sera of HIV-1-infected individuals. Using an antigen capture ELISA they detected soluble Nef in the sera of 27 of 32 HIV-infected individuals, but not in any of the 28 healthy individuals used as controls. The five sera in which Nef was not detected contained high titers (>1:5000) of anti-Nef antibodies, whereas the other sera contained no levels of anti-Nef antibodies. Importantly, in the sera of 21 of 27 individuals positive for Nef the concentrations were 5–10 ng/ml, which was close to the concentration at which Brigino *et al.*[105] were able to stimulate IL-10 release from normal human PBMCs. This concentration is also similar to that with which we have observed (Fig. 1) significant inhibition of anti-CD3-stimulated proliferation of normal human PBMCs with recombinant (*E. coli*) Nef. In those studies we found that recombinant Nef inhibits in a dose-dependent manner the proliferation of normal PBMCs in response to anti-CD3 monoclonal (OKT3; 50 ng/ml) antibody. Although Fujii *et al.*[106] and

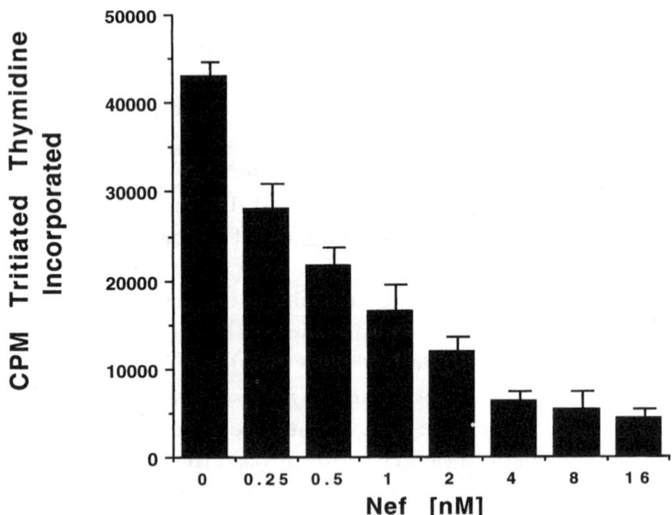

FIGURE 1. Effect of recombinant HIV-1 Nef on proliferation of human peripheral blood mononuclear cells in response to OKT3 stimulation. Recombinant Nef was expressed in *E. coli*, purified to homogeneity, and contaminating endotoxin removed by multiple passages over DeToxi-Gel (Pierce, Rockford, IL) until endotoxin levels were ≤0.05 ng/ml. Human PBMCs were isolated and cultured for 72 h at 2×10^5 total cells/well in 200 μl in 96-well, flat-bottom tissue culture plates as previously described.[61,62] Recombinant Nef was added at the indicated concentrations at the beginning of the cultures and OKT3 monoclonal antibody was added to each well at a final concentration of 50 ng/ml. One-half μCi of ^{3}H-thymidine (6.7 Ci/mmol; Amersham) was added to each well for the last 6 h of culture and the amount of incorporated radioactivity determined by harvesting the cells onto glass fiber filters and counting the filters in a liquid scintillation spectrophotometer. Results represent the means and standard errors of quadruplicate samples. Similar results were observed in at least two other experiments.

Okada *et al.*[107] have reported that soluble Nef is cytotoxic for CD4$^+$ cells from PBMC, they found that cytotoxicity required Nef cross-linking, which they effected with anti-Nef antibody. We have not observed cytotoxicity in our Nef-treated PBMCs, as determined by trypan blue dye exclusion, a result supported by our observation (data not shown) that soluble Nef had little or no inhibition on mitogen-stimulated (PHA, ConA) proliferation of PBMCs under the same conditions.

We have begun to examine the potential role of the CKS-17-related region of Nef in its potential immunosuppressive effects on human lymphoid cells. A peptide (MN10042) was synthesized to correspond to the region (amino acids 79–104) of HIV-1 Nef which was partially homologous to the immunosuppressive CKS-17 region of p15E TM of type C retroviruses. Although initial studies[61] with CKS-17 had indicated that biological activity required coupling to a protein carrier, subsequent studies revealed that dimerization of peptide monomers through a naturally occurring C-terminal cysteine was sufficient to generate biologically active molecules. For our studies examining the potential active site of Nef, we prepared peptide monomers (corresponding to HIV-1 Nef amino acid residues 79–104) with a C-terminal cysteine (not naturally occurring) and then either dimerized them (MN10042) or, as a control, prepared monomer in which the cysteine was blocked (MN10041.5). As shown in Fig. 2, MN10042 inhibited in a dose-dependent manner the tetanus toxoid-stimulated proliferation of human PBMCs, whereas the control peptide, MN10041.5, had no activity. This effect was not simply a result of the presence of an active disulfide, since we previously found (data not shown) that other peptide dimers with similar amino acid compositions had no effect on PBMC proliferative responses. Although the concentration of MN10042 required for 50% inhibition was approximately three orders of magnitude greater than that required with soluble recombinant Nef, that result was not unexpected. Our experience has been that peptides are generally much less active than the native proteins from which they are derived, probably because their conformations are not stably fixed in solution.

More recently, Haraguchi *et al.*[108] have shown that MN10042 inhibits the production of IL-2 and IFN-γ from human PBMCs stimulated with either OKT3 or staphylococcal enterotoxin A, but not from PBMCs stimulated with phorbol-12-myristate 13-acetate plus ionomycin. Expression of mRNAs for both cytokines was also inhibited in the presence of MN10042. This result would seem to be in agreement with the results reported by Beneviste *et al.*,[98] who described increased production of IL-2 and IFN-γ in macaques infected with Nef-deleted SIV compared to the wild-type virus. Although MN10042 suppresses IL-2 and IFN-γ as had been observed for CKS-17, there are mechanistic differences since CKS-17 has been shown to induce dramatic increases

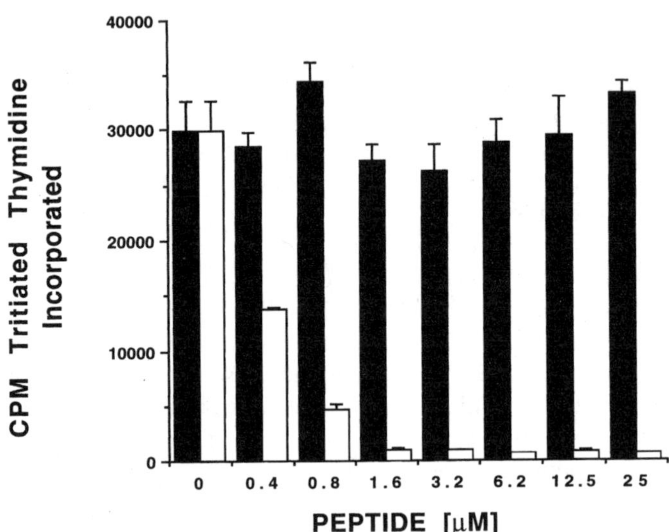

FIGURE 2. Effects of HIV-1 Nef-derived peptides on proliferation of human peripheral blood mononuclear cells in response to tetanus toxoid. Monomeric peptide corresponding to amino acid residues 79–104 of HIV-1 Nef was synthesized with a C-terminal cysteine (MTYKAAVDLSHFLKEKGGLEGLIHSQ-C) on a RAININ P3 peptide synthesizer and purified to >95% by reversed-phase HPLC. Homogeneity was confirmed by amino acid analysis and mass spectroscopy. Peptide dimers (MN10042) or peptides with a blocked C-terminal cysteine (MN10041.5) were prepared as previously described.[109] Purity and homogeneity of all peptides were confirmed by mass spectroscopy. Human PBMCs were isolated and cultured as described for Fig. 1 except that the cultures were incubated for 144 h. MN10042 (solid bars) or MN10041.5 (open bars) was added at the indicated concentrations at the beginning of the cultures. Tetanus toxoid was added to each well to a final concentration of 4 LF/ml. Thymidine incorporation was determined as for Fig. 1. Results represent the means and standard errors of quadruplicate samples. Similar results were observed in at least two other experiments.

in intracellular cAMP, while MN10042 appears to act through a cAMP-independent mechanism.[108,109]

Still another mechanism which has been proposed for Nef-associated immunosuppression is the effect of Nef on hematopoiesis. Calenda *et al.*[110] reported that supernatants from HIV-1 non-productivity infected cultures inhibited CFU-GM in clonogenic assays and that the active, growth-inhibitory factor was Nef. Anti-Nef antibodies were able to block the activity in the culture supernatants and recombinant Nef was able to mimic the activity from the culture supernatants. Furthermore, the authors confirmed the involvement of Nef by demonstrating that culture supernatants generated with Nef-deficient virus were inactive.

6. SUMMARY

It is clear from the numerous studies described above that a number of immunomodulating activities, mostly immunosuppressive, have been reported for the HIV-1 envelope proteins gp120/gp160 and gp41 and the regulatory proteins Tat and Nef on *in vitro* functions of lymphoid or hematopoietic cells. For at least some of these studies, the concentrations of soluble, extracellular viral proteins affecting cellular functions are comparable to those which have been reported or might be anticipated in the plasma of HIV-1-infected individuals. Nonetheless, with very little *in vivo* data and a relative paucity of animal models in which to test hypotheses regarding potential functions of these proteins as extracellular mediators, these studies can only serve to challenge our intellectual curiosity. The data described have offered a number of ways by which HIV-1 structural and regulatory proteins might play a significant role in the immunopathogenesis of HIV infection. We are now challenged to take the leads offered by these reports and design future experiments, particularly in animals or patients, to get more definitive answers.

Of the HIV-1 proteins discussed, Nef has certainly emerged as an interesting and exciting protein with potential for immunomodulatory activity. In part, this can be attributed to both studies using Nef-attenuated SIV and reports on certain long-term HIV nonprogressors. But Nef has also been implicated in downregulation of cytolytic T cell function and recent studies, including our own, suggest a potential role for Nef in the Th1 → Th2 cytokine shift characteristic of progression to AIDS. The most probable truth is that, as with so many other diseases, the mechanisms of pathogenesis of HIV infection are both numerous and diverse and a number of the potential immunosuppressive mechanisms outlined in this chapter will ultimately be determined to play at least some role at some time in some patients.

REFERENCES

1. Barre-Sinoussi, F., Chermann, J.-C., Rey, F., Nugeyre, M. T., Chamaret, S., Gruest, J., Dauguet, C., Axler-Blin, C., Vezinet-Brun, F., Rouzioux, C., Rozenbaum, W., and Montagnier, L., 1983, Isolation of a T-lymphotropic retrovirus from a patient at risk for acquired immune deficiency syndrome (AIDS), *Science* **220**:868–871.
2. Gallo, R. C., Sarin, P. S., Gelmann, E. P., Robert-Guroff, M., Richardson, E., Kalyanaraman, V. S., Mann, D., Sidhu, G. D., Stahl, R. E., Zolla-Pazner, S., Leibowitch, J., and Popovic, M., 1983, Isolation of human T-cell leukemia virus in acquired immune deficiency syndrome (AIDS), *Science* **220**:865–867.
3. Fauci, A. S., 1988, The human immunodeficiency virus: Infectivity and mechanisms of pathogenesis, *Science* **239**:617–622.

4. Lane, H. C., Depper, J. M., Greene, W. C., Whalen, G., Walkmann, T. A., and Fauci, A. S., 1985, Qualitative analysis of immune function in patients with the acquired immunodeficiency syndrome, *N. Engl. J. Med.* **313:**79–84.

5. Shearer, G. M., Bernstein, D. C., Tung, K. S., Via, C. S., Redfield, R., Salahuddin, S. Z., and Gallo, R. C., 1986, A model for the selective loss of major histocompatibility complex self-restricted T cell immune responses during the development of acquired immune deficiency syndrome (AIDS), *J. Immunol.* **137:**2514–2521.

6. Gurley, R. J., Ikeuchi, K., Byrn, R. A., Anderson, K., and Groopman, J. E., 1989, CD4[+] lymphocyte function with early immunodeficiency virus infection, *Proc. Natl. Acad. Sci. USA* **86:**1993–1997.

7. Viscidi, R. P., Mayur, K., Lederman, H. M., and Frankel, A. D., 1989, Inhibition of antigen-induced lymphocyte proliferation by tat protein from HIV-1, *Science* **246:**1606–1608.

8. Zauli, G., Davis, B. R., Re, M. C., Visani, G., Furlini, G., and LaPlaca, M., 1992, Tat protein stimulates production of TGF-β 1 by marrow macrophages: A potential mechanism for human immunodeficiency virus-1-induced hematopoietic suppression, *Blood* **80:**3036–3043.

9. Gibellini, D., Zauli, G., Re, M. C., Milani, D., Furlini, G., Caramelli, E., Capitani, S., and LaPlaca, M., 1994, Recombinant human immunodeficiency virus type-1 (HIV-1) Tat protein sequentially up-regulates IL-6 and TGF-β 1 mRNA expression and protein synthesis in peripheral blood monocytes, *Br. J. Haematol.* **88:**261–267.

10. Rautonen, J., Rautonen, N., Martin, N. L., and Wara, D. W., 1994, HIV type 1 Tat protein induces immunoglobulin and IL-6 synthesis by uninfected peripheral blood mononuclear cells, *AIDS Res. Hum. Retroviruses* **10:**781–785.

11. Hofman, F. M., Wright, A. D., Dohadwala, M. M., Wong-Stall, F., and Walker, S. M., 1993, Exogenous tat protein activates human endothelial cells, *Blood* **82:**2774–2780.

12. Meyaard, L., Otto, S. A., Schuitemaker, H., and Miedema, F., 1992, Effects of HIV-1 Tat protein on human T cell proliferation, *Eur. J. Immunol.* **22:**2729–2732.

13. Benjouad, A., Mabrouk, K., Moulard, M., Gluckman, J. C., Rochat, H., Van Rietschoten, J., and Sabatier, J. M., 1993, Cytotoxic effect on lymphocytes of Tat from human immunodeficiency virus (HIV-1), *FEBS Lett.* **319:**119–124.

14. Subramanyam, M., Gutheil, W. G., Bachovchin, W. W., and Huber, B. T., 1993, Mechanism of HIV-1 Tat induced inhibition of antigen-specific T cell responsiveness, *J. Immunol.* **150:**2544–2553.

15. Gutheil, W. G., Subramanyam, M., Flentke, G. R., Sanford, D. G., Munoz, E., Huber, B. T., and Bachovchin, W. W., 1994, Human immunodeficiency virus 1 Tat binds to dipeptidyl aminopeptidase IV (CD26): A possible mechanism for Tat's immunosuppressive activity, *Proc. Natl. Acad. Sci. USA* **91:**6594–6598.

16. Chirmule, N., Than, S., Khan, S. A., and Pabwa, S., 1995, Human immunodeficiency virus Tat induces a functional unresponsiveness in T cells, *J. Virol.* **69:**492–498.

17. Calenda, V., and Chermann, J. C., 1995, HIV Tat protein potentiates *in vitro* granulo-monocytic progenitor cell growth, *Eur. J. Haematol.* **54:**180–185.

18. Li, C. J., Friedman, D. J., Wang, C., Metelev, V., and Pardee, A. B., 1995, Induction of apoptosis in uninfected lymphocytes by HIV-1 Tat protein, *Science* **268:**429–431.

19. Zauli, G., Gibellini, D., Caputo, A., Bassini, A., Negrini, M., Monne, M., Mazzoni, M., and Capitani, S., 1995, The human immunodeficiency virus type-1 Tat protein upregulates Bcl-2 gene expression in Jurkat T-cell lines and primary peripheral blood mononuclear cells, *Blood* **86:**3823–3834.

20. Katsikis, P. D., Garcia-Ojeda, M. E., Torres-Roca, J. F., Greenwald, D. R., Herzenberg, L. A., and Herzenberg, L. A., 1997, HIV type 1 Tat protein enhances activation—but not Fas (CD95)-induced peripheral blood T cell apoptosis in health individuals, *Int. Immunol.* **9:**835–841.

21. McCloskey, T. W., Ott, M., Tribble, E., Khan, S. A., Teichberg, S., Paul, M. O., Pahwa, S., Verdin, E., and Chirmule, N., 1997, Dual role of HIV Tat in regulation of apoptosis in T cells, *J. Immunol.* **158**:1014–1019.
22. Lafrenie, R. M., Wahl, L. M., Epstein, J. S., Hewlett, I. K., Yamada, K. M., and Shawan, S., 1996, HIV-1-Tat protein promotes chemotaxis and invasive behavior by monocytes, *J. Immunol.* **157**:974–977.
23. Mitola, S., Sozzani, S., Luini, W., Primo, L., Borsatti, A., Weich, H., and Bussolino, F., 1997, Tat-human immunodeficiency virus-1 induces monocyte chemotaxis by activation of VEGF-1, *Blood* **90**:1365–1372.
24. Shalaby, M. R., Krowka, J. F., Gregory, T. J., Hirabayashi, S. E., McCabe, S. M., Kaufman, D. S., Stites, D. P., and Ammann, A. J., 1987, The effects of human immunodeficiency virus recombinant envelope glycoprotein on immune cell functions *in vitro, Cell. Immunol.* **110**:140–148.
25. Sandstrom, E. G., Andrews, C., Schooley, R. T., Byington, R., and Hirsch, M. S., 1986, Suppression of concanavalin A-induced blastogenesis by HTLV-III-infected H9 cells, *Clin. Immunol. Immunopathol.* **40**:253–258.
26. Mann, D. L., Lasane, F., Popovic, M., Arthur, L. O., Robey, W. G., Blattner, W. A., and Newman, M. J., 1987, HTLV-III large envelop protein (GP120) suppresses PHA-induced lymphocyte blastogenesis, *J. Immunol.* **138**:2640–2644.
27. Chirmule, N., Kalyanaraman, V., Oyaizu, N., and Pahwa, S., 1988, Inhibitory influences of envelope glycoproteins of HIV-1 on normal immune responses, *J. Acquired Immune Defic. Syndr.* **1**:425–430.
28. Krowka, J., Stites, D., Mills, J., Hollander, H., McHugh, T., Busch, M., Wilhelm, L., and Blackwood, L., 1988, Effects of IL-2 and large envelope glycoprotein (gp120) of human immunodeficiency virus (HIV) on lymphocyte proliferative responses to cytomegalovirus, *Clin. Exp. Immunol.* **72**:179–185.
29. Weinhold, K. J., Lyerly, H. K., Stanley, S. D., Austin, A. A., Matthews, T. J., and Bolognesi, D. P., 1989, HIV-1 gp120-mediated immune suppression and lymphocyte destruction in the absence of viral infection, *J. Immunol.* **142**:3091–3097.
30. Chirmule, N., Kalyanaraman, V. S., Oyaizu, N., Slade, H. B., and Pahwa, S., 1990, Inhibition of functional properties of tetanus antigen-specific T-cell clones by envelope glycoprotein gp120 of human immunodeficiency virus, *Blood* **75**:152–159.
31. Liegler, T. J., and Stites, D. P., 1994, HIV-1 gp120 and anti-gp120 induce reversible unresponsiveness in peripheral CD4 T lymphocytes, *J. Acquired Immune Defic. Syndr.* **7**:340–348.
32. Di Rienzo, A. M., Furlini, G., Olivier, R., Ferris, S., Heeney, J., and Montagnier, L., 1994, Different proliferative response of human and chimpanzee lymphocytes after contact with human immunodeficiency virus type 1 gp120, *Eur. J. Immunol.* **24**:34–40.
33. Wahl, L. M., Corcoran, M. L., Pyle, S. W., Arthur, L. O., Harel-Bellan, A., and Farrar, W. L., 1989, Human immunodeficiency virus glycoprotein (gp120) induction of monocyte arachidonic acid metabolites and IL-1, *Proc. Natl. Acad. Sci. USA* **86**:621–625.
34. Terai, C., Kornbluth, R. S., Pauza, C. D., Richman, D. D., and Carson, D. A., 1991, Apoptosis as a mechanism of cell death in cultured T lymphoblasts acutely infected with HIV-1, *J. Clin. Invest.* **87**:1710–1715.
35. Banda, N. K., Bernier, J., Kurahara, D. K., Kurrle, R., Haigwood, N., Sekaly, R. P., and Finkel, T. H., 1992, Crosslinking CD4 by human immunodeficiency virus gp120 primes T cells for activation-induced apoptosis, *J. Exp. Med.* **176**:1099–1106.
36. Amendola, A., Lombardi, G., Oliviero, S., Colizzi, V., and Piacentini, M., 1994, HIV-1 gp120-dependent induction of apoptosis in antigen-specific human T cell clones is characterized by "tissue" transglutaminase expression and prevented by cyclosporin A, *FEBS Lett.* **339**:258–264.
37. Oyaizu, N., McCloskey, T. W., Than, S., Hu, R., and Pahwa, S., 1995, Mechanism of

apoptosis in peripheral blood mononuclear cells of HIV-infected patients, *Adv. Exp. Med. Biol.* **374:**101–114.

38. Hashimoto, F., Oyaizu, N., Kalyanaraman, V. S., and Pahwa, S., 1997, Modulation of Bcl-2 protein by CD4 cross-linking: A possible mechanism for lymphocyte apoptosis in human immunodeficiency virus infection and for rescue of apoptosis by IL-2, *Blood* **90:**745–753.

39. Wahl, S. M., Allen, J. B., Gartner, S., Orenstein, J. M., Popovic, M., Chenoweth, D. E., Arthur, L. O., Farrar, W. L., and Wahl, L. M., 1989, HIV-1 and its envelope glycoprotein downregulate chemotactic ligand receptors and chemotactic function of peripheral blood monocytes, *J. Immunol.* **142:**3553–3559.

40. Durrbaum-Landmann, I., Kaltenhauser, E., Flad, H. D., and Ernst, M., 1994, HIV-1 envelope protein gp120 affects phenotype and function of monocytes *in vitro, J. Leuk. Biol.* **55:**545–551.

41. Kaneko, H., Hishikawa, T., Sekigawa, I., Hashimoto, H., Okumura, K., and Kaneko, Y., 1997, Role of tumor necrosis factor-alpha (TNF-alpha) in the induction of HIV-1 gp120-medicated CD4$^+$ cell anergy, *Clin. Exp. Immunol.* **109:**41–46.

42. Zembala, M., Pryjma, J., Plucienniczak, A., Szczepanek, A., Ruggiero, I., Jasinski, M., and Colizzi, V., 1994, Modulation of antigen-presenting capacity of human monocytes by HIV-1 gp120 molecule fragments, *Immunol. Invest.* **23:**189–199.

43. Chirmule, N., McCloskey, T. W., Hu, R., Kalyanaraman, V. S., and Pahwa, S, 1995, HIV gp120 inhibits T cell activation by interfering with expression of co-stimulatory molecules CD40 ligand and CD80 (B7-1), *J. Immunol.* **155:**917–924.

44. Capobianchi, M. R., Ankel, H., Amgelio, F., Paganelli, R., Pizzoli, P. M., and Dianzani, F., 1992, Recombinant glycoprotein 120 of human immunodeficiency virus is a potent interferon inducer, *AIDS Res. Hum. Retroviruses* **8:**575–579.

45. Ferbas, J. J., Toso, J. F., Logar, A. J., Navratil, J. S., and Rinaldo, C. R., Jr., 1994, CD4$^+$ blood dendritic cells are potent producers of IFN-α in response to *in vitro* HIV-1 infection, *J. Immunol.* **152:**4649–4662.

46. Hu, R., Oyaizu, N., Kalyanaraman, V. S., and Pahwa, S., 1994, HIV-1 gp160 as a modifier of TH1 and TH2 cytokine responses: gp160 suppresses IFN-γ and IL-2 production concomitantly with enhanced IL-4 production *in vitro, Clin. Immunol. Immunopathol.* **73:**245–251.

47. Ameglio, F., Capobianchi, M. R., Castiletti, C., Cordiali, F. P., Fais, S., Trento, E., and Dianzani, F., 1994, Recombinant gp120 induces IL-10 in resting peripheral blood mononuclear cells: Correlation with the induction of other cytokines, *Clin. Exp. Immunol.* **95:**455–458.

48. Borghi, P., Fantuzzi, L., Varano, B., Gessani, S., Puddu, P., Conti, L., Capobianchi, M. R., Ameglio, F., and Belardelli, F., 1995, Induction of IL-10 by human immunodeficiency virus type 1 and its gp120 protein in human monocytes/macrophages, *J. Virol.* **69:**1284–1287.

49. Ameglio, F., Cordiali, Fei, P., Solmone, M., Bonifati, C., Prignano, G., Giglio, A., Caprilli, F., Gentili, G., and Capobianchi, M. R., 1994, Serum II-10 levels in HIV-positive subjects: Correlation with CDC stages, *J. Biol. Regul. Homeostatic Agents* **8:**48–52.

50. Taoufik, Y., Lantz, O., Wallon, C., Charles, A., Dussaix, E., and Delfraissy, J. F., 1997, Human immunodeficiency virus gp120 inhibits IL-12 secretion by human monocytes: An indirect IL-10 mediated effect, *Blood* **89:**2842–2848.

51. Zauli, G., Re, M. C., Furlini, G., Giovannini, M., and LaPlaca, M., 1991, Evidence for an HIV-1 mediated suppression of *in vitro* growth of enriched (CD34$^+$) hematopoietic progenitors, *J. Acquired Immune Defic. Syndr.* **4:**1251–1253.

52. Zauli, G., Re, M. C., Furlini, G., Giovannini, M., and LaPlaca, M., 1992, Human immunodeficiency virus type 1 envelope glycoprotein gp120-mediated killing of human hematopoietic progenitors (CD34$^+$ cells), *J. Gen. Virol.* **73:**417–421.

53. Zauli, G., Re, M. C., Visani, G., Furlini, G., and LaPlaca, M., 1992, Inhibitory effect of HIV-1

envelope glycoproteins gp120 and gp160 on the *in vitro* growth of enriched (CD34$^+$) hematopoietic progenitor cells, *Arch. Virol.* **122:**271–280.

54. Sugiura, K., Oyaizu, N., Pahwa, R., Kalyanaraman, V. S., and Pahwa, S., 1992, Effect of human immunodeficiency virus-1 envelope glycoprotein on *in vitro* hematopoiesis of umbilical cord blood, *Blood* **80:**1463–1469.

55. Maciejewski, J. P., Weichold, F. F., and Young, N. S., 1994, HIV-1 suppression of hematopoiesis *in vitro* mediated by envelope glycoprotein and TNF-α, *J. Immunol.* **153:**4303–4310.

56. Zauli, G., Vitale, M., Gibellini, D., and Capitani, S., 1996, Inhibition of purified CD34$^+$ hematopoietic progenitor cells by human immunodeficiency virus 1 or gp120 mediated by endogenous TGF-β1, *J. Exp. Med.* **183:**99–108.

57. Wild, C. T., Shugars, D. C., Greenwell, T. K., McDanal, C. B., and Matthews, T. J., 1994, Peptides corresponding to a predictive alpha-helical domain of human immunodeficiency virus type 1 gp41 are potent inhibitors of virus infection, *Proc. Natl. Acad. Sci. USA* **91:**9770–9774.

58. Chen, C. H., Matthews, T. J., McDanal, C. B., Bolognesi, D. P., and Greenberg, M. L., 1995, A molecular clasp in the human immunodeficiency virus (HIV) type I TM protein determines the anti-HIV activity of gp41 derivatives: Implication for viral fusion, *J. Virol.* **69:**3771–3777.

59. Laurence, J., 1985, The immune system in AIDS, *Sci. Am.* **256:**84–93.

60. Cianciolo, G. J., Kipnis, R. J., and Snyderman, R., 1984, Similarity between p15E of murine and feline leukemia viruses and p21 of HTLV, *Nature* **311:**515.

61. Cianciolo, G. J., Copeland, T. D., Oroszlan, S., and Snyderman, R., 1985, Inhibition of lymphocyte proliferation by a synthetic peptide homologous to retroviral envelope proteins. *Science* **230:**453–455.

62. Cianciolo, G. J., Bogerd, H., and Snyderman, R., 1988, Human retrovirus-related synthetic peptides inhibit T lymphocyte proliferation, *Immunol. Lett.* **19:**7–14.

63. Ruegg, C. L., Monell, C. R., and Strand, M., 1989, Inhibition of lymphoproliferation by a synthetic peptide with sequence identity to gp41 of human immunodeficiency virus type 1, *J. Virol.* **63:**3257–3260.

64. Ruegg, C. L., and Strand, M., 1990, Inhibition of protein kinase C and anti-CD3-induced Ca^{2+} influx in Jurkat T cells by a synthetic peptide with sequence identity to HIV-1 gp41, *J. Immunol.* **144:**3928–3935.

65. Ruegg, C. L., and Strand, M., 1991, A synthetic peptide with sequence identity to the transmembrane protein gp41 of HIV-1 inhibits distinct lymphocyte activation pathways dependent on protein kinase C and intracellular calcium influx, *Cell. Immunol.* **137:**1–13.

66. Denner, J., Norley, S., and Kurth, R., 1994, The immunosuppressive peptide of HIV-1: Functional domains and immune response in AIDS patients, *AIDS* **8:**1063–1072.

67. Denner, J., Persin, C., Vogel, T., Haustein, D., Norley, S., and Kurth, R., 1996, The immunosuppressive peptide of HIV-1 inhibits T and B lymphocyte stimulation, *J. Acquired Immune Defic. Syndr.* **12:**442–450.

68. Wang, H., Nishanian, P., and Fahey, J. L., 1995, Characterization of immune suppression by a synthetic HIV gp41 peptide, *Cell. Immunol.* **16:**236–243.

69. Cauda, R., Tumbarello, M., Ortona, L., Kennedy, R. C., Shuler, K. R., Chanh, T. C., and Kanda, P., 1990, Inhibition of lymphokine-activated killer activity during HIV infection: Role of HIV-1 gp41 synthetic peptides, *Nat. Immun. Cell Growth Regul.* **9:**366–375.

70. Tyring, S. K., Cauda, R., Tumbarello, M., Ortona, L., Kennedy, R. C, Chanh, T. C., and Kanda, P., 1991, Synthetic peptides corresponding to sequences in HIV envelope gp41 and gp120 enhance *in vitro* production of IL-1 and TNF but depress production of IFN-α, IFN-γ, and IL-2, *Viral Immunol.* **4:**33–42.

71. Qureshi, N. M., Coy, D. H., Garry, R. F., and Henderson, L. A., 1990, Characterization of a

putative cellular receptor for HIV-1 transmembrane glycoprotein using synthetic peptides, *AIDS* **4**:553–558.

72. Ebenbichler, C. F., Roder, C., Vornhagen, R., Ratner, L., and Dierich, M. P., 1993, Cell surface proteins binding to recombinant soluble HIV-1 and HIV-2 transmembrane proteins, *AIDS* **7**:489–495.

73. Chen, Y. H., Bock, G., Vornhagen, R., Steindl, F., Katinger, H., and Dierich, M. P., 1993, HIV-1 gp41 binding to human peripheral blood mononuclear cells occurs preferentially to B lymphocytes and monocytes, *Immunobiology* **188**:323–329.

74. Oh, S. K., Cerda, S., Lee, B. G., Blanchard, G. C., Walker, J. E., and Hartshorn, K., 1993, Partial purification and identification of an immunosuppressive factor in AIDS sera, *AIDS Res. Hum. Retrovir.* **9**:365–373.

75. Chen, Y. H., Christiansen, A., and Dierich, M. P., 1995, HIV-1 gp41 selectively inhibits spontaneous cell proliferation of human cell lines and mitogen- and recall antigen-induced lymphocyte proliferation, *Immunol. Lett.* **48**:39–44.

76. Takeshita, S., Breen, E. C., Ivashchenko, M., Nishanian, P. G., Kishimoto, T., Vredevoe, D. L., and Martinez-Maza, O., 1995, Induction of IL-6 and IL-10 production by recombinant HIV-1 envelope glycoprotein 41 (gp41) in the THP-1 human monocytic cell line, *Cell. Immunol.* **165**:234–242.

77. Koutsonikolis, A., Haraguchi, S., Brigino, E. N., Owens, E. U., Good, R. A., and Day, N. K., 1997, HIV-1 recombinant gp41 induces IL-10 expression and production in peripheral blood monocytes but not in T-lymphocytes, *Immunol. Lett.* **55**:109–113.

78. Kestler, H. W., Ringler, D. J., Mori, K., Panicali, D. L., Sehgal, P. K., Daniel, M. D., and Desrosiers, R. C., 1991, Importance of the *nef* gene for maintenance of high virus loads and for development of AIDS, *Cell* **65**:651–662.

79. Deacon, N. J., Tsykin, A., Solomon, A., Smith, K., Ludford-Menting, M., Hooker, D. J., McPhee, D. A., Greenway, A. L., Ellet, A., Chatfield, C., Lawson, V. A., Crowe, S., Maerz, A., Sonza, S., Learmont, J., Sullivan, J. S., Cunningham, A., Dwyer, D., Dowton, D., and Mills, J., 1995, Genomic structure of an attenuated quasispecies of HIV-1 from a cohort of a blood transfusion donor and recipients, *Science* **270**:988–991.

80. Baba, T. W., Jeong, Y. S., Pennick, D., Bronson, R., Greene, M. F., and Ruprecht, R. M., 1995, Pathogenicity of live, attenuated SIV after mucosal infection of neonatal macaques, *Science* **267**:1820–1824.

81. Amiesen, J. C., Guy, B., Chamaret, S., Loche, M., Mouton, Y., Neyrinck, J. L., Khalife, J., Leprevost, C., Beaucaire, G., Boutillon, C., Gras-Masse, H., Maniez, M., Kieny, M.-P., Laustriat, D., Berthier, A., Mach, B., Montagnier, L., Lecocq, J.-P., and Capron, A., 1989, Antibodies to the nef protein and to nef peptides in HIV-1-infected seronegative individuals, *AIDS Res. Hum. Retrovir.* **5**:279–291.

82. Reiss, P., de Ronde, A., Lange, J. M., de Wolf, F., Dekker, J., Debouck, C., and Goudsmit, J., 1989, Antibody response to the viral negative factor (nef) in HIV-1 infection: A correlate of levels of HIV-1 expression, *AIDS* **3**:227–233.

83. Reiss, P., Lange, J. M., de Ronde, A., de Wolfe, F., Dekker, J., Debouck, C., and Goudsmit, J., 1990, Speed of progression to AIDS and degree of antibody response to accessory gene products of HIV-1, *J. Med. Virol.* **30**:163–168.

84. Wieland, U., Kuhn, J. E., Jassoy, C., Rubsamen-Waigmann, H., Wolber, V., and Braun, R. W., 1990, Antibodies to recombinant HIV-1 vif, tat, and nef proteins in human sera, *Med. Microbiol. Immunol.* **179**:1–11.

85. Rezza, G., Titti, F., Pezzotti, P., Sernicola, L., Lo Caputo, S., Angarano, G., Lazzarin, A., Sinicco, A., Rossi, G. B., and Verani, P., 1992, Anti-nef antibodies and other predictors of disease progression in HIV-1 seropositive injecting drug users, *J. Biol. Regul. Homeostatic Agents* **6**:15–20.

86. Fujii, Y., Ito, M., and Ikuta, K., 1993, Evidence for the role of human immunodeficiency

virus type 1 Nef protein as a growth inhibitor to CD4$^+$ T lymphocytes and for the blocking of the Nef function by anti-Nef antibodies, *Vaccine* **11**:837–847.

87. Fujii, Y., Nishino, Y., Nakaya, T., Tokunaga, K., and Ikuta, K., 1993, Expression of human immunodeficiency virus type 1 antigen on the surface of acutely and persistently infected human T cells, *Vaccine* **11**:1240–1246.

88. Otake, K., Fujii, Y., Nakaya, T., Nishino, Y., Zhong, Q., Fujinaga, K., Kameoka, M., Ohki, K., and Ikuta, K., 1994, The carboxy-terminal region of HIV-1 Nef protein is a cell surface domain that can interact with CD4$^+$ cells, *J. Immunol.* **153**:5826–5837.

89. Freund, J., Kellner, R., Konvalinka, J., Wolber, V., Kräusslich, H. G., and Kalbitzer, H. R., 1994, A possible regulation of negative factor (Nef) activity of human immunodeficiency virus type 1 by the viral protease, *Eur. J. Biochem.* **223**:589–593.

90. Gaedigk-Nitschko, K., Schön, A., Wachinger, G., Erfle, V., and Kohleisen, B., 1995, Cleavage of recombinant and cell derived human immunodeficiency virus 1 (HIV-1) Nef protein by HIV-1 protease, *FEBS Lett.* **357**:275–278.

91. Welker, R., Kottler, H., Kalbitzer, H. R., and Kräusslich, H.-G., 1996, Human immunodeficiency virus type 1 Nef protein is incorporated into virus particles and specifically cleaved by the viral proteinase, *Virology* **219**:228–236.

92. Bukovsky, A., Dorfman, T., Weiman, A., and Göttlinger, H. G., 1997, Nef association with human immunodeficiency virus type 1 virions and cleavage by the viral protease, *J. Virol.* **71**:1013–1018.

93. Chirmule, N., Oyaizu, N., Saxinger, C., and Pahwa, S., 1994, Nef protein of HIV-1 has B-cell stimulatory activity, *AIDS* **8**:733–734.

94. Torres, B. A., Tanabe, T., and Johnson, H. M., 1996, Characterization of Nef-induced CD4 T cell proliferation, *Biochem. Biophys. Res. Commun.* **225**:54–61.

95. Torres, B. A., Tanabe, T., Yamamoto, J. K., and Johnson, H. M., 1996, HIV encodes for its own CD4 T-cell superantigen mitogen, *Biochem. Biophys. Res. Commun.* **225**:672–678.

96. Tanabe, T., Torres, B. A., Subramaniam, P. S., and Johnson, H. M., 1997, V beta activation by HIV Nef protein: Detection by a simple amplification procedure, *Biochem. Biophys. Res. Commun.* **230**:509–513.

97. Novembre, F. J., Lewis, M. G., Saucier, M. M., Yalley-Ogunro, J., Brennan, T., McKinnon, K., Bellah, S., and McClure, H. M., 1996, Deletion of the Nef gene abrogates the ability of SIV smmPBj to induce acutely lethal disease in pigtail macaques, *AIDS Res. Hum. Retrovir.* **12**:727–736.

98. Benveniste, O., Vaslin, B., Le Grand, R., Cheret, A., Matheux, F., Theodoro, F., Cranage, M. P., and Dormont, D., 1996, Comparative interleukin (IL-2)/interferon (IFN)-gamma and IL-4/IL-10 responses during acute infection of macaques inoculated with attenuated nef-truncated or pathogenic SICmac251 virus, *Proc. Natl. Acad. Sci. USA* **93**:3658–3663.

99. Collette, Y., Dutartre, H., Benziane, A., and Olive, D., 1996, Similarity between Nef of primate lentiviruses and p15E of murine and feline leukaemia viruses, *AIDS* **10**:441–442.

100. Haraguchi, S., Good, R. A., Cianciolo, G. J., Engelman, R. W., and Day, N. K., 1997, Immunosuppressive retroviral peptides: Immunopathological implications for immuno-suppressive influences of retroviral infections, *J. Leuk. Biol.* **61**:654–666.

101. Clerici, M., and Shearer, G. M., 1993, A TH1 to TH2 switch is a critical step in the etiology of HIV infection, *Immunol. Today* **14**:107–111.

102. Clerici, M., and Shearer, G. M., 1994, The Th1/Th2 hypothesis of HIV infection: New insights, *Immunol. Today* **15**:575–581.

103. Clerici, M. A., Sarin, A., Coffman, R. L., Wynn, T. A., Blatt, S. P., Hendrix, C. W., Wolf, S. F., Shearer, G. M., and Henkart, P. A., 1994, Type 1/type 2 cytokine modulation of T cell programmed cell death as a model for HIV pathogenesis, *Proc. Natl. Acad. Sci. USA* **91**:11811–11815.

104. Clerici, M., Wynn, T. A., Berzofsky, J. A., Blatt, S. P., Hendrix, C. W., Sher, A., Coffman,

R. L., and Shearer, G. M., 1994, Role of IL-10 in T helper cell dysfunction in asymptomatic individuals infected with the human immunodeficiency virus, *J. Clin. Invest.* **93:**768–775.

105. Brigino, E., Haraguchi, S., Koutsonikolis, A., Cianciolo, G. J., Owens, U., Good, R. A., and Day, N. K., 1997, IL-10 is induced by recombinant HIV-1 Nef protein involving the calcium/calmodulin-dependent phosphodiesterase signal transduction pathway, *Proc. Natl. Acad. Sci. USA* **94:**3178–3182.

106. Fujii, Y., Otake, K., Tashiro, M., and Adachi, A., 1996, Soluble Nef antigen of HIV-1 is cytotoxic for human CD4$^+$ T cells, *FEBS Lett.* **393:**396.

107. Okada, H., Takei, R., and Tashiro, M., 1997, HIV-1 Nef protein-induced apoptotic cytolysis of a broad spectrum of uninfected human blood cells independently of CD95 (Fas), *FEBS Lett.* **414:**603–606.

108. Haraguchi, S., Cianciolo, G. J., Good, R. A., James-Yarish, M., Brigino, E., and Day, N. K., 1998, Inhibition of IL-2 and IFN-γ by a HIV-1 Nef-encoded synthetic peptide, *AIDS* **12:** 820–823.

109. Gottlieb, R. A., Kleinerman, E. S., O'Brian, C. A., Tsujimoto, S., Cianciolo, G. J., and Lennarz, W. J., 1990, Inhibition of protein kinase C by a peptide conjugate homologous to a domain of the retroviral protein p15E, *J. Immunol.* **145:**2566–2570.

110. Calenda, V., Graber, P., Delamarter, J. F., and Chermann, J. C., 1994, Involvement of HIV Nef protein in abnormal hematopoiesis in AIDS: *In vitro* study on bone marrow progenitor cells, *Eur. J. Haematol.* **52:**103–107.

4

Carbohydrate Interactions and HIV-1

THOMAS KIEBER-EMMONS

1. INTRODUCTION

More than 18 million people worldwide are estimated to be infected with the human immunodeficiency virus (HIV), the cause of acquired immunodeficiency syndrome (AIDS). There is accumulating evidence that HIV-1 can interact with target cells via high-mannose and/or biantennary structures on cell-surface-expressed glycolipids and glycoproteins. In this regard, HIV-1 recombinant envelope (Env) glycoprotein (gp) precursor gp160 (rgp160) behaves as a mannosyl/*N*-acetylglucosaminyl (GlcNAc)-binding protein.[1] Inhibition of HIV infection of CD4$^+$ cells *in vitro* by CD4-free glycopeptides or by carbohydrate forms has been shown.[1–4] The HIV envelope protein is also highly glycosylated. HIV-1 gp160 glycans are also representative of mannose, lacto-series, and sialo-syl carbohydrate residues. Characterized O- and N-linked carbohydrates on the HIV-1 envelope protein are also common saccharide subunits among bacterial, fungal, and tumor-associated antigens (Ag).[5,6] Glycosylation is necessary for HIV-1 gp120 to attain a functional conformation, and individual N-linked glycans of gp120 are important, but not essential, for replication of HIV-1 in cell culture.[6]

The pathogenic role associated with HIV glycosylation can take several forms which may be compared with the glycobiology of tumor cells or with other pathogens. For example, phenotypic alterations with respect to surface carbohydrate expression appear to provide a selective growth advantage for human adenocarcinoma cells in terms of facilitated extravasation

THOMAS KIEBER-EMMONS • Department of Pathology and Laboratory Medicine, University of Pennsylvania, Philadelphia, Pennsylvania 19104-6082.

Human Retroviral Infections, edited by Kenneth E. Ugen *et al.*
Kluwer Academic / Plenum Publishers, New York, 2000.

via selectins. Similarly, after HIV glycoproteins are glycosylated by host cell glycosyltransferases, the virus may use the carbohydrate determinants as ligands for expansion to a spectrum of target cells during spread of the infection through the body. Comparison of oligosaccharide profiles on the envelope glycoproteins of different virus isolates propagated in the same host cells yields very similar glycan patterns, whereas cultivation of an isolate in different host cells results in markedly divergent oligosaccharide maps.[7] Variations in the pattern concern the proportion of high-mannose-, hybrid-, and complex-type substituents, as well as the state of charge and structural parameters of the complex-type species. As a characteristic feature, complex-type glycans of macrophage-derived viral glycoprotein are almost exclusively substituted by lactosamine repeats.[7] Hence, glycosylation of the external envelope glycoprotein seems to be primarily governed by host cell-specific factors rather than by the amino acid sequence of the corresponding poly-peptide backbone. Changes in glycosylation might affect the neutralization ability of antibodies that otherwise neutralize primary isolates.[8,9] It has been observed that mutated virus lacking an N-linked glycan in the V1 loop of gp120 is more resistant to neutralization by monoclonal antibodies to the V3 loop and neutralization by soluble recombinant CD4 (sCD4).[6] Conversely, certain lectins and anticarbohydrate monoclonal antibodies (mAbs) have the capacity to neutralize a variety of different HIV field and laboratory isolates by interference with viral attachment.[10]

The primary molecular changes that lead to the development of AIDS are poorly understood. Recent data suggest that HIV may be using the glycosylation system of T lymphocytes to acquire glycans for its glycoproteins that enable it to disrupt carbohydrate-dependent immune cell interactions or induce aberrant immune reactions. These observations have led to viewing AIDS from a glycobiological perspective with potential linkages to the human fetoembryonic defense system hypothesis.[11] These findings sug-gest that mimicry or acquisition of glycans by pathogens or tumor cells may enable them to either subvert or misdirect the human immune response, thereby greatly increasing their pathogenicity, and that expression of car-bohydrates by normal cells and tissues may protect them from immune responses, especially in those cases where major histocompatibility recogni-tion is either absent or minimal.[11] A better understanding of this hypothesis and its corollaries may enable us to address the molecular mechanisms underlying not only AIDS, but also a host of other very serious pathological conditions.

Natural alteration of the glycosylation pattern of viral variants affects both the humoral and cellular immune response.[12] In this review, issues associated with the role carbohydrates play as targets in HIV vaccine devel-opment is contrasted with issues associated with carbohydrate vaccine development for bacterial and fungal pathogens in general. The unusually highly glycosylated state of the major envelope glycoprotein (gp160) of HIV has offered a challenge to both glycobiologists and virologists.[13] What is the functional significance of such a mass of glycans and how might they be manipulated to disadvantage HIV pathogenesis?

2. CARBOHYDRATE STRUCTURES ARE UBIQUITOUS IN NATURE

Carbohydrates that influence the patho/glycobiology of pathogens are ubiquitous in nature. Carbohydrates in general play an essential role in cell biology, being involved in cell–cell communication, cell proliferation, and differentiation (cell growth). Aberrant glycosylation is a basis for uncontrolled cell growth, invasiveness, and increased metastatic potential of tumor cells.[14] Glyconjugate forms found on normal cells mediate the adhesion of pathogens and toxins to host cells. Much of what we know about carbohydrates and pathogens comes from studies of Gram-negative and Gram-positive bacteria. The main targets of the protective immune response against bacterial infections are the capsular polysaccharide (PS) as well as the O-antigen carbohydrate moiety of the lipopolysaccharide (LPS). The epitopes recognized by anticarbohydrate mAbs range from one sugar unit up to ten sugar units. Although most anticarbohydrate mAbs are directed predominantly toward terminal sugar residues, a few mAbs are also reactive with internal sugar residues. LPS is usually the major glycolipid present in Gram-negative bacteria.[15] The structures of LPS have been reviewed.[16] Chemically, LPS, as characterized by enterobacterial LPS, is composed of a poly- or oligosaccharide covalently linked to a lipid component, termed lipid A (Fig. 1). High-molecular-weight (MW) LPS consists of an O-specific polysaccharide chain, which is a polymer of repeating oligosaccharide units, a core oligosaccharide, and lipid A; whereas the low-molecular weight rough form of LPS lacks the O-chain. Like wild-type strains of Enterobacteriaceae, *Helicobacter pylori* strains produce high-MW LPS (Fig. 1A). Wild-type strains of *Campylobacter jejuni* produce high-MW LPS, structurally distinct low-MW LPS resembling those of *Neisseria* and *Haemophilus* spp., or both (Fig. 1A, B). Lipooligosaccharides (LOS) are the major glycolipids expressed on mucosal Gram-negative bacteria, including members of the genera *Neisseria, Haemophilus, Bordetella,* and *Branhamella.* They can also be expressed on some enteric bacteria such as *Campylobacter jejuni* and *Campylobacter coli* strains. LOS is analogous to the LPS found in other Gram-negative families. Certain species of Gram-negative mucosal pathogens such as *Neisseria gonorrhoeae, Neisseria meningitidis,* and *Haemophilus influenzae* express a low-MW LPS (Fig. 1C), or what some investigators refer to as LOS.

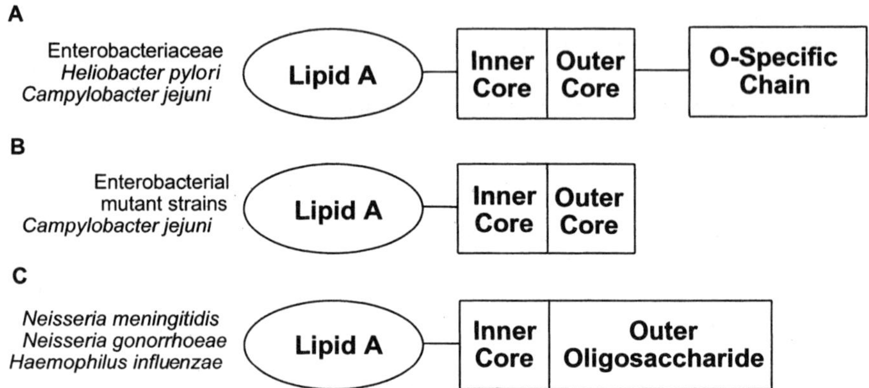

FIGURE 1. Schematic representation of the general organization of (A) high-MW LPS, (B) low-MW LPS, and (C) LOS.

Two characteristics of these LOS that make them different from rough-type LPS of enteric bacteria is their structural and antigenic similarity to human glycolipids and the potential for certain LOS to be modified *in vivo* by host substances or secretions. Therefore, modification of LOS in different environments of the host results in the synthesis of new LPS carbohydrates that probably benefit survival of the pathogen. Although the majority of information regarding LPS activity is related to its role as an endotoxin, the normal symbiotic relationship between many bacteria and their human hosts emphasizes not only that LPS is important in the survival of the bacteria, but also that the host must have mechanisms for neutralizing or regulating the potentially harmful effects of LPS. Variations in the structure of the saccharide moiety of LPS can contribute to the virulence of bacterial strains. Consequently, a large number of encapsulated bacteria cause human disease, and individual vaccines must be developed for each. It is possible, however, that the immune response to carbohydrates is restricted to only a few epitope types or structures. Consequently, the definition of immunogenic carbohydrate forms may be difficult to identify and the reproduction of appropriate immunogenic structures in vaccine formulations may also be hampered.

Bacteria with LOS that mimic structures of glycosphingolipids (GSL) and are also tumor-associated are shown in Table I. Immunochemical studies of sialylated LOS of the Gram-negative bacteria *Neisseria gonorrhoeae* and *Neisseria meningitidis* show them to be antigenically and/or chemically identical to lactoneoseries glycosphingolipids.[16] The terminal trisaccharide lactotriaose (Galβ1-4GlcNAcβ1-3Gal) is common among LOS of pathogenic

TABLE I
Bacterial Lipooligosaccharides That Mimic Host Glycosphingolipids

GSL series type	Structure	Bacterial species
Lacto	Galβ1→4GlcNAcβ1→3Galβ1→4Glcβ1→ Cer	*Neisseria gonorrhoeae* *N. meningitidis* *N. lactamica* *N. cinera* *Haemophilus influenzae* type b (S) H. influenzae NT (S) H. influenzae biotype aegyptius H. ducreyi (S)
Globo	Galα1→4Galβ1→4Glcβ1→Cer	*N. gonorrhoeae* *N. meningitidis* *H. influenzae* type b *H. influenzae* NT *Branhamella catarrhalis*
Gangilo	GalNAcβ1→4Galβ1→4Glcβ1→Cer Galβ1→3GalNAcβ1→4Galβ1→4Glcβ1 GalNAcβ1→3Galβ1→4GlcNAcβ1→ 3Galβ1→4Glcβ1→Cer	*N. gonorrhoeae*

Neisseria spp. since both serological and structural studies have shown that most serogroup B and C strains of meningococci and most gonococci strains possess this trisaccharide. Lactotriaose is a precursor of lacto-*N*-neotetraose, the precursor of lactoneoseries glycosphingolipids. In contrast, the O-chains of a number of *Helicobacter pylori* strains exhibit mimicry of LeX and Y blood group antigens.[17] Furthermore, *N. gonorrhoeae* 1291 has an LOS containing a paragloboside terminal oligosaccharide. Globotriaose (Galα1-4Galβ1-4Glc) has been identified as the terminal trisaccharide LOS of the pycocin-resistant mutant. An additional type of mimicry of *N. gonorrhoeae* involves a pentasaccharide with an *N*-acetylgalactosamine (GalNAc) residue β1-3 linked to a terminal galactose (Gal) of a lacto-*N*-neotetraose. Anti-lactoneo-series monoclonal antibodies bind to *Haemophilus ducreyi* and to nontypable and type b *H. influenzae*. The core oligosaccharides of LPS of *Campylobacter jejuni* serotypes exhibit mimicry of gangliosides.

These cancer-related mucin-type and bacterial-associated carbohydrate epitopes (Table I) are also expressed on HIV gp160. Since early evidence indicated that mannosyl residues are involved in HIV-1 pathogenesis, investigators have examined the ability of lectins and antimannan antibodies to neutralize HIV infection. The cross-reactivity of antibodies induced by microorganism preparations was thought a possibility to induce anticarbohydrate antibodies directed to HIV. For example, using a mannan preparation from *Saccharomyces cerevisiae* or from *Candida albicans*, neutralizing antiserum was raised in rabbits which prevents HIV-1 infection *in vitro* up to a titer of 1:128.[18] Antibodies reactive with gangliosides have also been identified to neutralize HIV-1 *in vitro*. Antigalactocerebroside (GalC) antibodies have been reported to inhibit myelin formation, cause demyelination, and block HIV-1 infection of neural cells.[19] These antibodies bound strongly to GM1 ganglioside, monogalactosyl diglyceride, asialo-GM1, or GD1b and psychosine.

Antibodies directed to the Tn (GalNAc-Ser/Thr) or sialo-syl-Tn (NeuAc-GalNAc-Ser/Thr) antigen, occurring as a surface antigen on most primary human breast carcinomas and their metastases, inhibit HIV infection and syncytium formation.[5,20,22] Antibody to histo-blood group A antigen neutralizes HIV produced by lymphocytes from blood group A donors, but not from blood group B or O donors.[23] As a characteristic feature, complex-type glycans of monocyte-derived, macrophage-derived viral glycoprotein are almost exclusively substituted by lactosamine repeats. One type of lactosoamine present on the envelope structure is the histo-blood-related Lewis Y (LeY) determinant. The blood group-related neolactoseries carbohydrate structures LeX, sialyl-LeX (sLeX), ABH, Lewis a (Lea), sialyl-Lea (sLea), and LeY, are examples of terminal carbohydrate structures related to tumor prognosis.[24] LeY is expressed on HIV-infected human T cells and monocytes[25–27] as well as released HIV particles, whereas uninfected counterparts are LeY-negative. Anti-LeY and anti-LeX antibodies and lectins exhibit HIV neutralization and block syncytium formation.[20,21,23,28]

LeY is particularly interesting. The incidence of LeY antigen expression on CD8$^+$ T cells increases as the disease progresses with the ongoing impairment of immune function. The phenotype change that occurs with LeY antigen expression might reflect the abnormal activation of T lymphocytes of some specific, but unknown population of CD8$^+$ T cells. Thus, carbohydrate changes on the cell surface may induce immunological abnormality and accelerate the damage within the CD4$^+$ T cell subset, resulting in an impairment of the antigen-specific immune system.[29] Thus the expression of the Lewis Y carbohydrate represents a unique feature associated with HIV infection since (1) other carbohydrate isomeric structures defined by specific monoclonal antibodies do not show significant differences between infected and noninfected cells, suggesting specific expression of Lewis Y carbohydrate, (2) Lewis Y expression is not influenced by the extensive genetic heterogeneity affecting immunogenic properties of viral polypeptides, and (3) the incidence of LeY expression increases progressively in patients with advanced stages of AIDS in which the CD4$^+$ population greatly decreases. The striking expression of LeY in HIV-infected cells and in lymphocytes of AIDS patients suggests that HIV infection alters the membrane phenotype of T cells[29] similarly to the expression of LeY, which is associated with tumor progression and appears at the early stages of tumor development. The cross-reactivity of anticarbohydrate antibodies for HIV and tumor cells indicates that the pathophysiology of infection and neoplasia is profoundly affected by the same or similar carbohydrate forms.

Anti-alpha-galactosyl (anti-Gal) is a natural human serum antibody that binds to the carbohydrate Galα1-3Galβ1-4GlcNAc-R (alpha-galactosyl epitope) and is synthesized by 1% of circulating B lymphocytes in response to immune stimulation by enteric bacteria.[30,31] However, the biological significance of this binding specificity in humans is unclear, since unsubstituted Galα1-3Galβ1-4GlcNAcβ sequences are not found in human tissues, due to suppression of the gene coding for the enzyme Gal beta 3-transferase. Interestingly, retroviruses from various mammalian species, excluding humans, are effectively inactivated with normal human serum (NHS). Studies have shown that NHS inactivation of retroviruses occurs through natural antibody recognition of the terminal alpha-galactosyl epitope on the viral envelope that is acquired during replication in the host cell. Recently, NHS sensitivity of HIV was assessed following viral propagation in human cells that were manipulated to express the alpha-galactosyl epitope.[32] HUT-78 cells were transduced with an exogenous alpha-1-3-galactosyl transferase gene which codes for the terminal glycosyl transferase responsible for generation of the alpha-galactosyl epitope. The transduced HUT-78 cells expressed high levels of the alpha-galactosyl epitope on their membrane surface, rendering them sensitive to killing in NHS. Similarly, HIV passaged through these cells acquired the alpha-galactosyl epitope in association with the envelope glycoprotein gp120 and was also effectively inactivated in NHS. These results demonstrate that, like other retroviruses bearing the alpha-galactosyl epitope, HIV modified to express this epitope is inactivated in NHS. Furthermore, these data suggest that expression of the alpha-galactosyl epitope on the surface of viruses may have implications in the interspecies transmission of such viruses to humans.

3. INFLUENCE OF CARBOHYDRATE MOIETIES ON THE IMMUNOGENICITY OF HIV

Containing most of the documented binding sites for virus-neutralizing antibodies, gp120 is a prime target for the humoral arm of the human immune response to HIV-1 function. Knowledge of the structure, function, and immunogenicity of gp120 is therefore central to several applied areas of HIV research: vaccine development, immunotherapeutics, and rational

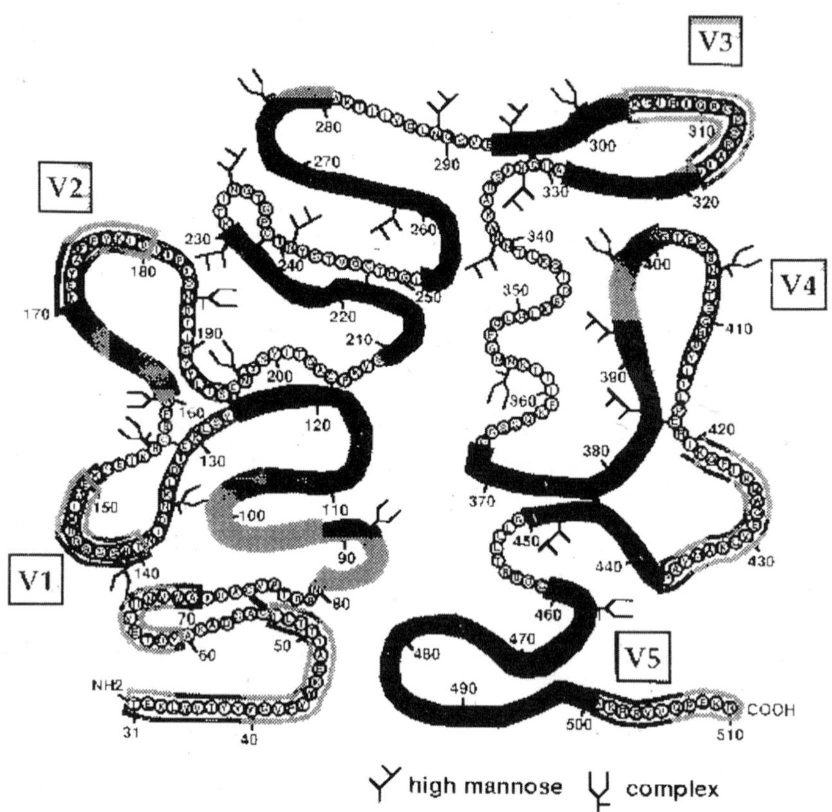

FIGURE 2. Schematic of exposure of continuous antibody epitopes on gp120. The darker the shading of blocked-in areas, the less accessible are the epitopes as determined from epitope mapping studies. The regions V1–V5 correspond to variable regions 1–5. Positions of high-mannose and complex carbohydrate sites will differ among isolates. Numbering scheme will also differ among isolates and is shown as illustration only. The figure is adapted from that published on a Web site by Dr. John P. Moore, Department of Microbiology, New York University, on HIV-1 gp120 Glycoprotein: Structure, Function, and Immunogenicity.

drug design (Fig. 2). The extensive glycosylation of gp120, coupled with its flexibility, has precluded crystallographic analysis of its full structure. To understand how the mature protein is folded, investigators have probed its surface with both murine and human monoclonal antibodies to defined or partially defined epitopes, using gp120 mutants to delineate the antibody epitopes.[33–35] To assist in the design of globally effective HIV vaccines, the antigenic variation of gp120 throughout the world has been explored.[36–38]

The glycans of gp160, part of which are highly sialylated, have been shown to influence gp160 immunogenicity. Recognition of viral Ag and of HIV gp120 in particular by human Th cells is critical in the immune response to the viral Ag, which includes antibody production and generation of cytotoxic cells. Procedures to increase antigenicity of gp120 are highly desirable from a vaccine perspective. Terminal sialic acid residues removed by neuraminidase treatment of gp120 were shown to enhance its uptake by exploiting the galactose receptors on antigen-presenting cells (APC).[39] Galactose residues were exposed and hence recognized by galactose receptors on APC. These findings and others indicate that the carbohydrate moieties of gp160 can modulate the specificity and perhaps the protective efficiency of the antibody response to the molecule.

Carbohydrate side chains of envelope glycoproteins of HIV-1 and other viruses have been postulated to interfere with binding of neutralizing antibodies. Consequently, the role played by glycans on the V3 domain of HIV-1 is of particular importance since the V3 region is a purported neutralization epitope. The effect of sialic acid removal on the antibody response to the third variable domain of HIV-1 envelope glycoprotein has been examined. Using a panel of synthetic V3 peptides, the anti-V3 antibodies generated in rabbits immunized by desialylated recombinant gp160LAI were characterized.[40] Amino acid residues flanking the GPGR tip of V3 (Fig. 2) were necessary for the recognition by anti-V3 antibodies raised against either native or desialylated gp160. Both types of antibodies reacted to V3 peptides of MN and SF2 strains and with a North American/European V3 consensus peptide, while anti-desialylated gp160LAI antibodies reacted in addition to the V3 and CDC4, WMJ2, and NY5 strains. The V3 peptides did not significantly differ in their secondary structure, as determined by circular dichroism. The titer and avidity for V3MN of anti-desialylated gp160LAI antibodies were significantly lower than those of anti-native gp160LAI, which likely accounts for the inability of anti-desialylated gp160LAI sera to neutralize HIV-1MN-induced syncytia. These results indicate that V3 immunogenicity may be influenced by subtle directed changes in the gp160 glycosylation pattern.[40]

In another investigation, four infectious HIV-1 molecular clones chimeric for their gp120 V3 domains were used to study the influence on HIV-1 neutralization of an N-glycan localized within the V3 loop.[22] Two clones lacking the 301N-glycan were at least eightfold more sensitive to neutralization by two V3-specific mAbs and 2- to 10-fold more sensitive to neutralization by a CD4-binding-site-specific human mAb than two HIV-1 clones glycosylated at this site. The affinity of the V3 mAbs for soluble gp120 of the four clones was similar. However, a decreased binding of these mAbs to the gp120 of the two 301N-glycosylated clones was observed when the majority of gp120 was virion-associated during the initial binding step. These findings indicate that the 301N-glycan may interfere with the binding of neutralizing antibodies by limiting the accessibility of neutralization sites or by inducing conformational changes in the HIV-1 gp120 molecule.

The role of glycans on V3 is further assessed by considering that HIV strains lacking a naturally conserved N-linked oligosaccharide (at position 306) within the V3 loop are highly sensitive to neutralization.[41] Sequencing of the V3 loop of resistant variants revealed that variants had become resistant at least partly by reacquisition of the 306N-glycan. Thus, HIV strains lacking the 306N-glycan primarily develop resistance to V3-directed neutralization through acquisition of the specific oligosaccharide. This demonstrates that protein glycosylation can be a primary modifier of virus antigenicity of possible importance for the interaction of HIV with the host immune response.

It is suggested that threonine or serine residues in the V3 loop of HIV-1 gp120 are glycosylated with the short-chain O-linked oligosaccharides Tn or sialosyl-Tn that function as epitopes for broadly neutralizing carbohydrate-specific antibodies. In a recent study, the effect of mutation of threonine or serine residues on the sensitivity to infectivity inhibition by Tn- or sialosyl-Tn-specific antibodies was examined.[10] All potentially O-glycosylated threonine and serine residues in the V3 loop of cloned HIV-1 BRU were mutagenized to alanine, thus abrogating any O-glycosylation at these sites. Additionally, one of these T-A mutants (T308A) also abrogated the signal for N-glycosylation at N306 inside the V3 loop. The mutant clones were compared with the wild-type virus as to sensitivity to neutralization with monoclonal and polyclonal antibodies specific for the tip of the V3 loop of BRU or for the O-linked oligosaccharides Tn or sialosyl-Tn. Deletion of the N-linked oligosaccharide at N306 increased the neutralization sensitivity to antibodies specific for a tip of the loop, which indicates that N-linked glycosylation modulates the accessibility to this immunodominant epitope as previously suggested.[22] However, none of the mutants with deletions of O-glycosylation signals in the V3 loop displayed any decrease in sensitivity to anti-Tn or anti-sialosyl-Tn antibody. This indicates that these broadly specific neutralization epitopes are located outside the V3 loop of gp120.

In related investigations the conformational properties of glycosylated analogues of the V3 neutralizing determinant were studied by NMR and circular dichroism spectroscopies. A 24-residue peptide from the HIV-1IIIB isolate (residues 308–331), designated RP135, was glycosylated with both N- and O-linked sugars.[42] The structures of two glycopeptides, one with an N-linked beta-glucosamine (RP135NG) and the other with two O-linked alpha-galactosamine units (RP135digal), were examined. The data showed that covalently linking a carbohydrate to the peptide has a major effect on the local conformation and imparts additional minor changes at more distant sites of partially defined secondary structure. In particular, the transient beta-type turn comprised of the -Gly-Pro-Gly-Arg- segment at the "tip" of the V3 loop is more highly populated in RP135digal than that in the native peptide and N-linked analogue. Binding data for the glycopeptides with 0.5beta, a monoclonal antibody mapped to the RP135 sequence, revealed a significant enhancement in binding for RP135digal as compared with the native peptide, whereas binding was reduced for the N-linked glycopeptide. These data show that glycosylation of V3-loop peptides can affect their conformations as well as their interactions with antibodies. These authors also showed that threonine glycosylation affects cytotoxic T targeting to the V3 region.[43]

The extent to which the peripheral carbohydrate structure of N-linked glycans influences the antigenic properties of HIV-1 gp120 has also been studied.[6,44] It was found that the carbohydrate structure NeuAc-Gal beta[1–4] of N-linked glycans, defined both by lectin reactivity and by specific glycosidases, is involved in modulating the binding of antibody to a number of epitopes of peptide nature. The binding of antibody to one class of epitopes, situated in a region between amino acids 200 and 230, was strongly increased by removal of NeuAc-Gal, whereas the binding to epitopes in the V3 region was decreased and the binding to epitopes in the far-N-terminal region was not altered by the treatment. These results suggested that peripheral structures of N-glycans are involved in modulating the overall conformation of gp120.

The glycosylation effect on neutralization sites was further illustrated using a mutant HIV-1 infectious clone lacking a signal for N-linked glycosylation in the V1 loop of HIV-1 gp120.[6] The mutated virus showed no differences in either gp120 content per infectious unit or infectivity, indicating that the N-linked glycan was neither essential nor affecting viral infectivity in cell culture. It was found that the mutated virus lacking an N-linked glycan in the V1 loop of gp120 was more resistant to neutralization by monoclonal antibodies to the V3 loop and neutralization by soluble recombinant CD4 (sCD4). Both viruses were equally well neutralized by concanavalin A (ConA) and a conformation-dependent human antibody IAM-2G12. These results suggest that the N-linked glycan in the V1 loop modulates the three-dimensional conformation of gp120 without changing the overall functional integrity of the molecule. This result is in contrast to neutralization effects upon glycosylation in the V1 region of the Env protein derived from simian immunodeficiency virus (SIV).[45] These data provide the first direct evidence that the carbohydrate profile of envelope protein can be distinct in SIV variants that evolve during infection of the host. Moreover, these studies showed that these changes in glycosylation in V1 were directly associated with changes in antigenicity. Specifically, serine and threonine changes in V1 allowed the virus to escape neutralization by macaque sera that contained antibodies that could neutralize the parental virus. The escape from antibody recognition appeared to be influenced by either O-linked or N-linked carbohydrate additions in V1. Moreover, when glycine residues were engineered at the positions where serine and threonine changes evolve in V1, there was no change in antigenicity compared to the parental form. This suggests that the amino acids in V1 are not part of the linear epitope recognized by neutralizing antibody. More likely, V1-associated carbohydrates mask the major neutralizing epitope of SIV. These experiments indicate that the selection of novel glycosylation sites in the V1 region of envelope during the course of disease is driven by humoral immune responses.

The correlation of N-linked glycosylation and neutralization resistance is further demonstrated in studies examining the biochemical and antigenic properties of the HIV-1 envelope produced in infected and radiolabeled primary peripheral blood mononuclear cell (PBMC) and monocyte-derived macrophage (MDM) cultures.[46] Comparison of the oligosaccharide profiles on the envelope protein reveals that different virus isolates, propagated in the same host cells, yield very similar glycan patterns, whereas cultivation of an isolate in different host cells result in markedly divergent oligosaccharide maps.[46] Variations concern the proportion of high-mannose-, hybrid-, and complex-type substituents, as well as the state of charge and structural parameters of the complex-type species. The gp120 produced in MDM migrates as a broad, diffuse band in sodium dodecyl sulfate–polyacrylamide gel electrophoresis gels compared with that of the more homogeneous gp120 released from PBMCs. Glycosidase analyses indicated that the diffuse appearance of the MDM gp120 is due to the presence of asparagine-linked carbohydrates containing lactosaminoglycans, a modification not observed with the gp120 produced in PBMCs. Neutralization experiments, using isogeneic PBMC and MDM-derived macrophage-tropic HIV-1 isolates, indicate that 8- to 10-fold more neutralizing antibody directed against the viral envelope is required to block virus produced from MDM. Consequently, cell types that are infected may alter the glycosylation pattern and subsequently neutralization capacity of antibodies.

4. ROLE OF CARBOHYDRATE IN VACCINE STRATEGIES TO HIV

Another mechanism for expanding the cellular tropism of HIV *in vitro* is through formation of phenotypically mixed particles (pseudotypes) with human T lymphotropic virus type I (HTLV-I).[47] It has been found that pseudotypes allow penetration of HIV particles into CD4-negative cells previously nonsusceptible to HIV infection. The infection of CD4-negative cells with pseudotypes could be blocked with anti-HTLV-I serum, but failed to be significantly inhibited with anti-HIV serum or a V3-neutralizing anti-gp120 monoclonal antibody. This may represent a possibility for pseudotypes to escape neutralization by the immune system *in vivo*. Importantly, the neutralizing capacity of lectins and anticarbohydrate monoclonal antibodies was found to block infection by cell-free pseudotypes in CD4-negative cells. These data suggest that although viral cofactors might expand the tropism of HIV *in vivo*, HIV and HTLV-I seem to induce common carbohydrate neutralization epitopes which might be conserved targets in a vaccine design strategy.

Targeting carbohydrates which are not structurally encoded by the viral genome as targets for neutralizing antibodies could present a new possibility for group-specific vaccine development because they are not sensitive to viral genetic variabilities present on a broader range of isolates. Early studies on consecutive escape virus isolates suggested that the majority of the change in neutralization sensitivity of variants is driven by the selective pressure of type-specific neutralizing antibodies.[48] No differences were observed in sensitivity to neutralization by anticarbohydrate neutralizing monoclonal antibodies or the lectin concanavalin A, indicating a conserved nature of certain carbohydrate neutralization epitopes during escape. The V3 sequence of three sets of consecutive virus isolates was analyzed revealing amino acid mutations in V3 sequences of all escape virus isolates. The biological significance of these variations was confirmed further by the demonstration of changes in sensitivity to neutralization by anti-V3 monoclonal antibodies. These results strongly suggest a participation of the neutralizing antibody response against the V3 loop in the immunoselection of escape virus.

The phenotypic and genotypic characteristics of HIV-1 from patients with AIDS in northern Thailand further demonstrate divergence of HIV-1 subtypes in an infected population that may have important implications for vaccine development.[49] In this study viral sequences were determined for 22 patients with AIDS, and all were subtype E HIV-1 on the basis of sequence analysis of a region from the envelope protein gp120. Syncytium-inducing (SI) viruses were detected for 16 of the 22 patients. All the SI viruses lost a potential N-linked glycosylation site in V3 which is highly conserved among previously described subtype E HIV-1 isolates from asymptomatic patients from Thailand. HIV-1 envelope sequences including V3 from some patients were significantly more divergent than viruses from asymptomatic patients in Thailand characterized earlier. These results suggest that emergence of subtype E SI HIV-1 variants is associated with the development of AIDS, as it is for subtype B HIV-1.

As discussed previously, changes in the glycosylation pattern of gp120 affect both titer and avidity of a humoral response.[40] Immunodominant epitopes are known to suppress a primary immune response to other antigenic determinants by a number of mechanisms. This concept was originally discussed as a means to induce a "biological smoke screen" thwarting the immune response.[50] Many pathogens have used this strategy to subvert the immune response and this may be a mechanism responsible for limited vaccine efficacies. HIV-1 vaccine efficacy appears to be complicated similarly by a limited, immunodominant, isolate-restricted immune response generally directed toward determinants in V3 of gp120. To overcome this problem, a cleaver approach based on masking the V3 domain through addition of N-linked carbohydrate and reduction in net positive charge has been investigated.[51] N-linked modified gp120s were expressed by recombinant vaccinia virus and used to immunize guinea pigs by infection and protein boosting. This modification resulted in variable site-specific glycosylation and antigenic dampening, without loss of gp120/CD4 binding or virus neutralization. Most importantly, V3 epitope dampening shifted the dominant type-specific neutralizing Ab response away from V3 to an epitope in V1 of gp120. Interestingly, in the presence of V3 dampening, V1 changes from an immunodominant nonneutralizing epitope to a primary neutralizing epitope with broader neutralizing properties.[51] In addition, Ab responses were also observed to conserved domains in C1 and C5. These results suggest that selective epitope dampening can lead to qualitative shifts in the immune response resulting in second-order neutralizing responses that may prove useful in the fine manipulation of the immune response and in the development of more broadly protective vaccines and therapeutic strategies.

Most carbohydrate antigens belong to the category of T cell-independent antigens that reflect their inability to stimulate major histocompatibility complex (MHC) class II-dependent T cell help.[52] As a consequence, carbohydrates are not capable of induction of a sufficient anamnestic or secondary immune response. Furthermore, antibodies produced in response to carbohydrate antigens usually are not of high affinity compared to those produced by responses to peptide or protein antigens. Consequently, new carbohydrate immunogens, formulations, and alternative vaccination strategies are constantly being evaluated. In this regard, novel approaches to refining a model to produce self-immunoadjuvanting vaccine constructs that will stimulate enhanced and persistent antibody responses to all of the components in a polysaccharide-based vaccine might prove useful. Genetically engineered protein–cytokine fusions have been conjugated to polysaccharides both in order to enhance the T cell-dependent response as well as to focus cytokine help onto polysaccharide-specific B cells. Interleukin-2 (IL-2) and granulocyte-macrophage colony-stimulating factor (GM-CSF) are two cytokine fusion partners shown to enhance Ig secretion to T cell-independent type 2 antigens *in vitro*. These combined approaches lay the groundwork for designing novel polysaccharide-based vaccines that stimulate high titer and persistent antibody in normal, neonatal, and immune-compromised hosts via systemic or mucosal routes.

In a further attempt to overcome the problems that arise from the T-independent immune response induced by carbohydrates, vaccine strategies have focused on development of either polysaccharide–protein conjugates or on antiidiotypic antibodies that mimic PS. The difficult steps in the former approach are the purification of the polysaccharide (especially when starting from LPS, which must be devoid of any residual lipid A-related endotoxin activity) and the loss of immunogenicity of the carbohydrate moiety during coupling to the protein carrier. Carbohydrate synthesis may diminish the problems associated with antigen purification, but remains a limited solution due to the overall difficulties of carbohydrate chemistry. The latter strategy, based upon the mimicry of carbohydrate antigens by antiidiotope antibodies, is not a simpler alternative, since obtaining these antibodies is relatively time-consuming, and discriminating among many possible antiidiotypes is also not straightforward.

An alternate approach to the development of T-dependent vaccines for carbohydrates is through the use of peptides that mimic capsular polysaccharides or LPS/LOS. We have developed peptides that induce immune responses to carbohydrate structures with *in vivo* and *in vitro* functionality.[53–56] Carbohydrate-mimicking peptides could revolutionize vaccines against infectious pathogens. Peptides that mimic carbohydrate structure have significant advantages as vaccines compared with carbohydrate–protein conjugates or antiidiotypic antibodies. First, the chemical composition and purity of synthesized peptides can be precisely defined. Second, the immunogenicity of the peptides can be significantly enhanced by polymerization or addition of relatively small carrier molecules that reduce the total amount of antigen required for immunization. Third, peptide synthesis may be more practical than synthesis of carbohydrate–protein conjugates or the production of antiidiotypes. Fourth, peptide-mimicking sequences can be engineered into DNA plasmids for DNA vaccination to further manipulate T cell responses.

Westerink and colleagues defined an antiidiotypic antibody called 6F9 that mimics the major C-polysaccharide (MCP) of *N. meningitidis*.[57] This antibody was sequenced and a YYRYD motif that was surface-exposed in the CDR3 of the heavy chain was identified as the putative MCP mimic.[56] We found the peptide sequence CARIYYRYDGTAY, when complexed to proteosomes, induces an anti-MCP antibody response in Balb/c mice that is protective when peptide-immunized mice were challenged with a lethal dose of bacteria.[56] We designed the peptide by placing a YGG spacer at the N-terminus, preceded by a cysteine. To induce the formation of dimers or cross-linking between the peptides, we placed cysteine just before the tyrosine. To facilitate hydrophobic complexing to the proteosomes, we placed a lauroyl group at the N-terminus.[56] Balb/c mice were hyperimmunized on a weekly basis with various peptide concentrations. The major Ig fraction upon immunization was found to be IgM, with IgG coming up later. The serum was found to be protective with a 100% survival rate across the dose range used. Immunization with proteosomes or saponin (QS21) gave the same result. Thus, immunization with either proteosomes or saponin provided the same level of protective immunity.

This peptide bears some homology with other peptides that mimic carbohydrates involving a planar–X–planar sequence motif (Table II). Peptide mimotopes for carbohydrates have also been defined containing a two-aromatic-amino acid repeat motif W/YXY found to bind to ConA (YPY),[58,59] in peptides that mimic the Lewis Y antigen (WLY),[60] and in peptides that bind to anti-*Crytococcous* polysaccharide antibodies.[61] Aromatic–aromatic residues have also been implicated as mimics for N-acetyl-beta-D-glucos-amine.[62–64] The immunological presentation of the putative motifs (i.e., short or longer peptides, presentation in a helix or beta bend) might mimic overlapping epitopes on otherwise different carbohydrate structures.

The homology displayed by the peptides in Table II suggests that the various carbohydrate forms mimicked by these peptides might display similar conformational features. This is further suggested by the fact that sera and monoclonals made to the YYRYD and YYPYD tracts cross-react with MCP on enzyme-linked immunosorbent assay (ELISA).[56] In addition, antibodies against these motifs might also cross-react with tumor cells that display various subunit forms reflective of these carbohydrate subunits. The sequence similarities among the putative motifs suggest that antibodies raised to this peptide set might cross-react with similar subunits expressed on what are otherwise dissimilar carbohydrate structures. Molecular modeling suggests that the Lewis Y (LeY) tetrasaccharide structure is similar to the core structure of MCP, suggesting that it is possible for antibodies to cross-react

TABLE II
Peptide Motifs That Mimic Carbohydrate Structures

Peptide	Carbohydrate	Structure
YYPY	Mannose	Methyl-α-D-mannopyranoside
WRY	Glucose	$\alpha(1-4)$glucose
PWLY	Lewis Y	Fucα1$\rightarrow$2Galβ1$\rightarrow$4(Fucα1$\rightarrow$3)GlcNAc
YYRYD	Group C polysaccharide	$\alpha(2-9)$sialic acid

with these moieties.[53] Similarly, reactivity with LeY might extend to Leb. Superposition of LeY and Leb structures indicates that in spite of the change of glycosidic linkage from $\beta1-3$ to $\beta1-4$ in the type 1 and 2 chains, resulting conformational features of the respective sugar moieties are still shared, forming a common topography.[65] The only effective difference is the position of the N-acetyl and hydroxymethyl groups projected on opposite sides of the type 1 and 2 difucosylated structures.

Antisera to peptides representative of the YXY motif found to mimic carbohydrate structures associated with the lipooligosaccharides were examined for their ability to bind to glycosylated HIV envelope protein and to neutralize HIV in syncytia neutralization assays.[53] Peptides containing motifs mimicking methyl-a-D-mannopyranoside (YYPYD) and sialo-syl (YYRYD) residues were synthesized and complexed to proteosomes. A third motif, YYRGD, which modifies the YXY sequence tract was also examined. We found that humoral immune responses can be induced in mice following immunization with respective peptide–proteosome complex that bound internal glycosylated glycoproteins gp140 and gp120 from two diverse HIV-1 isolates (MN and SF). Generation of antibodies was not Ir-gene-dependent, because at least two different strains of mice, Balb/c (H-2^d) and C57Bl/6 (H-2^b), responded equally to the peptides. Importantly, the same antisera did not bind nonglycosylated gp120 from HIV-1/SF. Anti-YYPYD and YYRYD, but not anti-YYRGD sera also displayed biological activity as shown by neutralization of HIV-1/MN and HIV-1/3B cell-free infection of target cells. This neutralization was as good as human anti-HIV sera. These results indicate that peptide–proteosome complexes are perhaps MHC-unrestricted and that these peptides potentially convert otherwise T cell-independent polysaccharide epitopes to T cell-dependent peptide epitopes which direct immune responses toward carbohydrates. This approach provides a novel strategy for the further development of an HIV vaccine.

Issues remain as to the relative roles played by humoral and cellular responses in affording either a therapeutic or a prophylactic vaccine. Induction of HIV-1 gp120-specific cytotoxic T lymphocyte responses in mice by recombinant CHO cell-derived gp120 is enhanced by enzymatic removal of N-linked glycans.[66] Some evidence suggests that T cells can recognize carbohydrate antigens directly.[67–71] This work postulates a new class of naturally occurring epitopes for T cells where branched-chain oligosaccharides linked to peptides with anchoring motifs for the major histocompatibility complex class I or class II pocket. Glycopeptide recognition by T cells is dependent on the chemical structure of the glycan as well as its position within the peptide. While analogous to the haptens trinitrophenyl and O-beta-linked acetyl-glucosamine, the potential implications of natural carbohydrates as antigenic epitopes for T cells in biology are considerable and understudied.

5. MUCOSAL IMMUNE RESPONSES

An important role of polysaccharides is attachment to mucosal surfaces. The control of pandemic infection of HIV-1 requires some means of developing mucosal immunity against HIV-1 because sexual transmission of the virus occurs mainly through the mucosal tissues. For an HIV vaccine to be effective, it will be essential that it protect against the virus variants to which individuals are most frequently exposed. HIV-1 is predominantly a sexually acquired virus, and thus variants in genital secretions are a potentially important reservoir of viruses that are transmitted. Recently, variants in the provirus population in the genital mucosa were analyzed and compared with proviruses in the blood of individuals chronically infected with HIV-1.[37] A major genetic difference between variants within a patient were insertions, apparently created by duplication of adjacent sequences, that resulted in acquisition of new potential glycosylation sites in V1 and V2. Comparisons of mucosal and PBMC variants suggest that these tissues harbor distinct, but related populations of HIV-1 variants. In two of three patients, the mucosal variants were most closely related to a minor variant genotype in blood. In a third individual, viruses in both tissues were surprisingly homogeneous, but the majority of variants in the cervix encoded a V1 sequence with a predicted glycosylation pattern similar to a minor variant in blood. The V3 sequence patterns of the mucosal isolates indicate they may be predominantly macrophage-tropic viruses.[37] The potential changes in the V1 glycosylation patterns of mucosal variants might imply that such variants could be hard to neutralize. This idea is based on the observation that such changes associated with V1 were observed in SIV variants in which glycosylation in V1 was directly associated with changes in antigenicity.[45]

There is, unfortunately, limited information on mucosal immunogenicity of polysaccharide conjugate vaccines even though mucosal immunization with polysaccharide–protein conjugates able to attach to and be taken up by the mucosa should have great potential usefulness. For capsular bacteria, it is likely that antibodies against capsular antigens, if available at the mucosal surface, should help prevent colonization and thus prevent subsequent disease. Secretory IgA (s-IgA) is presumed to be the mediator of mucosal immunity based on studies that show a correlation between protection and s-IgA titers. Thus, in the sexual transmission of HIV, the establishment of a genital mucosal immunity through s-IgA may be necessary to achieve protection.[72] The mucosal route might also control maternal mucosal infection and prevent HIV infection in infants.[73] Maternal antibodies may protect the fetus by reducing the quantity of infectious virus on the maternal circulation and/or by passive transfer of protective antibodies to the fetus.[74] The presence of s-IgA, however, does not ensure protection from HIV.

There is some evidence that the s-IgA antibody induced by immunization with antigens in experimental animals can neutralize HIV-1.[75] While IgA responses to HIV- and SIV-derived antigens were shown in early studies, attempts to vaccinate against infections of mucosal tissues have generally been less successful than vaccination against systemic infections, to a large extent reflecting incomplete knowledge about the most efficient means for inducing protective local immune responses at these sites. The large and repeated antigen doses often required to achieve a protective immune response make this vaccination approach impractical for many purified antigens. There is, therefore, a great need to develop strategies for enhancing delivery of antigen to the mucosal immune system as well as to identify mucosa-active immunostimulating agents (adjuvants).[76–79]

The function of IgA at mucosal surfaces appears to be one of cross-linking pathogens in the lumen and facilitating their clearance by peristalsis. This is significant in that it would suggest that IgA antibodies need not be directed to classical neutralizing epitopes on gp120 (V3 loop), for example. A secretory immune response against a single surface epitope of a virus protein might block entry into the host at a lumenal site safely removed from target cells. Since neutralizing antibodies might not be required in this system, it is conceivable that IgAs against conserved epitopes might be protective. Mucosal responses against carbohydrate epitopes that are presented on the envelope protein of mucosal variants might prove to be a very effective means for vaccination. Targeting carbohydrates on HIV is not easy. The advantage to targeting such determinants is that whatever changes occur in the glycosylation patterns of viral variants that appear to affect neutralizing antibodies that are targeting protein elements, such variations will not affect antibodies targeting carbohydrates. The issue with anticarbohydrate antibodies is their avidity/affinity. Nevertheless, specific S-IgA to HIV-1-expressed carbohydrates could be relevant in decreasing infectivity of HIV-1 for the long-term maintenance of mucosal immune responses.

6. SUMMARY

The HIV-1 envelope glycoprotein demonstrates an unusual degree of complex posttranslational glycosylation with both high-mannose and complex-type N-linked oligosaccharides. The relative conservation of these sites on gp120/41 may indicate an evolutionarily important function for carbohydrate moieties. These functions may involve the protection of the viral protein from nonspecific proteolysis. In addition, carbohydrate expression affects the tertiary and quaternary structure of monomers and oligomers. These changes may affect cell tropism and/or immune escape. Glycans play a critical role in both humoral and cellular immune responses to viruses. During mutation and selection, sites can be added or deleted, resulting in antigenic variation and effectively masking or directing antibody responses to nonneutralizing sites. Carbohydrates have been shown to interfere with peptide presentation by antigen-presenting cells. Therefore it is also plausible that the heavy glycosylation of HIV-1 interferes with (1) proteolytic degradation into peptides, (2) peptide binding to MHC, and (3) recognition of the MHC peptide complex by T lymphocytes. Glycosylation should also result in tolerance for HIV-1. Consequently, following exposure, there is an inability to evoke a specific immune response to viral epitopes that are protective in nature. These aspects point out the importance of targeting carbohydrate antigens on HIV-1 in vaccine strategies or redirecting immune responses. In vaccine design for HIV-1, carbohydrates, while having profound biological and immunological roles in the pathophysiology of HIV infection, have been understudied and basically forgotten entities.

REFERENCES

1. Mbemba, E., Carre, V., Atemezem, A., Saffar, L., Gluckman, J. C., and Gattegno, L., 1996, Inhibition of human immunodeficiency virus infection of CD4$^+$ cells by CD4-free glycopeptides from monocytic U937 cells, *AIDS Res. Hum. Retrovir.* **12**:47–53.
2. Bhat, S., Mettus, R. V., Reddy, E. P., Ugen, K. E., Srikanthan, V., Williams, W. V., and Weiner, D. B., 1993, The galactosyl ceramide/sulfatide receptor binding region of HIV-1 gp120 maps to amino acids 206-275, *AIDS Res. Hum. Retrovir.* **9**:175–181.
3. Pearce, P. R., and Phillips, D. M., 1996, Sulfated polysaccharides inhibit lymphocyte-to-epithelial transmission of human immunodeficiency virus-1, *Biol. Reprod.* **54**:173–182.
4. Leydet, A., Jeantet, S. C., Bouchitte, C., Moullet, C., Boyer, B., Roque, J. P., Witvrouw, M., Este, J., Snoeck, R., Andrei, G., and De, C. E., 1997, Polyanion inhibitors of human immunodeficiency virus and other viruses. 6. Mecille-like anti-HIV polyanionic compounds based on a carbohydrate core, *J. Med. Chem.* **40**:350–356.
5. Bernstein, H. B., Tucker, S. P., Hunter, E., Schutzbach, J. S., and Compans, R. W., 1994, Human immunodeficiency virus type 1 envelope glycoprotein is modified by O-linked oligosaccharides, *J. Virol.* **68**:463–468.
6. Gram, G. J., Hemming, A., Bolmstedt, A., Jansson, B., Olofsson, S., Akerblom, L., Nielsen, J. O., and Hansen, J. E., 1994, Identification of an N-linked glycan in the V1-loop of HIV-1 gp120 influencing neutralization by anti-V3 antibodies and soluble CD4, *Arch. Virol.* **139**:253–261.
7. Liedtke, S., Adamski, M., Geyer, R., Pfutzner, A., Rubsamen, W. H., and Geyer, H., 1994, Oligosaccharide profiles of HIV-2 external envelope glycoprotein: Dependence on host cells and virus isolates, *Glycobiology* **4**:477–484.
8. Trkola, A., Purtscher, M., Muster, T., Ballaun, C., Buchacher, A., Sullivan, N., Srinivasan, K., Sodroski, J., Moore, J. P., and Katinger, H., 1996, Human monoclonal antibody 2G12 defines a distinctive neutralization epitope on the gp120 glycoprotein of human immunodeficiency virus type 1, *J. Virol.* **70**:1100–1108.
9. Moore, J. P., Cao, Y., Qing, L., Sattentau, Q. J., Pyati, J., Koduri, R., Robinson, J., Barbas, C.,

Burton, D. R., and Ho, D. D., 1995, Primary isolates of human immunodeficiency virus type 1 are relatively resistant to neutralization by monoclonal antibodies to gp120, and their neutralization is not predicted by studies with monomeric gp120, *J. Virol.* **69:**101–109.

10. Hansen, J. E., Jansson, B., Gram, G. J., Clausen, H., Nielsen, J. O., and Olofsson, S., 1996, Sensitivity of HIV-1 to neutralization by antibodies against O-linked carbohydrate epitopes despite deletion of O-glycosylation signals in the V3 loop, *Arch. Virol.* **141:**291–300.

11. Clark, G. F., Dell, A., Morris, H. R., Patankar, M., Oehninger, S., and Seppala, M., 1997, Viewing AIDS from a glycobiological perspective: Potential linkages to the human feto-embryonic defence system hypotheses, *Mol. Hum. Reprod.* **3:**5–13.

12. Nara, P., 1996, Humoral immunity to HIV-1: Lethal force or trojan horse, in: *Immunology of HIV Infection* (S. Gupta, ed.), Plenum Press, New York, pp. 243–276.

13. Fenouillet, E., Gluckman, J. C., and Jones, I. M., 1994, Functions of HIV envelope glycans, *Trends Biochem. Sci.* **19:**65–70.

14. Zanetta, J. P., Badache, A., Maschke, S., Marschal, P., and Kuchler, S., 1994, Carbohydrates and soluble lectins in the regulation of cell adhesion and proliferation, *Histol. Histopathol.* **9:**385–412.

15. Preston, A., Mandrell, R. E., Gibson, B. W., and Apicella, M. A., 1996, The lipooligosaccharides of pathogenic Gram-negative bacteria, *Crit. Rev. Microbiol.* **22:**139–180.

16. Moran, A. P., Prendergast, M. M., and Appelmelk, B. J., 1996, Molecular mimicry of host structures by bacterial lipopolysaccharides and its contribution to disease, *FEMS Immunol. Med. Microbiol.* **16:**105–115.

17. Appelmelk, B. J., Simoons, S. I., Negrini, R., Moran, A. P., Aspinall, G. O., Forte, J. G., De, V. T., Quan, H., Verboom, T., Maaskant, J. J., Ghiara, P., Kuipers, E. J., Bloemena, E., Tadema, T. M., Townsend, R. R., *et al.*, 1996, Potential role of molecular mimicry between *Helicobacter pylori* lipopolysaccharide and host Lewis blood group antigens in autoimmunity, *Infection Immunity* **64:**2031–2040.

18. Muller, W. E., Bachmann, M., Weiler, B. E., Schroder, H. C., Uhlenbruck, G., Shinoda, T., Shimizu, H., and Ushjima, H., 1991, Antibodies against defined carbohydrate structures of *Candida albicans* protect H9 cells against infection with human immunodeficiency virus-1 *in vitro, J. Acquired Immune Defic. Syndr.* **4:**694–703.

19. McAlarney, T., Ogino, M., Apostolski, S., and Latov, N., 1995, Specificity and cross-reactivity of anti-galactocerebroside antibodies, *Immunol. Inv.* **24:**595–606.

20. Hansen, J. E., Clausen, H., Hu, S. L., Nielsen, J. O., and Olofsson, S., 1992, An O-linked carbohydrate neutralization epitope of HIV-1 gp120 is expressed by HIV-1 env gene recombinant vaccinia virus, *Arch. Virol.* **126:**11–20.

21. Hansen, J. E., Nielsen, C., Arendrup, M., Olofsson, S., Mathiesen, L., Nielsen, J. O., and Clausen, H., 1991, Broadly neutralizing antibodies targeted to mucin-type carbohydrate epitopes of human immunodeficiency virus, *J. Virol.* **65:**6461–6467.

22. Back, N. K., Smit, L., De, J. J., Keulen, W., Schutten, M., Goudsmit, J., and Tersmette, M., 1994, An N-glycan within the human immunodeficiency virus type 1 gp120 V3 loop affects virus neutralization, *Virology* **199:**431–438.

23. Arendrup, M., Hansen, J. E., Clausen, H., Nielsen, C., Mathiesen, L. R., and Nielsen, J. O., 1991, Antibody to histo-blood group A antigen neutralizes HIV produced by lymphocytes from blood group A donors but not from blood group B or O donors, *AIDS* **5:**441–444.

24. Dabelsteen, E., 1996, Cell surface carbohydrates as prognostic markers in human carcinomas, *J. Pathol.* **179:**358–369.

25. Lefebvre, J. C., Giordanengo, V., Doglio, A., Cagnon, L., Breittmayer, J. P., Peyron, J. F., and Lesimple, J., 1994, Altered sialylation of CD45 in HIV-1-infected T lymphocytes, *Virology* **199:**265–274.

26. Lefebvre, J. C., Giordanengo, V., Limouse, M., Doglio, A., Cucchiarini, M., Monpoux, F.,

Mariani, R., and Peyron, J. F., 1994, Altered glycosylation of leukosialin, CD43, in HIV-1-infected cells of the CEM line, *J. Exp. Med.* **180:**1609–1617.

27. Adachi, M., Hayami, M., Kashiwagi, N., Mizuta, T., Ohta, Y., Gill, M. J., Matheson, D. S., Tamaoki, T., Shiozawa, C., and Hakomori, S., 1988, Expression of Le^Y antigen in human immunodeficiency virus-infected human T cell lines and in peripheral lymphocytes of patients with acquired immune deficiency syndrome (AIDS) and AIDS-related complex (ARC), *J. Exp. Med.* **167:**323–331.

28. Gattegno, L., Ramdani, A., Joualt, T., Saffar, L., and Gluckman, J. C., 1992, Lectin–carbohydrate interactions and infectivity of human immunodeficiency virus type 1 (HIV-1), *AIDS Res. Hum. Retrovir.* **8:**27–37.

29. Kashiwagi, N., Gill, M. J., Adachi, M., Church, D., Wong, S. J., Poon, M. C., Hakomori, S., Tamaoki, T., and Shiozawa, C., 1994, Lymphocyte membrane modifications induced by HIV infection, *Tohoku J. Exp. Med.* **173:**115–131.

30. Hamadeh, R. M., Galili, U., Zhou, P., and Griffiss, J. M., 1995, Anti-alpha-galactosyl immuno-globulin A (IgA), IgG, and IgM in human secretions, *Clin. Diag. Lab. Immunol.* **2:**125–131.

31. Teneberg, S., Lonnroth, I., Torres, L. J., Galili, U., Halvarsson, M. O., Angstrom, J., and Karlsson, K. A., 1996, Molecular mimicry in the recognition of glycosphingolipids by Gal alpha 3 Gal beta 4 GlcNAc beta-binding *Clostridium difficile* toxin A, human natural anti alpha-galactosyl IgG and the monoclonal antibody Gal-13: Characterization of a binding-active human glycosphingolipid, non-identical with the animal receptor, *Glycobiology* **6:** 599–609.

32. Reed, D. J., Lin, X., Thomas, T. D., Birks, C. W., Tang, J., and Rother, R. P., 1997, Alteration of glycosylation renders HIV sensitive to inactivation by normal human serum, *J. Immunol.* **159:**4356–4361.

33. Moore, J. P., Sattentau, Q. J., Wyatt, R., and Sodroski, J., 1994, Probing the structure of the human immunodeficiency virus surface glycoprotein gp120 with a panel of monoclonal antibodies, *J. Virol.* **68:**469–484.

34. Ditzel, H. J., Parren, P. W., Binley, J. M., Sodroski, J., Moore, J. P., Barbas, C., and Burton, D. R., 1997, Mapping the protein surface of human immunodeficiency virus type 1 gp120 using human monoclonal antibodies from phage display libraries, *J. Mol. Biol.* **267:** 684–695.

35. Sattentau, Q. J., and Moore, J. P., 1995, Human immunodeficiency virus type 1 neutralization is determined by epitope exposure on the gp120 oligomer, *J. Exp. Med.* **182:**185–196.

36. Zhang, Y. M., Dawson, S. C., Landsman, D., Lane, H. C., and Salzman, N. P., 1994, Persistence of four related human immunodeficiency virus subtypes during the course of zidovudine therapy: Relationship between virion RNA and proviral DNA, *J. Virol.* **68:** 425–432.

37. Overbaugh, J., Anderson, R. J., Ndinya, A. J., and Kreiss, J. K., 1996, Distinct but related human immunodeficiency virus type 1 variant populations in genital secretions and blood, *AIDS Res. Hum. Retrovir.* **12:**107–115.

38. Gao, F., Morrison, S. G., Robertson, D. L., Thornton, C. L., Craig, S., Karlsson, G., Sodroski, J., Morgado, M., Galvao, C. B., von Briesen, H., *et al.*, 1996, Molecular cloning and analysis of functional envelope genes from human immunodeficiency virus type 1 sequence subtypes A through G. The WHO and NIAID Networks for HIV Isolation and Characterization, *J. Virol.* **70:**1651–1667.

39. Manca, F., 1992, Galactose receptors and presentation of HIV envelope glycoprotein to specific human T cells, *J. Immunol.* **148:**2278–2282.

40. Benjouad, A., Mabrouk, K., Gluckman, J. C., and Fenouillet, E., 1994, Effect of sialic acid removal on the antibody response to the third variable domain of human immunodeficiency virus type-1 envelope glycoprotein, FEBS Lett. **341:**244–250.

41. Schonning, K., Jansson, B., Olofsson, S., and Hansen, J. E., 1996, Rapid selection for an . N-linked oligosaccharide by monoclonal antibodies directed against the V3 loop of human immunodeficiency virus type 1, *J. Gen. Virol.* **77**:753–758.

42. Huang, X., Barchi, J. J., Lung, F. D., Roller, P. P., Nara, P. L., Muschik, J., and Garrity, R. R., 1997, Glycosylation affects both the three-dimensional structure and antibody binding properties of the HIV-1IIIB Gp120 peptide RP135, *Biochemistry* **36**:10846–10856.

43. Huang, X., Smith, M. C., Berzofsky, J. A., and Barchi, J. J., 1996, Structural comparison of a 15 residue peptide from the V3 loop of HIV-1IIIb and an O-glycosylated analogue, *FEBS Lett.* **393**:280–286.

44. Bolmstedt, A., Olofsson, S., Sjogren, J. E., Jeansson, S., Sjoblom, I., Akerblom, L., Hansen, J. E., and Hu, S. L., 1992, Carbohydrate determinant NeuAc-Gal beta (1–4) of N-linked glycans modulates the antigenic activity of human immunodeficiency virus type 1 glycoprotein gp120, *J. Gen. Virol.* **73**:3099–3105.

45. Chackerian, B., Rudensey, L. M., and Overbaugh, J., 1997, Specific N-linked and O-linked glycosylation modifications in the envelope V1 domain of simian immunodeficiency virus variants that evolve in the host alter recognition by neutralizing antibodies, *J. Virol.* **71**: 7719–7727.

46. Willey, R. L., Shibata, R., Freed, E. O., Cho, M. W., and Martin, M. A., 1996, Differential glycosylation, virion incorporation, and sensitivity to neutralizing antibodies of human immunodeficiency virus type 1 envelope produced from infected primary T-lymphocyte and macrophage cultures, *J. Virol.* **70**:6431–6436.

47. Sorensen, A. M., Nielsen, C., Arendrup, M., Clausen, H., Nielsen, J. O., Osinaga, E., Roseto, A., and Hansen, J. E., 1994, Neutralization epitopes on HIV pseudotyped with HTLV-I: Conservation of carbohydrate epitopes, *J. Acquired Immune Defic. Syndr.* **7**:116–123; Erratum, *J. Acquired Immune Defic. Syndr.* **7**:740.

48. Arendrup, M., Sonnerborg, A., Svennerholm, B., Akerblom, L., Nielsen, C., Clausen, H., Olofsson, S., Nielsen, J. O., and Hansen, J. E., 1993, Neutralizing antibody response during human immunodeficiency virus type 1 infection: Type and group specificity and viral escape, *J. Gen. Virol.* **74**:855–863.

49. Yu, X. F., Wang, Z., Beyrer, C., Celentano, D. D., Khamboonruang, C., Allen, E., and Nelson, K., 1995, Phenotypic and genotypic characteristics of human immunodeficiency virus type 1 from patients with AIDS in northern Thailand, *J. Virol.* **69**:4649–4655.

50. Kieber-Emmons, T., Jameson, B. A., and Morrow, W. J., 1989, The gp120-CD4 interface: Structural, immunological and pathological considerations, *Biochim. Biophys. Acta* **989**: 281–300.

51. Garrity, R. R., Rimmelzwaan, G., Minassian, A., Tsai, W. P., Lin, G., de Jong, J. J., Goudsmit, J., and Nara, P. L., 1997, Refocusing neutralizing antibody response by targeted dampening of an immunodominant epitope, *J. Immunol.* **159**:279–289.

52. Mond, J. J., Lees, A., and Snapper, C. M., 1995, T cell-independent antigens type 2, *Annu. Rev. Immunol.* **13**:655–692.

53. Agadjanyan, M., Luo, P., Westerink, M. A. J., Carey, L. A., Hutchins, W., Steplewski, Z., Weiner, D. B., and Kieber-Emmons, T., 1997, Peptide mimicry of carbohydrate epitopes on human immunodeficiency virus, *Nature Biotechnol.* **15**:547–551.

54. Hutchins, W., Adkins, A., Kieber-Emmons, T., and Westerink, M. A. J., 1996, Molecular characterization of a monoclonal antibody produced in response to a group-C meningococcal polysaccharide peptide mimic, *Mol. Immunol.* **33**:503–510.

55. Kieber-Emmons, T., Luo, P., Qiu, J., Agadjanyan, M., Carey, L., Hutchins, W., Westerink, M. A. J., and Steplewski, Z., 1997, Peptide mimicry of adenocarcinoma-associated carbohydrate antigens, *Hybridoma* **16**:3–10.

56. Westerink, M. A. J., Giardina, P. C., Apicella, M. A., and Kieber-Emmons, T., 1995, Peptide

mimicry of the meningococcal group C capsular polysaccharide, *Proc. Natl. Acad. Sci. USA* **92**:4021–4025.

57. Westerink, M. A. J., Campagnari, A. A., Giardina, P., and Apicella, M. A., 1994, Antiidiotype antibodies as surrogates for polysaccharide vaccines, *Ann. N.Y. Acad. Sci.* **730**:209–216.

58. Scott, J. K., Loganathan, D., Easley, R. B., Gong, X., and Goldstein, I. J., 1992, A family of concanavalin A-binding peptides from a hexapeptide epitope library, *Proc. Natl. Acad. Sci. USA* **89**:5398–5402.

59. Oldenburg, K. R., Loganathan, D., Goldstein, I. J., Schultz, P. G., and Gallop, M. A., 1992, Peptide ligands for a sugar-binding protein isolated from a random peptide library, *Proc. Natl. Acad. Sci. USA* **89**:5393–5397.

60. Hoess, R., Brinkmann, U., Handel, T., and Pastan, I., 1993, Identification of a peptide which binds to the carbohydrate-specific monoclonal antibody B3, *Gene* **128**:43–49.

61. Valadon, P., Nussbaum, G., Boyd, L. F., Margulies, D. H., and Scharff, M. D., 1996, Peptide libraries define the fine specificity of anti-polysaccharide antibodies to *Cryptococcus neoformans*, *J. Mol. Biol.* **261**:11–22.

62. Shikhman, A. R., and Cunningham, M. W., 1994, Immunological mimicry between *N*-acetyl-beta-D-glucosamine and cytokeratin peptides. Evidence for a microbially driven anti-keratin antibody response, *J. Immunol.* **152**:4375–4387.

63. Shikhman, A. R., Greenspan, N. S., and Cunningham, M. W., 1994, Cytokeratin peptide SFGSGFGGGY mimics *N*-acetyl-beta-D-glucosamine in reaction with antibodies and lectins, and induces *in vivo* anti-carbohydrate antibody response, *J. Immunol.* **153**:5593–5606.

64. Shikhman, A. R., Greenspan, N. S., and Cunningham, M. W., 1993, A subset of mouse monoclonal antibodies cross-reactive with cytoskeletal proteins and group A streptococcal M proteins recognizes *N*-acetyl-beta-D-glucosamine, *J. Immunol.* **151**:3902–3913.

65. Thurin-Blaszczyk, M., Murali, R., Westerink, M. A. J., Steplewski, Z., Co, M.-S., and Kieber-Emmons, T., 1996, Molecular recognition of the Lewis Y antigen by monoclonal antibodies, *Protein Eng.* **9**:101–113.

66. Doe, B., Steimer, K. S., and Walker, C. M., 1994, Induction of HIV-1 envelope (gp120)-specific cytotoxic T lymphocyte responses in mice by recombinant CHO cell-derived gp120 is enhanced by enzymatic removal of N-linked glycans, *Eur. J. Immunol.* **24**:2369–2376.

67. Haurum, J. S., Arsequell, G., Lellouch, A. C., Wong, S. Y., Dwek, R. A., McMichael, A. J., and Elliott, T., 1994, Recognition of carbohydrate by major histocompatibility complex class I-restricted, glycopeptide-specific cytotoxic T lymphocytes, *J. Exp. Med.* **180**:739–744.

68. Haurum, J. S., Tan., L., Arsequell, G., Frodsham, P., Lellouch, A. C., Moss, P. A., Dwek, R. A., McMichael, A. J., and Elliott, T., 1995, Peptide anchor residue glycosylation: Effect on class I major histocompatibility complex binding and cytotoxic T lymphocyte recognition, *Eur. J. Immunol.* **25**:3270–3276.

69. Jensen, T., Galli, S. L., Mouritsen, S., Frische, K., Peters, S., Meldal, M., and Werdelin, O., 1996, T cell recognition of Tn-glycosylated peptide antigens, *Eur. J. Immunol.* **26**:1342–1349.

70. Jensen, T., Hansen, P., Galli, S. L., Mouritsen, S., Frische, K., Meinjohanns, E., Meldal, M., and Werdelin, O., 1997, Carbohydrate and peptide specificity of MHC class II-restricted T cell hybridomas raised against an O-glycosylated self peptide, *J. Immunol.* **158**:3769–3778.

71. Michaelsson, E., Broddefalk, J., Engstrom, A., Kihlberg, J., and Holmdahl, R., 1996, Antigen processing and presentation of a naturally glycosylated protein elicits major histocompatibility complex class II-restricted, carbohydrate-specific T cells, *Eur. J. Immunol.* **26**:1906–1910.

72. Rosenthal, K. L., and Gallichan, W. S., 1997, Challenges for vaccination against sexually-transmitted diseases: Induction and long-term maintenance of mucosal immune responses in the female genital tract, *Semin. Immunol.* **9**:303–314.

73. Re, M. C., Furlini, G., Vignoli, M., Ricchi, E., Ramazzotti, E., Bianchi, S., Guerra, B.,

Costigliola, P., and La, P. M., 1992, Vertical transmission of human immunodeficiency virus type 1. Prognostic value of IgA antibody to HIV-1 polypeptides during pregnancy, *Diagn. Microbiol. Infect. Dis.* **15:**553–556.

74. Hocini, H., Belec, L., Iscaki, S., Garin, B., Pillot, J., Becquart, P., and Bomsel, M., 1997, High-level ability of secretory IgA to block HIV type 1 transcytosis: Contrasting secretory IgA and IgG responses to glycoprotein 160, *AIDS Res. Hum. Retrovir.* **13:**1179–1185.

75. Bukawa, H., Sekigawa, K., Hamajima, K., Fukushima, J., Yamada, Y., Kiyono, H., and Okuda, K., 1995, Neutralization of HIV-1 by secretory IgA induced by oral immunization with a new macromolecular multicomponent peptide vaccine candidate, *Nature Med.* **1:** 681–685.

76. Holmgren, J., Czerkinsky, C., Lycke, N., and Svennerholm, A. M., 1994, Strategies for the induction of immune responses at mucosal surfaces making use of cholera toxin B subunit as immunogen, carrier, and adjuvant, *Am. J. Trop. Med. Hyg.* **50:**42–54.

77. Staats, H. F., Montgomery, S. P., and Palker, T. J., 1997, Intranasal immunization is superior to vaginal, gastric, or rectal immunization for the induction of systemic and mucosal anti-HIV antibody responses, *AIDS Res. Hum. Retrovir.* **13:**945–952.

78. Wang, B., Dang, K., Agadjanyan, M. G., Srikantan, V., Li, F., Ugen, K. E., Boyer, J., Merva, M., Williams, W. V., and Weiner, D. B., 1997, Mucosal immunization with a DNA vaccine induces immune responses against HIV-1 at a mucosal site, *Vaccine* **15:**821–825.

79. Okada, E., Sasaki, S., Ishii, N., Aoki, I., Yasuda, T., Nishioka, K., Fukushima, J., Miyazaki, J., Wahren, B., and Okuda, K., 1997, Intranasal immunization of a DNA vaccine with IL-12- and granulocyte-macrophage colony-stimulating factor (GM-CSF)-expressing plasmids in liposomes induces strong mucosal and cell-mediated immune responses against HIV-1 antigens, *J. Immunol.* **159:**3638–3647.

5

HTLV-I and HTLV-II Infection
Immunological and Molecular Aspects

MOSI K. BENNETT and MICHAEL G. AGADJANYAN

1. INTRODUCTION

Human T lymphotropic virus type I (HTLV-I) holds a place in history as the first human retrovirus linked to a specific disease. The isolation of HTLV-I in 1980 was the culmination of a long, intensive search for human retroviruses. HTLV-I is the causal agent of adult T cell leukemia (ATL) and infects at least 10 million people worldwide. Since its initial association with ATL, HTLV-I has also been associated with HTLV-I associated myelopathy (HAM).

No disease etiology has been linked to a second human T lymphotropic virus (HTLV-II). However, this virus has been isolated from several patients with unusual lymphocytic leukemias. The limited number of individuals shown to harbor HTLV-II in association with specific diseases has complicated the determination of a specific disease etiology.

In addition to its historical importance, research on HTLV has played a significant role in the characterization of another retrovirus, human immunodeficiency virus (HIV). HTLV research advances have aided efforts to identify and further understand HIV. Unlike HIV, the cellular receptor for HTLV has not been characterized and the search for this receptor remains a main focus of HTLV research. In this chapter, we will discuss the molecular

MOSI K. BENNETT • Department of Pathology and Laboratory Medicine, University of Pennsylvania School of Medicine, Philadelphia, Pennsylvania 19104.   MICHAEL G. AGADJANYAN • Department of Pathology and Laboratory Medicine, University of Pennsylvania School of Medicine, Philadelphia, Pennsylvania 19104, and Institute of Viral Preparations, Russian Academy of Medical Science, Moscow, Russia 129028.

Human Retroviral Infections, edited by Kenneth E. Ugen *et al.*
Kluwer Academic / Plenum Publishers, New York, 2000.

biology of HTLV including recent efforts to identify the HTLV-I receptor and specific adhesion molecules important in HTLV-I cell fusion. Our goal is to further understand the nature of the immunological and molecular aspects of HTLV-I and HTLV-II infection.

2. BIOLOGY OF HTLV

Human T lymphotropic virus was first isolated by Robert Gallo's group[1,2] and was the first human retrovirus for which a disease association was made. This new virus was classified as human T cell lymphotrophic/leukemic virus type 1, or HTLV-I. Shortly after the discovery of HTLV-I, Kalyanaraman *et al.* isolated another human T cell leukemia virus, designated HTLV-II.[3] HTLV-I is the causal agent of adult T cell leukemia and HTLV-I associated myelopathy/tropical spastic paraparesis (HAM/TSP). Although a clear disease association for HTLV-II infection has not been established, some researchers suggest that it may be associated with a neurodegenerative disorder similar to HAM/TSP.[4]

Both HTLV-I and HTLV-II are members of the oncovirinae subfamily of retroviruses and have recently been reclassified as members of the group of viruses which includes bovine leukemia virus (BLV).[1] The discovery and characterization of HTLV-I and II greatly facilitated the characterization of the etiological agents for another retrovirus, human immunodeficiency virus types 1 and 2 (HIV-1 and HIV-2), a member of the subfamily of lentiviruses.[4] HTLV shares many biological and molecular characteristics of HIV including routes of transmission, a general T-cell tropism, syncytia induction, and the presence of viral-encoded core proteins that function at both the transcriptional and posttranscriptional level.

The HTLV-I proviral genome consists of 9032 nucleotides and includes regions coding for the structural proteins (*gag* and *env*) and the viral protease and polymerase (*pol*), similar to HIV-1 (Fig. 1). The genome also contains an additional sequence designated pX. This region is adjacent to the envelope (*env*) gene and contains three overlapping regulatory genes that encode the transactivator protein (Tax p40), transmodulator protein (Rex p27), and a third gene which encodes for the protein p21. The *tax* and *rex* genes are two regulatory genes that positively and negatively control viral gene expression and replication. The function of p21 is not known.[5–7]

HTLV-I and HTLV-II share a 66% sequence homology. HTLV also shares high sequence homologies with other retroviruses such as bovine leukemia virus (BLV) and the simian T cell leukemia virus (STLV). Unlike other oncoviruses, HTLV-I does not express an analogue of any known

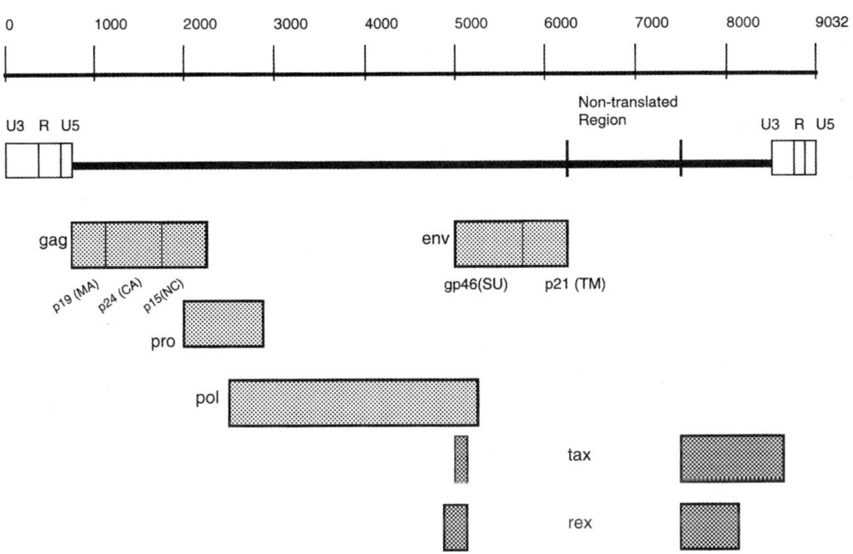

FIGURE 1. The HTLV genome.

cellular oncogene, nor does it insert its proviral genome into a specific regulatory region of the host DNA. Rather, the provirus acts randomly, taking up residence in any open region provided by the host cell's transcriptional activity.

The *env* gene of HTLV encodes a glycoprotein precursor, gp61, which is cleaved to the gp46 external and gp21 transmembrane glycoproteins. These glycoproteins are present on mature virions and are expressed on the infected cells.[8,9] Similar to other viruses, HTLV infects target cells by fusion with the cell membrane mediated by way of binding through its envelope glycoprotein to a cell surface. HTLV is unusual in its necessity for direct cell contact for efficient infection, possibly a result of the interactions of the Env proteins with currently unidentified cellular receptor or receptors. Culture media from HTLV-transformed cells have been shown to contain large amounts of free gp46 which are shed from the surface of cells.[9]

Although the mechanisms involved in the oncogenic transformation of infected cells are poorly understood, it is thought that the HTLV-I transactivation protein Tax may play an indirect role in cell transformation. The *tax* gene is capable of positively regulating HTLV gene expression and can also affect a number of transcription factors. As mentioned earlier, *tax* is located in the pX region. Tax activates the nuclear factor (NF)-κB-binding site of the interleukin-2 (IL-2) receptor alpha gene, the serum-responsive element (SRE) of the c-*fos* and c-*egr* protooncogenes, and the 21-bp enhancer of HTLV-I. The mechanism of the activation was shown to be the binding of Tax to transcription factors that bind to specific enhancers[10–13] (Fig. 2). Tax has the ability to associate with host cell enhancer-binding proteins such as the cyclic AMP-responsive element (CRE)-binding protein (CREB), the CRE modulator protein (CREM), the p50 subunit of NF-κB, and the serum-responsive factor (SRF).[13] Expression of Tax alone has been shown to transform a number of cell lines *in vitro* and cause these primary T cells to exhibit many of the features similar to those of T cells transformed by HTLV-I.

In addition, the *tax* gene can activate the expression of several other cellular genes such as those for the IL-2 receptor (IL-2R) γ chain, granulocyte-macrophage colony-stimulating factor (GM-CSF), and IL-6.[14] This characteristic may explain how Tax is responsible for making HTLV-I-infected cells

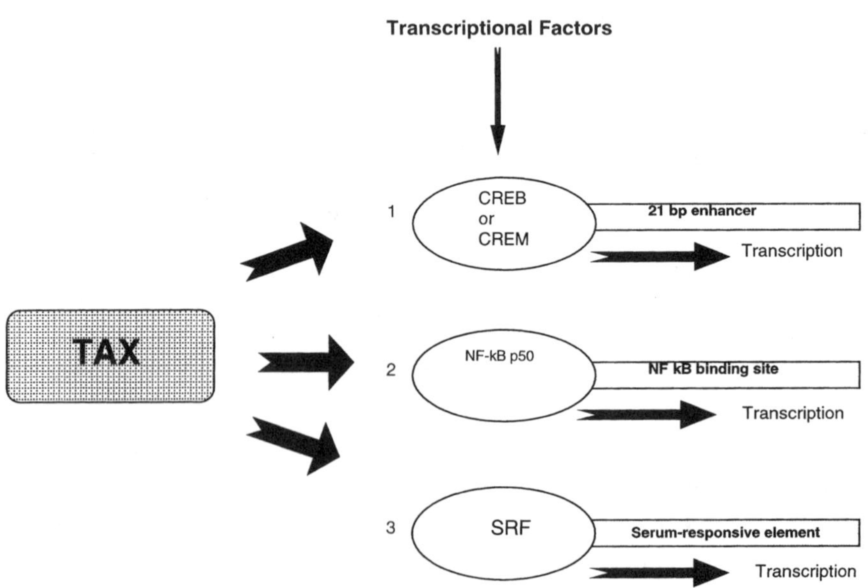

FIGURE 2. Binding of Tax to the enhancer-binding proteins CREB, CREM, NF-κB p50, and SRF. Indirect association of Tax protein to the enhancer DNA through each enhancer-binding protein, or increase of the active NF-κB p50 through binding to the precursor, seems to be the activation mechanism of Tax. Indirect binding of Tax to the three different classes of transcriptional enhancers promotes transcription.

more susceptible to IL-2-dependent growth. Typically, HTLV-I-infected cells upregulate the expression of not only IL-2R, but also other molecules involved in T cell activation such as major histocompatibility class (MHC) class II antigens, adhesion molecules, and costimulatory molecules. Infected cells also have the ability to stimulate the proliferation of uninfected T cells.[15] Recently, researchers have found that HTLV-I Tax induces expression of gp34 and OX40.[16] OX40 is expressed on activated T cells and is involved in T-cell proliferation. A member of the tumor necrosis factor (TNF) receptor family, OX40 is a ligand for gp34. Since Tax-positive cell lines express higher levels of both gp34 and OX40, this system may possibly play an important role in the process of malignant transformation of HTLV-I-infected cells.[16]

The *rex* gene is another unique gene found in the pX sequence of the HTLV genome. Like Tax, the exact mechanism of Rex action is not clear. The *rex* gene encodes proteins of 27/21 kDa for HTLV-I and 26/24 kDa for HTLV-II. The Rex protein is essential for HTLV replication and control of gene expression. While Rex does not directly regulate RNA transcription, it has been shown to act at the posttranscriptional level to regulate viral gene expression by processing viral RNA.[17] It is also believed that Rex can play a role in the process of cell transformation. Researchers have shown that Rex, like Tax, increases the expression of IL-2R by prolonging the IL-2R mRNA half-life.[18,19]

3. HTLV VACCINE DEVELOPMENT

Epidemiological studies demonstrate that HTLV-I is endemic in Japan, the Caribbean, Central America, and certain parts of the African continent. The number of people worldwide infected with HTLV-I alone has been estimated at between 10 and 20 million, with 1–5% ultimately developing disease. A vaccine against HTLV-I is desperately needed to control the spread of this serious illness.

In order to design a vaccine for HTLV-I, it is first necessary to document and understand the immune responses elicited against this retrovirus. Like other retroviruses, the envelope glycoproteins of HTLV-I can act as major antigens that induce serum antibodies in infected individuals. It has been demonstrated that both monkeys and rabbits can be productively infected with this retrovirus, thus providing animal models for HTLV-I infection and vaccine development.[20,21]

Research from several groups suggests that humoral immune responses alone may be sufficient to protect against infection, and that a vaccine based on the HTLV-I envelope glycoproteins may be feasible. It has been shown that antibodies against the HTLV-I envelope glycoprotein have the ability to protect monkeys from challenge with cell-associated HTLV-I.[22] Anti-envelope antibodies in patient sera have been shown to cross-neutralize HTLV-I strains originating from disparate geographic regions around the world. Therefore, a vaccine developed against one strain may be effective in neutralizing other strains.[23] Additionally, maternal antibodies delivered through breast-feeding are able to provide newborns with immunity and protection from congenital transmission.[24]

Generally, traditional vaccination has relied on either inactivated/ subunit preparations or live attenuated infectious material for the generation of a protective immune response. Inactivated or subunit vaccines provide protection through the induction of protective T helper (Th) cell and humoral immunity, but do not induce significant cytotoxic T lymphocyte (CTL) immunity. In contrast, live attenuated vaccines induce protective CTLs as well as Th cells and humoral immunity through a nonpathogenic infection of the host. This induction of MHC class I-restricted CTL immunity is thought to be crucial for protection from many viral infections. The importance of cellular immune responses in protection from HTLV-I infection is not clearly identified. It is likely that cellular responses do play a positive role in protection along with humoral responses.

A novel immunization strategy, DNA or nucleic acid vaccination, elicits both humoral and cellular immune responses *in vivo*. Since DNA vaccines are nonreplicating and the vaccine components are produced within the host cells, they can be constructed to function with all the safety features as well as the specificity of a subunit vaccine. However, DNA vaccine cassettes could produce immunological responses that are more similar to live vaccine preparations. By directly introducing DNA into the host cell, the host cell is essentially directed to produce the antigenic protein inside the cell thus mimicking viral replication in host cells.[25–28] Accordingly, this process has been demonstrated to generate both antibody and cell-mediated, particularly killer T cell-mediated, protective immunity. Unlike the attenuated vaccine, there is a little risk for pathogenesis by reversion to a disease-causing form with the injected DNA.[29,30]

We have investigated the ability of plasmid DNA inoculation to elicit protective immune responses against retroviruses in mice, rats, rabbits, monkeys, and chimpanzees.[25–28] Our work on the HTLV-I-envelope DNA plasmid vaccines in the rabbit and rat models for this retrovirus demonstrates the induction of humoral and cellular immune responses in the animal systems. Our data and the work of others suggest the potential utility of the DNA plasmid vaccine as well as other HTLV vaccine candidates.

Another possible vaccine strategy is the generation of protein subunits that will block the binding and fusion of HTLV to target cells. In order to block binding and fusion successfully, we must first identify the HTLV receptor. Knowledge of the cellular receptor could aid attempts to understand HTLV-I infection and help focus efforts to develop therapeutic and prophylactic agents for HTLV and related viruses.

4. RECEPTORS FOR HTLV

The cellular receptor(s) for HTLV-I and HTLV-II have not been identified. For years, one of the major goals in understanding HTLV biology has been the identification and characterization of its binding and/or fusogenic receptors.[31–33] The identification of the HTLV receptor is particularly important not only in understanding the viral replication, tropism, and pathogenesis of the virus, but also in generating therapeutic and prophylactic agents against HTLV. Several studies have reported that HTLV-I and HTLV-II share the same receptor.[33–35] However, attempts to identify this receptor using various techniques such as syncytium formation inhibition, fluorescence-activated cell sorter (FACS) analysis, coprecipitation, and immunoprecipitation have not been successful.

The task of identifying the HTLV receptor is further complicated by the fact that there may be more than one cellular receptor and possibly several cofactors and coreceptors involved in HTLV infection and cell fusion. For example, HTLV-I may have a cellular binding receptor and a separate fusion receptor. In addition, adhesion molecules could be also involved in infection of target cells by HTLV. Research on another retrovirus, HIV-1, has revealed that in addition to the specific cellular receptor CD4, coreceptors such as CXCR4 and CCR5 and other seven-transmembrane molecules[36–38] are important for viral fusion and infection. In the case of HIV, the identification and understanding of these important coreceptors lagged years behind the discovery of the CD4 molecule as the primary receptor for HIV.

The first event in viral infection of the host cell is the binding of the viral protein to molecules expressed on the target cells. In retroviral infections, cell fusion and syncytium formation has been attributed to the interaction of viral envelope expressed on the surface of infected cells and the viral receptor on the surface of neighboring cells. In HTLV, syncytium formation is dependent on the interaction between gp46, the envelope glycoprotein expressed on the surface of infected cells, and cellular receptors on the membrane of target cells. Fusion of infected and uninfected target cells takes place following this interaction.

TABLE I
Cells That Have Ability to Form Syncytia with HTLV-I-Infected MT2 Cells[a]

Cell	Type	Syncytia formation with HTLV-I cells
BJAB-WH	Human B-lymphosarcoma	++++
Jurkat	Human T-lymphosarcoma	+
SUPT-1	Non-Hodgkins human T-lymphosarcoma	+++
Molt-4	Human T-lymphoblastoid	++
HUT78	Human cutaneous T-lymphoma, African green monkey	++
HOS	Human osteogenic TE85, clone F-5	+
HeLa	Human cervical carcinoma, cervix	++
MDBK	Bovine kidney cells	+
PBL	Peripheral blood lymphocytes, phytohemagglutin-activated	++

[a]These data confirmed not only that HTLV receptor is present and biologically active on a wide variety of cell lines, but also may be a common conserved biochemical structure in a wide variety of species.

HTLV-I and HTLV-II have demonstrated a general T lymphocyte tropism,[3,39–41] preferentially infecting human CD4[+] T cells, although other cell populations such as B cells, fibroblasts, endothelial cells, and glial cells[42–44] are known to be infected. Most interesting is the ability of HTLV to infect target cells of several species[44] (Table I). These data suggest that the HTLV receptor is present and biologically active on a wide variety of cell lines and also may be a common conserved biochemical structure in a wide variety of species. Importantly, HTLV-I and HTLV-II infection *in vitro* generally requires contact between target and infected cells in order to transmit virus, since the cell-free transmission of HTLV is generally known to be inefficient. Therefore close contact between infected and target cells is very important for HTLV infection. Adhesion molecules expressed on the surface of target cells enhance this cell–cell contact.

Thus, in an attempt to identify and characterize the HTLV receptor, some investigators have focused on generating monoclonal antibodies that have the ability to inhibit the contact between target and infected cells that is necessary for HTLV-I-induced syncytium formation. One group generated and identified four monoclonal antibodies (mAbs) that inhibited syncytium formation in a culture of HTLV-I/MT2-infected cells and MolT-4, but not HOS or human PBMC target cells; however, these mAbs did not inhibit the binding of purified HTLV-I to target cells. Three of the four mAbs (C33, M101, and M104) recognized the same 28-kDa core protein and one mAb, M38, bound to a monomeric 26-kDa protein. Molecular cloning of cDNA encoding C33 and M38 antigens determined that both molecules are members of the transmembrane-4 superfamily, which includes CD9, CD37, CD53, CD63, L6, CO-029, T11, Sm23, and TAPA-1.[45,46] Interestingly, the M38 antigen was identical to TAPA-1, and along with C33 antigen, was encoded by human chromosome 11. These data clearly indicate that these two membrane proteins do not represent an HTLV-I-binding receptor itself; however, it is likely that both of these antigens are indirectly involved in HTLV-I syncytia formation. In the following section, we will discuss in detail several molecules that could be important for HTLV-I infection and our attempts to characterize a specific adhesion molecule involved in HTLV-I-induced cell fusion.

5. ADHESION MOLECULES AND HTLV INFECTION

Our group has utilized a variety of approaches in order to identify a novel cell membrane antigen important for HTLV-I and HTLV-II infection and syncytium formation. We reported that a membrane antigen expressed on human B and T cells is involved in HTLV-I and HTLV-II cell-free infection and syncytium formation.[45] It was observed that polyclonal monospecific sera generated against the fusogenic subline (WH) of the human B-cell line BJAB recognized an approximately 80- to 90-kDa antigen that was not detectable on the nonfusogenic subline of BJAB-CC/84.[47] Importantly, we demonstrated that the BJAB-WH subline could also be infected with cells infected with different HTLV-I and HTLV-II isolates (FLW, MoT, and MT2). Confirming results of others, we demonstrated that infection by cell-free virus was generally not as efficient as cell-contact-mediated infection. Only in half of the experiments were we able to infect target cells with cell-free virus. This process generally required more than 3 weeks of incubation with concentrated cell-free HTLV-I and HTLV-II. As expected, there was no detectable cell-free infection with the nonfusogenic BJAB-CC/84 cells.

We generated monoclonal antibodies against this molecule in an attempt to further characterize this membrane glycoprotein that plays an important role in syncytium formation.[47] These mAbs, as determined by FACS, were able to bind to the fusogenic/infectible WH cells, but not to the nonfusogenic CC/84 cells of the BJAB cell line. MAbs not only recognized the 80- to 90-kDa antigen on the surface of target cells, but also inhibited syncytium formation between target cells and HTLV I and HTLV II infected cells.

Subsequently, our attempts to identify this molecule using screening of human T-cell cDNA libraries as well as purification and sequencing of this molecule were not successful. Therefore, we used FACS analysis, immunoprecipitation with monospecific sera, and Western blotting in order to determine the identity of the 80- to 90-kDa glycoprotein. We focused on a range of molecules that may be relevant in HTLV biology and infection. Studies indicate that different cell surface antigens including receptors and ligands can be altered after HTLV-I and/or HTLV-II infection, including CD9, CD11a, CD13, CD18, CD33, CD40, CD49e, CD54, CD58, and CD80. First, antibodies against a range of known cell surface antigens were tested for their ability to bind BJAB-WH and not BJAB-CC84 cells. FACS analysis showed that anti-ICAM-3 mAbs bound to fusogenic BJAB-WH cells and not to BJAB-CC84 cells (Fig. 3). All other anti-CD mAbs exhibited similar bind-

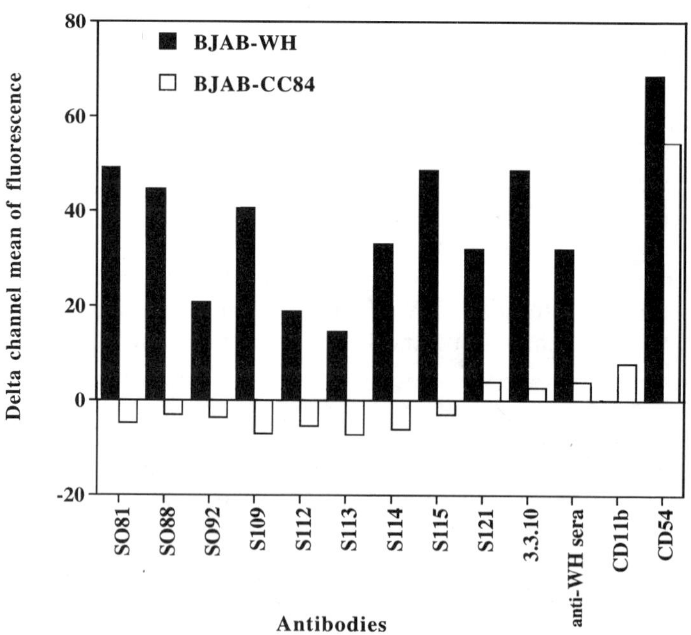

FIGURE 3. ICAM-3 mAbs bind to WH, not CC84 sublines of BJAB cells. Binding of various anti-CD50 mAbs to BJAB-WH and BJAB-CC84 cells. Binding activity was measured by FACS analysis and is shown as mean fluorescence for one representative experiment. Monospecific antisera and mAbs (3.3.10) against BJAB-WH cells were used as a positive control. For FACS, 10 μl of each mAb (1 μg/ml) or polyclonal sera (1:100 dilution) was incubated with 0.2×10^6 WH or CC84 sublines of BJAB cells for 1 h on ice. After incubation cells were washed and stained with anti-mouse Ig–FITC conjugate as described earlier.[47]

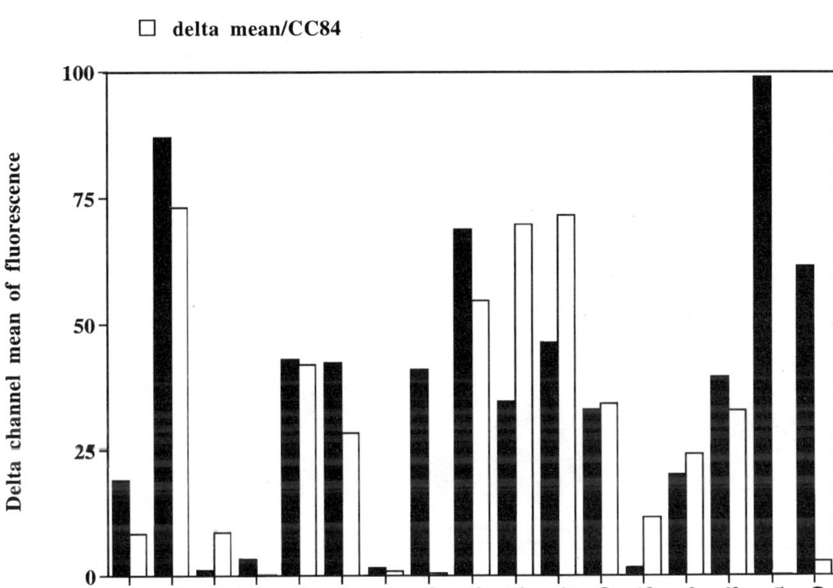

FIGURE 4. FACS comparison of mAbs to various known cell surface antigens. Binding of various mAbs to BJAB-WH and BJAB-CC84 cells. Results are shown as change in mean fluorescence as determined by FACS analysis. Monospecific antisera and mAbs (3.3.10) against BJAB-WH cells were used as a positive control. Ten μl of each mAb (1 μg/ml) or polyclonal sera (1:100 dilution) was incubated with 0.2×10^6 WH or CC84 sublines of BJAB cells for 1 h on ice. After incubation, cells were washed and stained with anti-mouse Ig–FITC conjugate as described previously.[47] Data are shown for one representative experiment.

ing characteristics on both BJAB-WH and BJAB-CC84 cells (Fig. 4). Several anti-ICAM-3 mAbs inhibited over 86% of HTLV-I-induced syncytia formation (Fig. 5). Syncytia inhibition by anti-ICAM-3 was comparable to inhibition by the anti-BJAB-WH monospecific sera as well as mAbs against BJAB-WH. The mAbs against surface antigens did not inhibit syncytia formation as effectively as the anti-ICAM-3 mAbs (Fig. 6). Importantly, anti-BJAB-WH monospecific sera, anti-BJAB-WH mAb, and anti-ICAM-3 mAb immunoprecipitated the same size molecule (80–90 kDa) from the lysates of metabolically labeled BJAB-WH, but not BJAB-CC/84 cells (Fig. 7). Additionally,

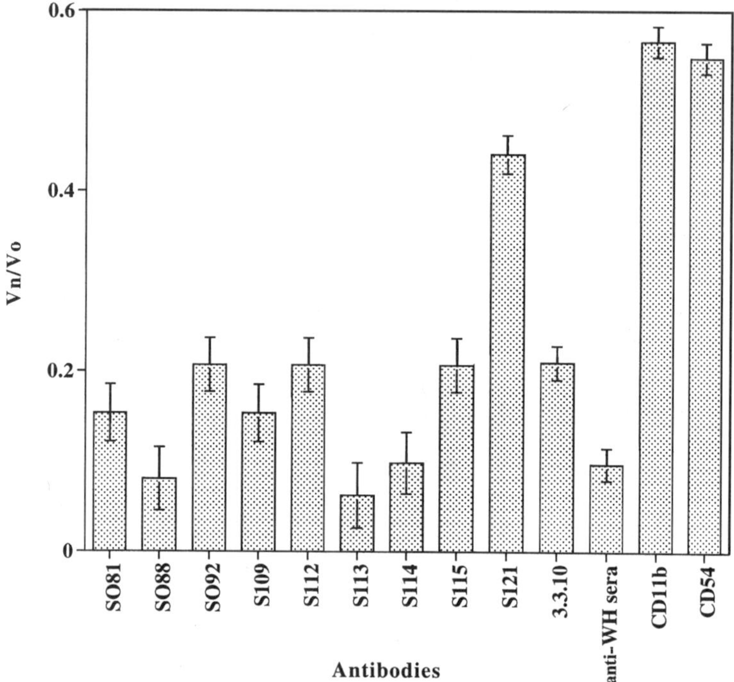

FIGURE 5. Syncytium inhibition with different anti-ICAM-3 mAbs. Syncytium formation inhibition ability of various anti-CD50 mAbs. Syncytium inhibition was performed as described.[47] A mixture of BJAB-WH target cells (50×10^3/well) and HTLV-I/MT2 infected cells (50×10^3/well) was used for the fusion assay. For analysis of syncytium inhibition, we incubated a 1:100 dilution of anti-BJAB-WH monospecific polyclonal sera or 1 µg/ml mAb antibodies first with target cells followed by the addition of infected cells. The quantity of syncytia was calculated and data were presented as V_n/V_o, where V_n is the number of syncytia in the control and V_o is the number of syncytia in experimental wells. Negative control wells contained media only. Data represent the average of two experiments.

anti-BJAB-WH sera blocked binding of fluorescein isothiocyanate (FITC)-conjugated anti-ICAM-3 mAb with target cells (Fig. 8). Therefore, we demonstrated that monoclonal anti-ICAM-3, our monospecific antisera, and mAb against BJAB-WH cells bound the same cellular molecule, ICAM-3, which is involved in HTLV infection.

ICAM-3 is a member of the immunoglobulin supergene family. This is a five-immunoglobulin-domain, 120-kDa adhesion molecule expressed on a variety of cells including lymphocytes and professional antigen-presenting cells (APC). ICAM-3 is a costimulatory molecule for both resting and acti-

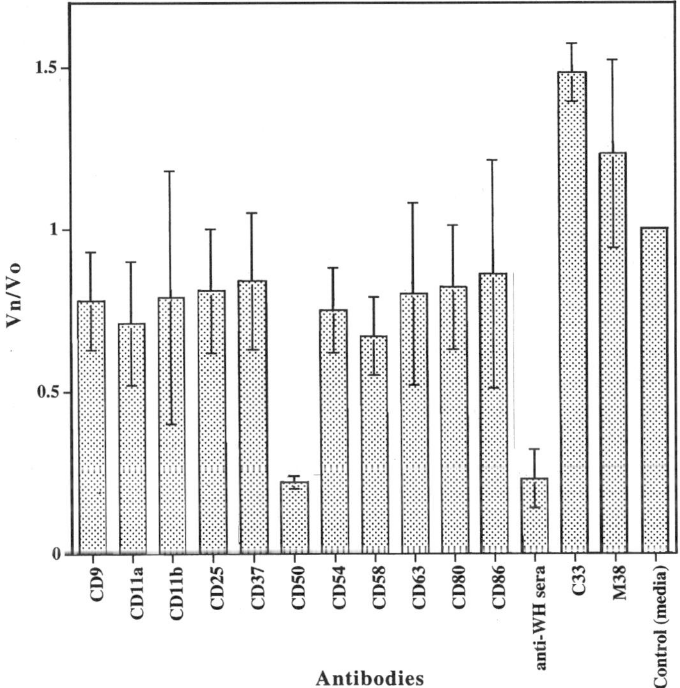

FIGURE 6. Syncytium inhibition with mAbs to various known cell surface antigens. Experiments were performed exactly as discussed in Fig. 5. Data represent the average of two experiments.

vated T lymphocytes and plays an essential role in the initiation of the immune response.[48] ICAM-3 is closely related to ICAM-1 and binds LFA-1 (lymphocyte function-associated molecule) through its two N-terminal domains.[49] The constitutive high expression of ICAM-3 on resting leukocytes coupled with the observation that ICAM-3 is the primary LFA-1 ligand on resting T cells suggests that ICAM-3 may be most important LFA-1 ligand in the initiation of the immune/inflammatory response.[50] The ability of ICAM-3 to mediate syncytium formation indicates that this molecule could represent a cellular binding/fusogenic receptor or could be an adhesion molecule involved in cell-to-cell spreading of HTLV. As mentioned above, ICAM-3 has molecular weight 120 kDa; interestingly, our identified antigen has a molecular lower weight of 80–90 kDa. It is possible that ICAM-3 exists in a truncated form on BJAB-WH cells, accounting for the lower molecular weight.

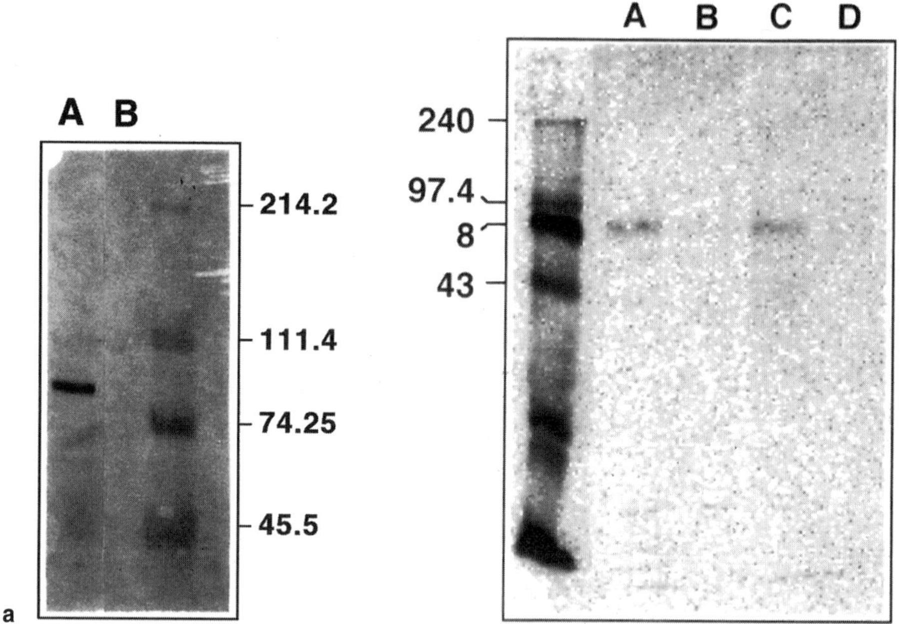

FIGURE 7. Gel immunoprecipitation. (a) Anti-BJAB-WH monospecific sera immunoprecipitated 80–90 kDa molecule from the lysate of metabolically labeled BJAB-WH (lane A), but not BJAB-CC84 (lane B) cells. (b) Both anti-BJAB-WH (lane A) and anti-ICAM-3 (lane C) mAb immunoprecipitated the same size molecule (80–90 kDa) from the lysates of metabolically labeled BJAB-WH (lanes A, C), but not BJAB-CC84 (lanes B, D) cells.

Earlier, Sommerfelt *et al.*[35] determined through the use of human–mouse somatic hybrids that HTLV uses a receptor coded by human chromosome 17. In these experiments, the only human chromosome common to all cell hybrids susceptible to HTLV infection was human chromosome 17.[35] More recently, researchers have found that the HTLV receptor gene may reside on the long arm of chromosome 17 localized between q21 and q23.[4] The gene encoding ICAM-3 has been mapped to chromosome 19. We feel that it is quite possible that chromosome 17 actually encodes binding receptor, whereas ICAM-3 act as a fusogenic receptor or adhesion molecule for viral binding, fusion, and infection. Alternatively, the gene on chromosome 17 may not encode an actual receptor, but may be important for the elaboration of a cofactor to the primary receptor or possibly the expression of a coreceptor.

There is other evidence that adhesion molecules can play a variety of roles in virus infection. The role of ICAMs as receptors or coreceptors for

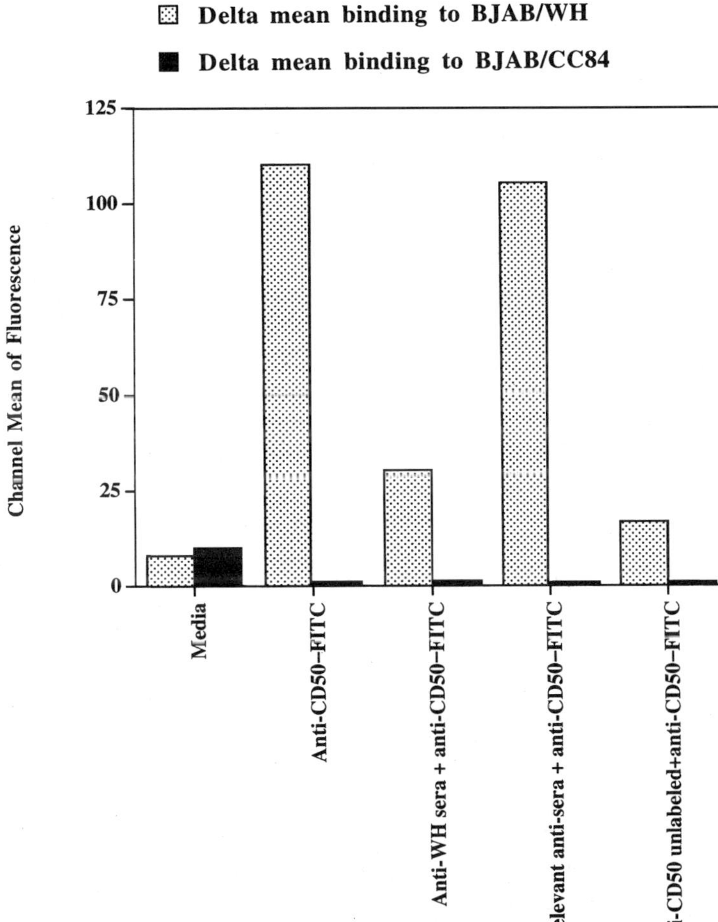

FIGURE 8. FACS blocking with labeled anti-ICAM-3. Anti-BJAB-WH sera blocked binding of FITC conjugated anti-ICAM-3 mAb with target cells. Blocking of anti-ICAM-3 binding with anti-WH sera as determined by FACS analysis. Antisera at a dilution up to 1:100 had the ability to block binding of anti-CD50–FITC with BJAB-WH cells. We used anti-CD50 MoAb labeled with FITC and unlabeled in the blocking experiments. Cells were incubated (30 min, 4°C) first with anti-WH monospecific sera (1:5 diluted, 5ul) or with anti-CD50 (1:10 diluted, 5ul) and than with anti-CD50–FITC (1:10 diluted, 5ul). Data is shown for one representative experiment.

other viruses has been documented. It has been shown that ICAM-1, ICAM-2, and ICAM-3 function as ligands for LFA-1 in HIV-1-mediated syncytia formation.[52] Results suggest that certain epitopes of ICAM-3 may be involved in mediating HIV-1-specific entry into lymphoid and monocytoid cells.[52] Certain Abs to ICAM-3 significantly inhibited HIV-1-specific entry, but not syncytium formation. These data are consistent with our earlier results demonstrating that monospecific antibodies do not inhibit syncytia formation between HIV-1/MN-infected cells and SUPT-1 target cells.[47] Additionally, it has been shown that antibodies against the cell adhesion molecule LFA-1 can block HIV-1-induced syncytium formation.[53] Others have also shown that LFA-1 plays an important role as an accessory protein in the biology of HIV.[54,55]

Other researchers have indicated that adhesion molecules may be central to HTLV infection. Fukudome *et al.*[56] showed that antibodies against ICAM-1 (CD54) and LFA-I (CD11a) also partially inhibit fusion of HTLV-infected cells with target cells. In a search for the cellular receptor for HTLV, Hildreth *et al.*[57] found that vascular cell adhesion molecule-1 (VCAM-1), another member of the Ig supergene family, can mediate HTLV-I-induced syncytium formation. VCAM-1 is the major cellular ligand for the integrin VLA-4 (CD29/CD49d).[57] Syncytium formation between HTLV-I-expressing cells and VCAM-1-positive cells could be blocked with antiserum against HTLV-I gp46 and with a monoclonal antibody against VCAM-1. Also, antibodies against the VCAM-1 counter-receptor VLA-4 failed to block VCAM-1-mediated HTLV-I syncytium formation.

These data along with our results suggested that several different adhesion molecules (VCAM-1, ICAM-1, and ICAM-3) could be involved in the HTLV infection process. Whether or not they serve as viral binding or fusion receptors or simply accessory molecules for HTLV-I-induced cell fusion remains to be investigated. Molecules such as VCAM-1 and ICAM-3 could increase the binding energy between HTLV-I-infected cells and uninfected cells, thereby increasing fusion efficiency. Since integrins have also been shown to be signaling molecules, adhesion molecules may support HTLV-I syncytium formation through signal transduction after interacting with its ligand.[57] Therefore, these adhesion molecules may be important in this type of signaling event. Such events may be necessary to trigger fusion.

Another possible function of adhesion molecules is simply to bring HTLV-I-infected cells and target cells closer together. After the initial association of the adhesion molecules and their ligands, the cellular receptor for HTLV may interact with the viral envelope expressed on the surface of infected cells. Another possibility is that adhesion molecules may function as the viral receptor. In this case, this molecule will interact directly with viral

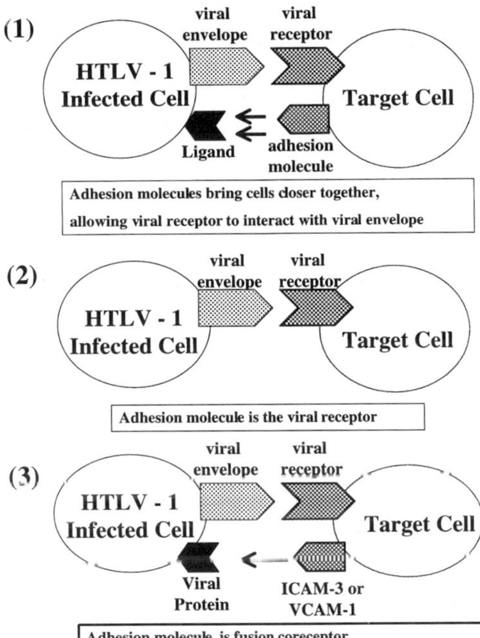

FIGURE 9. The three possible roles of adhesion molecules in HTLV-I cell–cell infection.

protein, thereby triggering fusion of target cells with infected cells. Finally, adhesion molecules may be the HTLV-I fusion receptor (Fig. 9). In such a model, gp46 should bind to its principal viral receptor first to bridge the two membranes, followed by the interaction between gp46/gp21 and the fusion receptor, causing viral infection. This scenario, with the adhesion molecule as the fusion coreceptor of HTLV-I, brings to mind the role of chemokine receptors as coreceptors for HIV-1, since chemokine receptors can regulate expression and function of integrins. Our data and the work of Hildreth *et al.*[57] serve as direct evidence that adhesion molecules are involved in the biology of HTLV-I.

REFERENCES

1. Poiesz, B. J., Ruscetti, F., Gazdar, A. F., Bunn, P. A., Minna, J. D., and Gallo, R. C., 1980, Detection and isolation of new type-C retrovirus particles from fresh and cultured lympho- cytes of a patient with cutaneous T-cell lymphoma, *Proc. Natl. Acad. Sci. USA* **77**:6815–6819.
2. Poiesz, B. J., Ruscetti, F. W., Reitz, M. S., Kalyanaraman, V. S., and Gallo, R. C., 1981,

Isolation a new type-C retrovirus (HTLV) in primary uncultured cells of patient with Sezary T-cell leukemia and evidence for virus nucleic acids and antigens in fresh leukemic cells, *Nature* **294**:268–271.

3. Kalyanaraman, V. C., Sarngadharan, M. G., Robert-Guroff, M., Miyoshi, I., Blayney, D., Golde, D., and Gallo, R. C., 1982, A new subtype of human T-cell leukemia virus (HTLV-II) associated with a T-cell variant of hairy cell leukemia, *Science* **218**:571–573.

4. Chen, I. S. Y., and Rosenblatt, J. D., 1991, Human T-cell leukemia virus type II, in: *The Human Retroviruses* (R. C. Gallo and G. Jay, eds.), Academic Press, New York, pp. 21–34.

5. Nagashima, K., Yoshida, M., and Seiki, M., 1986, A single species of pX mRNA of HTLV-I encodes trans-activator p40x and two other phosphoproteins, *J. Virol.* **60**:394–399.

6. Sagata, N., Yasunaga, T., Ohishi, K., Tszuku-Kawamura, J., Onuma, M., and Ikawa, Y., 1984, Comparison of the entire genomes of bovine leukemia virus and human T-cell leukemia virus and characterization of their unidentified open reading frames, *EMBO J.* **3**:323–327.

7. Seiki, M., Hatton, S., Hirayama, Y., and Yoshida, M., 1983, Human adult T-cell leukemia virus complete nucleotide sequence of the provirus genome integrated in leukemia cell DNA, *Proc. Natl. Acad. Sci. USA* **80**:3618–3622.

8. Schneider, J., Yamamoto, N., Hinuma, Y., and Hunsmann, G., 1984, Sera from adult T-cell leukemia patients react with envelope and core polypeptides of adult T-cell leukemia virus, *Virology* **132**:1–11.

9. Yamamoto, N., Schneider, J., Hinuma, Y., and Hunsmann, G., 1982, Adult T cell leukemia-associated antigen (ATLA): Detection of a glycoprotein in cell- and virus-free supernatant, *Z. Naturforsch. C* **37c**:731–732.

10. Leung, K., and Nabel, G. J., 1988, HTLV-I transactivator induces interleukin-2 receptor expression through an NF-κB-like factor, *Nature* **333**:776–778.

11. Fujii, M., Tsuchiya, H., Chuhjo, T., *et al.*, 1992, Interaction of HTLV-I Tax1 with p67SRF causes the aberrant induction of cellular immediate early genes through CarG boxes, *Genes Dev.* **6**:2066–2076.

12. Zhao, L.-J., and Giam, C.-Z., 1992, Human T cell lymphotrophic virus type I (HTLV-I) transcriptional activator, Tax, enhances CREB binding to HTLV-I 21-base-pair repeats by protein–protein interaction, *Proc. Natl. Acad. Sci. USA* **89**:7070–7074.

13. Suzuki, T., Hirai, H., Fujisawa, J., *et al.*, 1993, A *trans*-activator Tax of human T-cell leukemia virus type I (HTLV-I) interacts with cAMP-responsive element (CRE) binding and CRE modulator proteins that bind to the 21-base-pair enhancer of HTLV-I, *Proc. Natl. Acad. Sci. USA* **90**:610–614.

14. Yoshida, M., 1996, Molecular biology of HTLV-I: Recent progress, *J. AIDS* **13**(Suppl. 1):S63–S68.

15. Lal, R. B., Rudolph, D. L., Dezzutti, C. S., Linsley, P. S., and Prince, H. E., 1996, Costimulatory effects of T cell proliferation during infection with human T lymphotrophic virus types I and II are mediated through CD80 and CD86 ligands, *J. Immunol.* **157**:1288–1296.

16. Nakamura, M., Takasawa, N., Ohbo, K., Higashimura, N., Ohtni, K., Tanaka, Y., and Sugamura, K., 1997, HTLV-I Tax *trans*-activation and cell growth signaling, *Leukemia* **11**(Suppl. 3):7–9.

17. Hidaka, M., Inoue, J., Yoshida, M., and Seiki, M., 1988, Post-transcriptional regulator (rex) of HTLV-I initiates expression of viral structural proteins but suppresses expression of regulatory proteins, *EMBO J.* **7**:519–523.

18. Kanamori, H., Suzuki, N., Siomi, H., *et al.*, 1990, HTLV-I p27(rex) stabilizes human interleukin-2 receptor alpha chain mRNA, *EMBO J.* **9**:4161–4166.

19. White, K. N., Nosaka, T., Kanamori, H., Hatanaka, M., and Honjo, T., 1991, The nucleolar localisation signal of the HTLV protein p27rex is important for stablisation of IL-2 receptor alpha subunit mRNA by p27rex, *Biochem. Biophys. Res. Commun.* **175**:98–103.

20. Nacamura, H., Hatami, M., Ohta, Y., Ishikawa, K., Tsumoto, H., Kiyokawa, T., Yoshida, M., Sasagawa, A., and Hongo, S. 1987, Protection of cynomolgus monkeys against infection by human T-cell leukemia virus type-I by immunization with viral *env* gene products produced by *Escherichia coli, Int. J. Cancer* **40**:403.

21. Akagi, T., Takeda, I., Oka, T., Ohtsuki, Y., Yano, S., and Miyoshi, I., 1985, Experimental infection of rabbits with human T-cell leukemia virus type I, *Jpn. J. Cancer Res.* **76**:86.

22. Dezzutti, C. S., Frazier, D. E., Lafrado, L. J., and Olsen, R. G., 1990, Evaluation of a HTLV-I subunit vaccine in prevention of experimental STLV-I infection in *Macacca nemestrina, J. Med. Primatol.* **19**:305–316.

23. Clapham, P., Nagy, K., Cheingsong-Popov, R., Exley, M., and Weiss, R. A., 1983, Productive infection and cell-free transmission of human T-cell leukemia virus in a nonlymphoid cell line, *Science* **222**:1125–1127.

24. Ichimaru, M., Ikeda, S., Kinoshita, K., Hino, S., and Tsuji, Y., 1991, Mother-to-child transmission of HTLV-I, *Cancer Detect. Prev.* **15**:177–181.

25. Agadjanyan, M. G., Traverdi, N., Levine, W., Kuchkadar, S., and Weiner, D. B., 1997, Characterization of immune responses against HIV-2 and SIVmac after inoculation of mice with HIV-2 expression vector, *AIDS Hum. Retrovir.*

26. Boyer, J., Ugen, K., Wang, B., Agadjanyan, M., Gilbert, I., Bagarazzi, M., Chattergoon, M., Frost, P., Javadian, A., Ciccareli, R., Coney, L., and Weiner, D., 1997, *Nature Med.* **3**:526–532.

27. Wang, B., Ugen, K. E., Srikantan, V., Agadjanyan, M. G., Dang, K., Refaeli, Y., Sato, A., Boyer, J., Williams, W. V., and Weiner, D. B., 1993, Gene inoculation generates immune responses against human immunodeficiency virus type 1, *Proc. Natl. Acad. Sci. USA* **90**:4156–4160.

28. Wang, B., Boyer, J. D., Ugen, K. E., Srikantan, V., Ayyavoo, V., Agadjanyan, M. G., Williams, W. V., Newman, M., Coney, L., Carrano, R., and Weiner, D. B., 1995, Nucleic acid-based immunization against HIV-1: Induction of protective *in vivo* immune responses, *AIDS* **9**:S159–S170.

29. Moss, B., Fuerst, T. R., Flexner, C., and Hugin, A., 1988, Roles of vaccinia virus in the development of new vaccines, *Vaccine* **6**:161–163.

30. Ellis, R. W., 1996, New technologies for making vaccines, in: *Vaccines* (S. A. Plotkin and E. A. Mortimer, eds.), Saunders, Philadelphia, pp. 867–887.

31. Kido, H., Fukutomi, A., and Katanuma, N., 1990, A novel membrane-bound serine esterase in human T4$^+$ lymphocytes immunologically reactive with antibody inhibiting syncytia induced by HIV-1, *J. Biol. Chem.* **265**:21978.

32. Sommerfelt, M. A., and Weiss, R. A., 1990, Receptor interference groups of 20 retroviruses plating on human cells, *Virology* **176**:58–69.

33. Weiss, R. A., 1991, Receptors for human retroviruses, in: *The Human Retroviruses* (R. C. Gallo and G. Jay, eds.), Academic Press, New York, pp. 127–136.

34. Cann, A. J., and Chen, I. S. Y., 1990, Human T-cell leukemia virus types II, in: *Virology* (B. N. Fields, D. M. Knipe, *et al.*, eds.), Raven Press, New York, pp. 1501–1527.

35. Sommerfelt, M., Williams, B., Clapham, P. R., Solomon, E., Goodfellow, P. N., and Weiss, R. A., 1988, Human T-cell leukemia viruses use a receptor determined by human chromosome 17, *Science* **242**:1557–1559.

36. Alkhatib, G., Combadiere, C., Broder, C. C., Feng, Y., Kennedy, P. E., Murohy, P. M., and Berger, E. A., 1996, CC CKR5: A RANTES, MIP-1α, MIP-1β receptor as a fusion cofactor for macrophage-tropic HIV-1, *Science* **272**:1955–1956.

37. Cocchi, F., DeVico, A. I., Garzino-Demo, A., Arya, S. K., Gallo, R. C., and Lusso, O., 1995, Identification of RANTES, MIP-1_, MIP-1_ as the major HIV-suppressive factors produced by CD8$^+$ T cells, *Science* **270**:1811–1815.

38. Feng, Y., Broder, C. C., Kennedy, P. E., and Berger, E. A., 1996, HIV-1 entry cofactor:

Functional cDNA cloning of a seven-transmembrane, G-protein-coupled receptor, *Science* **272**:872–877.

39. Ehrlich, G. D., Dave, F., Kishner, J., Sninsky, J., Kwok, S., Slamon, R., Kalish, D., and Poiesz, B. J., 1989, A polyclonal CD4$^+$ and CD8$^+$ lymphocytosis in a patient doubly infected with HTLV-I and HIV-I: A clinical and molecular analysis, *Am. J. Hematol.* **30**:128–139.

40. Kalyanaraman, V. S., Narynan, P., Foeorino, P., Ramsey, R., Palmer, E., McDougal, T., Getchell, J. P., Holloway, B., Harrison, A. K., Cabradola, C., Telfer, M., and Evett, B., 1985, Isolation and characterization of a human T-cell leukemia virus type II from a hemophilia A patient with pancytopenia, *EMBO J.* **4**:1455–1460.

41. Rosenblatt, J. D., Giorgi, J. V., Golde, D. W., Ben-Ezra, J., Wu, A., Winberg, C. D., Glaspy, J., Wachsman, W., and Chen, I. S. Y., 1988, Integrated HTLV-II genome in CD8$^+$ T-cells from a patient with "atypical" hairy-cell leukemia: Evidence for distinct T- and B-cell lymphoproliferative disorders, *Blood* **30**:128–139.

42. Fan, N., Gavalchin, J., Paul, B., Wells, K., Lane, M. J., and Poiesz, B. J., 1992, Infection of peripheral blood mononuclear cells and cell lines by cell free human T-cell lymphoma/leukemia virus type I, *J. Clin. Microbiol.* **30**:905–910.

43. Agadjanyan, M. G., Chattergoon, M. A., Petrushina, I., Bennett, M., Kim, J., Ugen, K. E., Kieber-Emmons, T., and Weiner, D. B., 1998, Monoclonal antibodies define a cellular antigen involved in HTLV-I infection, *Hybridoma* **17**:9–19.

44. Li, Q. X., Camerini, D., Xie, Y., Greenwald, M., Kuritzkes, D. R., and Chen, I. S., 1996, Syncytium formation by recombinant HTLV-II envelope glycoprotein, *Virology* **218**:279–284.

45. Imai, T., Fukudome, K., Takagi, S., Nagira, M., Furuse, M., Fukuhara, N., Nashimura, M., Hinuma, Y., and Yoshie, O., 1992, C33 antigen recognized by monoclonal antibodies inhibitory to human T-cell leukemia virus type-I induced syncytium formation is a member of a new family of transmembrane proteins including CD9, CD37, CD53, and CD63, *J. Immunol.* **149**:2879–2886.

46. Imai, T., and Yoshie, O., 1993, C33 antigen and M38 antigen recognized by monoclonal antibodies inhibitory to syncytium formation by human T-cell leukemia virus type I are both members of the transmembrane 4 superfamily and associate with each other and with CD4 or CD8 in T cells, *J. Immunol.* **151**:6470–6482.

47. Agadjanyan, M. G., Ugen, K. E., Wang, B., Williams, W. V., and Weiner, D. B., 1994, Identification of an 90-kilodalton membrane glycoprotein important for human T-cell leukemia virus type I and type II syncytium formation and infection, *J. Virol.* **68**:485–493.

48. Hernandez-Caselles, T., Rubio, G., Camanero, M. R., del Pozo, M. A., Muro, M., Sanchez-Madrid, F., and Aparicio, P., 1993, ICAM-3 the third LFA-1 counterreceptor, is a costimulatory molecule for both resting and activated T lymphocytes, *Eur. J. Immunol.* **23**:2799–2807.

49. Fawcett, J., Holness, C. L., Needham, L. A., Turley, H., Gatter, K. C., Mason, D. Y., and Simmons, D. L., 1992, Molecular cloning of ICAM-3, a third ligand for LFA-1, constitutively expressed on resting leukocytes, *Nature* **360**:481–484.

50. Skubitz, K. M., Campbell, K. D., and Skubitz, A. P., 1997, CD50 monoclonal antibodies inhibit neutrophil activation, *J. Immunol.* **159**:820–828.

51. Tajima, Y., Tashiro, K., and Camerini, D., 1997, Assignment of the possible HTLV receptor gene to chromosome 17q21–q23, *Somat. Cell Mol. Genet.* **23**:225–227.

52. Sommerfelt, M. A., and Asjo, B., 1995, Intercellular adhesion molecule 3, a candidate human immunodeficiency virus type 1 co-receptor on lymphoid and monocytoid cells, *J. Gen. Virol.* **76**:1345–1352.

53. Hildreth, J. E., and Orentas, R. J., 1989, Involvement of a leukocyte adhesion receptor (LFA-1) in HIV induced syncytium formation, *Science* **244**:1075–1078.

54. Fecondo, J. V., Pavuk, N. C., Siburn, K. A., Read, D. M., Mansell, A. S., Boyd, A. W., and

McPhee, D. A., 1993, Synthetic peptide analogs of intercellular adhesion molecule 1 (ICAM-1) inhibit HIV-1 replication in MT-2 cells, *AIDS Res. Hum. Retrovir.* **9:**733–740.

55. Pantaleo, G., Butini, L., Graziosi, C., Poli, G., Scnittman, S. M., Greenhouse, J. J., Gallin, J. I., and Fauci, A. S., 1991, Human immunodeficiency virus (HIV) infection in CD4$^+$ T lymphocytes genetically deficient in LFA-1: LFA-1 is required for HIV-mediated cell fusion but not for viral transmission, *J. Exp. Med.* **173:**511–514.

56. Fukudome, K., Furuse, M., Imai, T., Nishimura, M., Takagi, S., Hinuma, Y., and Yoshie, O., 1992, Identification of membrane antigen C33 recognized by monoclonal antibodies inhibitory to human T-cell leukemia virus type 1 (HTLV-I)-induced syncytium formation: Altered glycosylation of C33 antigen in HTLV-I-positive T cells, *J. Virol.* **66:**1394–1401.

57. Hildreth, J., Subramanium, A., and Hampton, R., 1997, Human T-cell lymphocyte virus type 1 (HTLV-I)-induced syncytium formation mediated by vascular cell adhesion molecule-1: Evidence for involvement of cell adhesion molecules in HTLV-I biology, *J. Virol.* **71:**1173–1180.

6

Vaccine Approaches for Human T-Cell Lymphotropic Virus Type I

G. A. DEKABAN, A. PETERS, J. ARP, and G. FRANCHINI

1. INTRODUCTION

The first exogenous pathogenic human retrovirus, human T-lymphotropic virus type I (HTLV-I), was discovered in the late 1970s.[1] Soon after, HTLV-I was etiologically linked to a human cancer, adult T-cell leukemia/lymphoma (ATLL).[2] Epidemiological studies have demonstrated that HTLV-I is geographically distributed throughout the world, but is particularly endemic in certain regions of Japan, the southeastern United States, the Caribbean basin, several countries in Africa, Central and South America, and along the Pacific Rim.[3,4] While HTLV-I is clearly linked to the development of ATLL in these regions of the world, there is also an association between HTLV-I infection and a neurological condition known as HTLV-I-associated myelopathy/tropical spastic paraparesis (HAM/TSP).[5-7] In addition, HTLV-I-infected asymptomatic individuals are immunosuppressed to varying degrees, resulting in concomitant increases in their susceptibility to opportunistic infections and the development of other malignancies.[8-14] HTLV-I infection has also been linked to infective dermatitis in children,[15] arthropathy,[16] and uveitis.[17-19] Thus, the initial belief that ATLL was the only consequence of

G. A. DEKABAN, A. PETERS, and J. ARP • Gene Therapy and Molecular Virology Group, The John P. Robarts Research Institute, and the Department of Microbiology and Immunology, University of Western Ontario, London, Ontario, Canada, N6A 5K8. G. FRANCHINI • Basic Research Laboratory, National Cancer Institute, National Institutes of Health, Bethesda, Maryland 20892.

Human Retroviral Infections, edited by Kenneth E. Ugen *et al.* Kluwer Academic / Plenum Publishers, New York, 2000.

HTLV-I infection is now being reassessed to include a much broader spectrum of diseases in which HTLV-I can play a direct or indirect role. The cumulative risk among the 10–25 million HTLV-I-infected individuals of developing an HTLV-I-associated disease is between 2% and 5%.[7,20–24] The infectious nature of ATLL, HAM/TSP, and other HTLV-I-associated diseases suggests that it should be possible to develop proper and effective prophylactic measures to prevent HTLV-I infection, or at least to prevent the appearance of disease. The prevention of HTLV-I disease would be a substantial achievement on its own. Accounting for both the invariable fatality of ATLL and HAM/TSP and the approximately one million people at risk for the progression to all HTLV-I-associated diseases, the justification for the development of an HTLV-I vaccine is clearly evident.

Transmission of HTLV-I occurs by sexual contact, transfusion of whole blood, sharing of intravenous drug equipment, and from mother to child, principally via breast milk,[25] with occasional transplacental transmission.[25–27] The prevalence of HTLV-I infection and associated diseases varies considerably from one geographic location to the next. The prevalence of HTLV-I infection in the United States and Canada is low as determined by U.S. and Canadian Red Cross serological surveys[28] (Dr. Peter Gill, Head of the Canadian Red Cross, personal communication); however, more extensive studies are identifying new endemic areas in North America.[29] Unlike human immunodeficiency virus type 1 (HIV-1), HTLV-I is predominantly associated in the United States with heterosexual individuals and, to a lesser extent, with intravenous drug abusers, but not with homosexual individuals.[30,31] This appears to be true in other regions of the world as well.[32] Thus, with HTLV-I being transmitted largely by sexual contact, the heterosexual population as a whole would appear to be at greatest risk of HTLV-I infection. In Canada, it is not clear which risk groups are most highly associated with HTLV-I infection. However, medical observations suggest HTLV-I-associated diseases (ATLL and HAM/TSP) occur almost entirely in recent immigrants from HTLV-I-endemic areas of the Caribbean or South America,[33–35] or in the coastal Amerindians of British Columbia.[36] HTLV-I-induced disease remains a significant problem in southern Japan, and the prevalence of HTLV-I and its associated diseases in Africa, the Caribbean, South America, and Pacific Rim countries is likely to remain the same or increase with time. A recent European multinational study indicates that HTLV-I infection is spreading from immigrants of HTLV-I-endemic areas to native Europeans.[37] One of its conclusions is that an effective vaccine strategy would be useful in preventing the further spread of this virus within the European community. Although HTLV-I and HIV-1 coexist in many regions of the world, whether this dual infection has an impact on disease progression for either virus remains controversial.[38,39] Thus, the development of a successful HTLV-I vaccine through the understanding of HTLV-I biology and its interaction with the human immune system may be the only way to prevent the spread of the virus and its associated diseases.

An HTLV-I vaccine would be of particular benefit to underdeveloped countries. In addition, appropriate coupling of the vaccine with a sensitive and specific diagnostic assay would help countries in North America and Europe, which appear to be largely free of HTLV-I, to remain so. It is the purpose of this chapter to review the reasons why it may be possible to successfully develop an HTLV-I vaccine, to discuss which animal models are available for testing the efficacy of potential vaccine candidates, and to examine the vaccine approaches that will be most appropriate for the development of such a therapy.

2. ISSUES RELEVANT TO HTLV-I VACCINE DEVELOPMENT

2.1. HTLV-I Tropism

Compared to other retroviruses, including HIV-1, HTLV-I virions are not highly transmissible.[40,41] HTLV-I virions are produced in low titer and, as a result, the *in vitro* transmission of HTLV-I to various cell types, including lymphocytes and monocytes, with cell-free virus is difficult. Efficient cell-free infection can only take place following the concentration of virus from infected cell supernatants.[42,43] Thus, efficient *in vitro* infection of susceptible cells usually requires cocultivation with HTLV-I-infected cells.[40] HTLV-I infects various human cells, with a preference for CD4$^+$ T-cells.[40,44–46] Endothelial and neural-derived cell lines have also been successfully infected.[47,48] Similarly, under normal circumstances, successful *in vivo* transmission of HTLV-I to humans and to animals appears to require infected cells,[34,49,50] although some exceptions may exist. HTLV-I infects cells from different animal species, including nonhuman primate, rabbit, rat, and hamster lymphocytes. This has allowed several animal models of human HTLV-I infection to be developed.[24,51–55] Several studies suggest that, next to the primate models of HTLV-I infection, New Zealand rabbits and various inbred rat strains may be the most appropriate small-animal models of HTLV-I infection for immunological and vaccine studies.[24,56,57] Regardless of the animal model employed, however, it would be important that vaccine designs take into account that the primary target of the vaccine is HTLV-I-infected cells. Furthermore, since the virus is predominantly sexually transmitted, considerable effort should go into developing a vaccine capable of eliciting strong mucosal immunity. As will be discussed later, little effort has been made to address these issues.

2.2. The Natural Immune Response to HTLV-I Infection

Individuals infected with HTLV-I have both cell-mediated and humoral immune responses against the virus.[58–64] HTLV-I-seropositive asymptomatic individuals possess antibodies capable of blocking the formation of syncytia between HTLV-I-infected and HTLV-I-noninfected cells and antibodies that neutralize the infectivity of vesicular stomatitis virus (VSV)/HTLV-I pseudotypes.[61,62] Also present in infected individuals are antibodies involved in antibody-dependent cell-mediated cytotoxicity (ADCC).[60,65,66] Antibodies to the envelope transmembrane protein gp21 usually appear first,[59] followed closely by antibodies to the core proteins p19 and p24.[59] T-cell proliferation and cytotoxic T-cell responses have been documented in asymptomatic and HAM/TSP patients.[63,64] A critical issue that remains unresolved is whether it is necessary for an HTLV-I vaccine candidate to induce sterilizing immunity, since it has not yet been documented that an individual contracting HTLV-I infection can subsequently clear all traces of HTLV-I infection.

2.3. Feasibility of Generating an HTLV-I Vaccine

Several successful viral vaccines have been devised that have prevented significant disease morbidity and mortality throughout the world. These include vaccines for polio, small pox, measles, and the world's first genetically engineered vaccine: the recombinant hepatitis B virus vaccine. In all of these cases, the vaccines have been successful because they were designed to prevent only the appearance of disease and because the target viruses are intrinsically genetically stable. In contrast, there are many other viral infections for which vaccines have been only partially successful, such as for influenza viruses and herpes simplex virus, or are difficult to develop, as with HIV-1 and hepatitis C virus. With the latter viruses, issues such as high genetic variability, viral integration into the host genome, viral latency, the need for mucosal immunity, and/or a requirement for long-lasting sterilizing immunity need to be addressed. HTLV-I vaccine prospects most closely resemble those of the latter group of viruses, which require particular attention to the issues of latency and the potential need for sterilizing immunity.

2.4. Genetic Stability

Despite these apparent obstacles, there are several reasons why a vaccine for HTLV-I might be successfully developed. First, HTLV-I shows very limited degrees of genetic variability among isolates from geographically distant areas of the world.[67–69] This is in sharp contrast to HIV-1, where there are at least ten different phylogenetic clades recognized so far. Within HIV-1 clades, genetic variation up to 20% is observed, and as much as 40–50% variation can exist between clades.[70] There are only four phylogenetically distinct HTLV-I clades known, differing within themselves by 0.2–2% and by no more than 5–10% between clades. Furthermore, in contrast to HIV-1, where it is difficult to obtain antibody cross-neutralization between different clades by different HIV-1 vaccine candidates, this does not appear to be the case with respect to HTLV-I. Cross-neutralizing antibodies have been demonstrated even between the most divergent HTLV-I isolates, suggesting that genetic variability may not be an impediment to the development of an effective HTLV-I vaccine.[71,72] Protective vaccines have been developed for the relatively genetically stable feline leukemia virus.[73,74] Friend murine leukemia virus,[75,76] and the SAIDS retrovirus (a simian type D retrovirus).[77,78] Success with vaccine development against bovine leukemia virus (BLV) appears highly relevant to HTLV-I vaccine design since BLV is structurally and phylogenetically related to HTLV-I.[79] Importantly, BLV is transmitted in much the same manner as HTLV-I and shares the property of high genetic stability.[79,80] Small-scale efficacy studies using recombinant pox viruses expressing the BLV envelope protein have demonstrated varying degrees of protection, depending on the vaccinia construct used. Nevertheless, those studies indicate the critical contribution of cell-mediated immunity in protection.[81–85] Collectively, these studies suggest that when genetic variability is not a factor, the chances of creating a protective vaccine for a retrovirus are greatly increased.

3. AVAILABLE ANIMAL MODELS

HTLV-I vaccine testing is potentiated by the availability of several useful animal models such as rabbit, rat, mouse, and nonhuman primate.[51–56] Each of the models has its own unique attributes and usefulness. The choice of the appropriate model must be based on the nature of the scientific hypothesis to be tested. Correlates of protection, the availability of syngeneic HTLV-I-infected cell lines, differential immune responses, disease progression, and infectivity may all dictate which model should be used for a particular study.

3.1. Rodent Models

In theory, HTLV-I vaccine studies in inbred rats will have the advantage that correlates of protection with regard to cell-mediated immunity can be more easily determined through adoptive transfer experiments. The availability of syngeneic HTLV-I-infected cell lines has enabled the detection of anti-HTLV-I cytotoxic T-cell responses.[86,87] Furthermore, mother-to-offspring transmission of HTLV-I has also been demonstrated in rats and rabbits, thus offering two potential models that enable the testing of the protective efficacy of passive antibody immunization in newborns.[88,89] In vaccine studies conducted thus far, greater success has been obtained with the rabbit model of HTLV-I infection than the rat model. This may be due in part to the rat's normally robust immune system, which in general seems to be capable of dealing with infections very efficiently.

Successful HTLV-I infection of rabbits and rats in the laboratory appears to be variable with respect to the virus strain used, as well as the breed of rabbit and rat employed. Different strains of rabbits and rats have been evaluated immunologically on a scale of low or poor responders to high responders to HTLV-I.[50,56,90–92] Poor or atypical responses to the surface glycoprotein gp46 and sometimes to the capsid proteins p19 or p24 are apparent[50,93] (G. A. Dekaban and J. Arp, manuscript in preparation). Interestingly, the immune response seen in certain strains of rabbits and rats can be quite similar to human responses observed in asymptomatic HTLV-I carriers that exhibit indeterminate serological profiles on Western blot analysis.[50,94] However, some caution in the interpretation of these results is warranted. In studies claiming observations of either poor or no response to gp46, the assay for antibody detection was by Western blot.[92,93] While we concur with the Western blot results, data from our experiments in the Fisher F344 rat and other research performed in the rabbit suggest that antienvelope antibody is indeed produced, but can be detected only by radioimmunoprecipitation assays using a conformationally correct gp46 produced by the vaccinia/T7 recombinant expression system (G. A. Dekaban and J. Arp, manuscript in preparation) or from a lysate of an HTLV-I-infected cell line.[50] This finding suggests that conformation-sensitive antibodies may be preferentially produced as opposed to antibodies to linear epitopes. This type of antibody response is not unusual, as most antibody responses to hepatitis B surface antigens in humans are often of the conformational type and cannot be detected in Western blot assays.[95–97] Thus, in testing for antibody responses in rat and rabbit vaccine trials, it may be crucial that appropriate HTLV-I antibody diagnostic assays be employed to provide accurate results.

Reports of ATLL-like disease induction in an inbred rabbit strain following injection with an HTLV-I-infected rabbit cell line have appeared, but this is not a model that has been pursued by others.[98,99] Immunosuppressed adult rats also develop ATLL-like disease; however, a compromised host is not particularly relevant in a vaccine context.[53] Some authors have reported the induction of neurological disease in the WKA strain of rats that is observed 18–24 months after HTLV-I infection.[100] Unfortunately, no one has been able to reproduce these findings and the pathology of the HTLV-I-associated neurological disease witnessed in these WKA rats is not the same as that observed in HTLV-I-infected humans with HAM/TSP. In the related rat strain, WKY, postmortem examination of HTLV-I-infected rats revealed neurological abnormalities; however, the WKY rats were reported to be clinically normal.[101] Since the WKA rat strain is not readily available for general research use, it would be unlikely to serve as a model to determine whether candidate HTLV-I vaccines are effective in preventing HTLV-I-associated neurological disease. The rat model is also the most economical of all the animal challenge models available when a large number of animals is required. Indeed it might serve as an initial testing ground for HTLV-I vaccine immunogenicity before testing is expanded to the more expensive rabbit and simian models of HTLV-I infection.

The use of the hu-SCID mouse may also prove to be useful. This model, with its ability to support human adoptive transfer experiments, would allow the study of various aspects of vaccine-induced immunity to HTLV-I, and thus allow the examination of the correlates of HTLV-I protective immunity and disease pathogenesis in humans. [102,103] However, the short duration with which human cells are viable in hu-SCID mice and the difficulty in inducing primary immune responses in these mice make the use of this model problematic.[104] Nonimmunocompromised mice cannot be used in HTLV-I vaccine challenge studies since they are naturally resistant to HTLV-I infection due to the lack of expression of the HTLV-I receptor.[105] Mice with intact immune systems can only be useful in the preliminary HTLV-I vaccine experiments to study the antigenicity of the vaccine candidate.[106,107]

3.2. Nonhuman Primate Models

Unlike the rabbit and rat animal models, which are limited to testing the effectiveness of a vaccine to prevent infection, the use of simians may expand the role of animal models since they can manifest disease under normal circumstances. The simian models of HTLV-I usually involve the use of a human isolate as the challenge virus. However, phylogenetically highly related simian retroviruses can cause a naturally occurring disease similar, if not identical, to ATLL[108–111] and could serve as a challenge strain.

Recently, squirrel monkeys have been assessed for their ability to be a useful vaccine model for HTLV-I.[94] This monkey species is not naturally infected with simian T-lymphotropic virus type-I (STLV-I). Importantly, HTLV-I-transformed allogeneic squirrel monkey cell lines generated by cocultivation with irradiated MT-2 cells were shown to be infectious for squirrel monkeys, despite the fact that MT-2 cells themselves were found not to be. These results are in contrast to the aforementioned results obtained in rabbit and rat infection studies that demonstrated that xenogeneic MT-2 infection was possible. Similar to the results of the rat model, injection of squirrel monkeys with syngeneic HTLV-I-infected cells did not result in a substantive virus infection as compared to that observed when allogeneic HTLV-I-infected cells were used. From this report it appears that the squirrel monkey model could be effectively used to test HTLV-I vaccine candidates in a manner that more naturally mimics HTLV-I infection in humans due to the availability of an allogeneic challenge model. This may thus serve as a better animal challenge model for an HTLV-I vaccine.

Perhaps the best opportunity to evaluate HTLV-I vaccine candidates would be in the use of captive nonhuman primate colonies where STLV-I is naturally endemic. For example, at the Southwest Foundation in New Mexico, up to 80% seroprevalence can be seen in sexually mature captive baboons, accompanied by a high incidence of lymphoma. The colony thus provides an excellent model to test vaccine efficacy in a natural setting of STLV-I transmission. Such studies are underway (G. Franchini, J. Allen, G. A. Dekaban, and J. Tartaglia, unpublished).

4. MODES OF CHALLENGE

In most cases of simian, rabbit, and rat vaccine challenge studies, human infected lymphocytes were the source of challenge virus[54,57,72,93,112] and in only some cases were minimum infectious doses determined[57] (J. Arp and G. A. Dekaban, manuscript in preparation). HTLV-I-infected simian, rabbit, and rat cell lines have been successfully derived; however, not until recently have they been used in vaccine challenge studies.[94] Blood from HTLV-I-infected rabbits has been directly transfused in some passive antibody immunization experiments as a means of challenge of virus.[113,114] Passive immunization of rabbits with immunoglobulin obtained from HTLV-I-infected asymptomatic humans or rabbits was protective in rabbits challenged with live HTLV-I-infected cells.[72,115,116] Similarly, passive immunization of macaques with immunoglobulin from HTLV-I asymptomatic humans was protective.[117] Some experimental evidence also indicates that it may be possible to protect rabbits and monkeys from HTLV-I infection using subunit proteins/peptides consisting of *Escherichia coli*-produced envelope protein,[54] inactivated whole-virus preparations,[118,119] or synthetic peptides.[57] However, in these studies the immunological correlates associated with protection were either not sought or were not found.

5. VACCINE APPROACHES RELEVANT TO HTLV-I

In designing a vaccine for any infectious agent, it is necessary to define the antigens, the method of delivery capable of inducing protective immunity, and any safety concerns. We will first consider several vaccine approaches relevant to HTLV-I and will then assess the HTLV-I antigens most appropriate for use in one of the vaccine strategies. Finally, a review of the HTLV-I vaccine animal studies will be presented and the future directions of HTLV-I vaccine research will be discussed.

5.1. Inactivated and Live Attenuated Viruses

Two vaccine approaches that have been successful in combating other virus infections are the use of inactivated and live attenuated vaccines. For several reasons, however, neither the inactivated nor the live attenuated vaccine approaches are appropriate for an HTLV-I vaccine. With HTLV-I clearly being a human oncogenic virus, it is problematic to ensure that a viral preparation is entirely inactivated. Although inactivated HTLV-I vaccines have been tested in animal models,[113,119] the problems of ensuring the complete safety of this type of vaccine would likely lead to its rejection by regulatory authorities. The creation of a live attenuated HTLV-I vaccine faces the same oncogenic concerns. However, it may be theoretically possible to remove from the live virus genome the two HTLV-I genes that have demonstrated oncogenic potential, those for Tax and p12^I.[120,121] Another confounding aspect of HTLV-I biology makes the inactivated and live attenuated vaccine approaches even less feasible. As would be expected for a virus that is mainly cell-associated, HTLV-I virions are produced in only very low quantities by infected cell lines. Therefore, the expense of growing and purifying sufficient quantities of HTLV-I for vaccine purposes would be great. Similarly, creating noninfectious virus-like HTLV-I particles with the genes for Tax and p12^I deleted would not allow the generation of enough virus-like particles to make such a vaccine approach economically feasible. The molecular aspects of HTLV-I virus assembly that result in the low yields of virus particles are not yet understood. A potential alternative is the creation of a retroviral pseudotype in which the HTLV-I envelope is used to coat another retroviral core, such as the one for the Moloney murine leukemia virus or Gibbon ape leukemia virus. However, even though the parental Moloney and Gibbon ape leukemia viruses are produced in high titer, prior attempts have failed to generate high titers of HTLV-I pseudotypes.[122,123] These data suggest that the HTLV-I envelope protein possesses inherent properties that limit virus particle production.[124]

5.2. Subunit Approach

In order to generate a safe HTLV-I vaccine, a subunit approach would be the most appropriate choice. Several different subunit vaccine approaches appear to offer real possibilities for creating a suitable HTLV-I vaccine. These approaches (summarized in Table I) include the conventional subunit approach, in which purified HTLV-I proteins would be combined with a suitable adjuvant. However, the isolation of large amounts of viral proteins from HTLV-I particles or infected cells is not feasible. Due to these limitations, recombinant sources of the different HTLV-I proteins should be investigated and pursued to provide protein for vaccine use. Alternatively, other attractive vaccine approaches include live viral or bacterial vectors that are capable of expressing one or more of the HTLV-I viral proteins. In addition, synthetic peptides spanning immunologically important regions of HTLV-I proteins can also serve as the basis for an HTLV-I subunit vaccine. The synthetic peptide approach is a viable option due to the genetic stability of HTLV-I; however, a long-term protective immunity against HTLV-I may require a broad-based immune response that is difficult to achieve with a peptide-based vaccine.

5.3. Live Vectors

The use of live attenuated poxviruses, adenoviruses, or bacteria, such as attenuated *Salmonella*, has been proposed and tested as vaccine delivery vehicles for HTLV-I proteins. These live vectors are attractive as vaccine candidates since naturally occurring infections are mimicked. For example, in a recombinant viral vector, the produced HTLV-I antigens will be expressed, processed, and presented to the immune system in a fashion very similar to that occurring in a natural HTLV-I infection. Furthermore, any of

TABLE I

HTLV-I Vaccine Approaches

1. Inactivated (whole-killed) virus: Natural or recombinant source
2. Live attenuated variants: Natural or engineered (e.g., Tax- and p12^L-deleted)
3. Subunit vaccine: Natural or engineered envelope glycoproteins (gp63, gp46, p21), Gag proteins (p24, p19)
4. HTLV-I proteins in live vectors: E.g., vaccinia virus, avipox, adenovirus, various bacteria, herpes simplex, poliovirus
5. Sequence-derived peptides of HTLV-I: B-/Th-/CTL-epitopes
6. DNA gene transfer: Genetic immunization
7. Prime-boost regimes with combinations of the various aforementioned vaccine candidates

the live vectors can be modified to also express cytokine genes, whose presence might serve as an immune stimulant or modulator. These vectors also offer the opportunity to deliver antigen to mucosal surfaces. However, a disadvantage of this approach is that prior exposure to any of the live vectors, particularly adenoviruses or poxviruses, may serve to dampen the immune response to the recombinant HTLV-I protein expressed by the vector. To circumvent this possibility, vectors expressing heterologous proteins, independent of their replication, have been constructed. In addition, prime–boost immunization regimens have been proposed in which the live vector delivering HIV-1 or HTLV-I antigens would serve to prime the immune system and the injections of purified recombinant protein would boost the initial response.[125–127] Another issue to be addressed is the observation that in certain individuals with preexisting medical conditions, such as immunosuppression, the use of live attenuated virus or bacterial vectors for vaccine purposes is contraindicated. Fortunately, however, more re cently derived replication-defective adenovirus and poxvirus vectors allow at least partial circumvention of this problem.[104,126]

5.4. Immunization by DNA Gene Transfer

DNA or genetic immunization is the latest type of subunit vaccine approach to be designed. In this method, eukaryotic plasmid DNA-based vectors which are capable of expressing the gene of interest are injected directly into the vaccinee. This results in a small, but sufficient number of the surrounding cells taking up the plasmid DNA, which in turn provide a platform for short-term recombinant gene expression. As with recombinant virus vectors, the plasmid-encoded recombinant protein is synthesized, posttranslationally modified, and then processed and presented via the major histocompatibility complex (MHC) class I pathways to the immune system, all in a manner similar to that of the native viral protein. Usually intramuscular immunization is conducted with 50–100 μg of the plasmid DNA; other routes of injection have been tried, but with much less success.[128,129] It has been suggested that injury of the muscle by the injection of an anesthetic, such as bupivacane, followed by an injection with the immunizing plasmid DNA into the same muscle site enhances the immune response. It is proposed that the regenerating muscle cells are more efficient at taking up DNA than mature, fully differentiated myotubes. Furthermore, the injury may set up a mild inflammatory reaction that serves to attract immune effector cells, thus enhancing the immunization process similar to the effect of an adjuvant. This DNA injection approach has recently been tested in HTLV-I vaccine trials in both the rat and rabbit models.[93,130,131]

Alternatively, the plasmid DNA can be delivered with the use of a biolistic device such as the Accell gene gun.[129] This apparatus can deliver plasmid DNA-coated gold particles into the dermis such that keratinocytes and some dendritic cells are directly transfected. One of the advantages of this DNA immunization route is that it requires no more than 1–2 μg of plasmid DNA; as little as 0.05 μg has been found to be effective.[129] This immunization strategy is currently being explored as a vaccine strategy for HTLV-I in the rat challenge model (A. Peters and G. A. Dekaban, unpublished data).

As with live recombinant viral and bacterial vectors, genetic vectors permit the incorporation of additional genes that can act as immune stimulants. The immunomodulating genes could be encoded with the same immunization vector or be coadministered on a separate DNA vector.[128] Advantages of the DNA immunization approach are its low cost and lack of inherent immunogenecity, and the apparent absence of traditional vaccine side effects. Anti-DNA responses and potential oncogenicity issues are theoretical concerns about this vaccine approach; however, no evidence has been found to validate these concerns. Interestingly, there is some evidence that intramuscular and gene gun immunization do not elicit equivalent immune responses.[132] In general, it appears that intramuscular immunization more readily induces a cell-mediated Th1-type of immune response, whereas gene gun immunization of the skin is more effective in inducing an antibody or Th2-type of immune response. Prime–boost immunization regimens in which DNA is used in combination with recombinant poxviruses, adenoviruses, or purified recombinant proteins are also potential strategies to maximize the effectiveness of the immune response generated. Since HTLV-I is a mainly cell-associated virus, it may be more important to have a strong cell-mediated response to clear an HTLV-I infection. However, whether this is the case remains to be seen; this may eventually need to be addressed when choosing the appropriate genetic immunization strategy.

6. HTLV-I VACCINE CANDIDATES

HTLV-I proteins are currently being assessed for their ability to induce protective immune responses. Several HTLV-I proteins are being tested individually and/or in various combinations in a polyvalent vaccine approach. These include the envelope gp46 (surface protein) associated with gp21 (transmembrane anchor protein), the gp46 surface envelope protein alone, the major core protein p24, and the matrix protein p19. It appears that the envelope proteins represent the major targets of the immune system during HTLV-I infection,[28] as is observed with other retroviruses, where immune responses to retroviral envelope proteins can be protective.[76] While any HTLV-I-based envelope vaccine would probably require gp46 to play a role, the inclusion of all or part of gp21 requires some consideration as gp21 is known to contain an immunosuppressive region.[133,134] However, other retroviral proteins may provide additional protection. For instance, the HIV-1 matrix protein p17 is capable of eliciting neutralizing antibodies,[136,137] although a recombinant HIV-1 *gag* precursor protein did not induce neutralizing antibodies and was not protective in a chimpanzee.[138] In addition, immunization with a recombinant vaccinia virus (RVV) expressing HIV-1 p24 core protein has been shown to elicit cytotoxic T-cells in both humans[139] and mice.[128] Furthermore, it has also been demonstrated that the murine Friend leukemia virus *gag* proteins expressed by an RVV protected mice from a live virus challenge.[76] So far, no evidence exists on whether the HTLV-I *gag* proteins p19 and p24 are capable of eliciting a protective response individually or in combination.

6.1. *Tax*

The HTLV-I *tax* protein could also be considered as a component of a polyvalent vaccine candidate. However, some biological properties of the HTLV-I *tax* protein may be of concern. First, Tax has been demonstrated to have oncogenic and pathogenic potential on its own.[120,140,141] It has also been suggested that immune responses to the Tax protein, particularly CD8$^+$ cytotoxic T-cell responses, may play a role in the pathogenesis of HAM/TSP.[63] There are, however, other interpretations that have been proposed for the role of anti-Tax cytotoxic T-cells in HAM/TSP; thus, the issue remains unresolved.[64,142] Furthermore, while Tax transactivates the HTLV-I viral promoter, it also influences the expression of a number of cellular genes that regulate cell proliferation.[143,144] Tax is also secreted from infected cells and is taken up by noninfected cells and transported to the nucleus, where it may affect the transcription of cellular genes.[145] For all these reasons, the use of an intact Tax protein as a vaccine component may not be desirable.

6.2. *Pol* and pX Region Proteins

Other HTLV-I proteins could be used, such as the protease, the reverse transcriptase, p27 (Rex), and other proteins from the X region such as p12^I, p13, and p27 (Rof). However, there has been very little analysis of T cell,[146] and antibody[147] responses to these HTLV-I proteins and thus little is known about whether they may be important to protective HTLV-I immunity.

6.3. Envelope Glycoprotein

In contrast, the immunogenicity and potential usefulness of the HTLV-I envelope protein as a vaccine candidate has been extensively studied. A number of efforts have been made to express the HTLV-I envelope precursor glycoprotein gp63[106,107,126,135,148] and the cleaved gp46 surface glycoprotein.[106,149] However, the generation of a conformationally correct gp46 recombinant protein able to induce neutralizing antibodies, as defined by syncytia inhibition assays, has been proven difficult in most expression systems. While both a direct neutralization of infectivity assay and an VSV/HTLV-I pseudotype neutralization assay have been described, they are both difficult to perform and are thus not routinely used.[61,62] The presence of neutralizing antibodies, described in the studies reviewed here, has been determined by the more commonly used syncytia inhibition assay. Interpretation of these results must be performed with some degree of caution since direct inhibition of infectivity is not being measured.

In assessment of the many recombinant systems designed and implemented for expression of the HTLV-I envelope protein, it appears that only the gp46 expressed by recombinant vaccinia viruses is capable of reproducibly inducing neutralizing antibodies *in vivo*.[106,107,134] Furthermore, vaccinia virus-derived recombinant gp46 appears to be the only recombinant version of gp46 that is capable of binding human conformation-sensitive monoclonal antibodies capable of neutralization.[149,150] This suggests that proper folding and posttranslational modification, with respect to glycosylation, of the envelope protein may be critical for the generation of effective neutralizing antibody responses in humans.[149] Neutralizing monoclonal and polyclonal antienvelope antibodies have been isolated and characterized from HTLV-I-infected humans. On the other hand, neutralizing antibodies have been induced by synthetic envelope peptide immunization of mice, rats, and rabbits,[151–154] suggesting that linear envelope epitopes are also able to elicit neutralizing antibody responses. With regard to the antienvelope cellular response, it has been demonstrated that the context within which the HTLV-I envelope protein is presented to human T-cells can yield different responses. Human T-cells educated *in vitro* to proliferate to envelope synthetic peptides, recombinant *E. coli*-derived truncated versions of gp46, and a baculovirus-derived gp63 do not cross-stimulate each other.[155] Thus, because human T-cells educated to proliferate in the presence of recombinant baculovirus-derived gp63 do not proliferate in the presence of HTLV-I envelope synthetic peptides and vice versa, individuals immunized with a synthetic peptide might not be protected from HTLV-I infection. Taking this into account, one would predict that regimens involving recombinant poxviruses or DNA immunization strategies which allow for the proper processing and presentation of native HTLV-I envelope protein would be more effective than the use of synthetic peptides or denatured and/or incompletely posttranslationally modified recombinant envelope proteins.

Several studies have focused on mapping potential B-cell epitopes through the use of synthetic peptides and human HTLV-I-infected sera[151] or human monoclonal antibodies.[156–158] Synthetic peptides correlated with human antibody binding have also been tested for their ability to induce neutralizing antibody responses in different species, including mice, rabbits, and goats.[151,159] Using this method, several epitopes important for virus neutralization have been identified.[151,154] In the rabbit challenge model, putative B-cell epitopes were presented in a synthetic peptide vaccine and tested for their ability to induce a protective response against HTLV-I challenge.[57,113] In another study, murine and rat monoclonal antibodies able to neutralize in a syncytia inhibition assay were shown to bind to predicted B-cell epitopes.[57,160] Epitopes involved in antibody-dependent cellular cytotoxicity have also been identified using sera from HTLV-I envelope-peptide immunized rabbits.[66] Thus, based on this information it is possible to formulate correlates of protection with respect to the role of antibody.

In contrast to the extensive characterization of B-cell responses to HTLV-I envelope proteins, virtually no systematic mapping has been carried out to identify helper or cytotoxic T-cell epitopes. In only one case has a cytotoxic T-cell envelope epitope been identified, and only for humans.[161] A helper T-cell envelope epitope that stimulates T-cell proliferation has also been identified.[156] Thus, much work remains to be done in this area if helper and cytotoxic T-cell responses are to be characterized and vaccine responses optimized to induce T-cell-mediated immunity toward HTLV-I.

6.4. *Gag*

As outlined above, despite recent and past immunization studies demonstrating that HTLV-I envelope proteins can induce protective responses, no effort has been made to determine if other HTLV-I proteins, particularly the *gag* proteins, can also elicit a protective immune response. Mapping the HTLV-I *gag* region for immunogenic B-cell epitopes through the use of human-infected sera has identified several B-cell epitopes especially in p19.[162] This information has resulted in the development of diagnostic tests for HTLV-I, but has not lead to any synthetic *gag* peptide approaches being tested in animal models. There also has not been any attempt to identify T-helper or cytotoxic epitopes within any of the *gag* proteins. There have been studies of HTLV-I-infected rats describing anti-*gag* T-cell responses.[86,87] Another study demonstrated that incorporation of a *gag*–envelope fusion protein into a mannan-coated liposome is able to prime predominantly *gag*, but also envelope cytotoxic T-cell responses.[163] Whether any of these cytotoxic T-lymphocyte (CTL) responses could prevent HTLV-I infection remains unclear because none of the animals were challenged with HTLV-I.

The development of recombinant viral, DNA, and *gag*-protein immunization strategies is hampered by apparent *cis*-acting negative regulatory sequences present in the *gag* region of HTLV-I that possibly require the presence of Rex to overcome.[164] Similar sequences have been described for HIV-1.[165] Though recombinant adenoviruses containing the HTLV-I *gag–protease* region have been constructed, these recombinant adenoviruses failed to generate detectable levels of *gag* protein in infected cultured cells and were unable to induce anti-*gag* antibody responses in mice (G. A. Dekaban and F. Graham, unpublished data). Recombinant poxviruses, on the other hand, are able to effectively express the HTLV-I *gag* and *env* proteins due to the cytoplasmic nature of poxvirus replication. These viruses are currently being used in rat, rabbit, and baboon immunogenicity studies (G. A. Dekaban, J. Allen, J. Tartaglia, and G. Franchini, unpublished data). In addition, DNA immunization vectors have been successfully constructed to express the HTLV-I p24; however, no expression of p19 has yet been observed (G. A. Dekaban, unpublished data). Lack of successful p19 expression suggests that the *cis*-acting negative regulatory sequences may be present in the p19 region of the *gag* gene; however, further experimentation, currently underway, is required to resolve this issue. Thus, if adenovirus or DNA immunization vectors are to be used as vaccines capable of expressing HTLV-I *gag* proteins, Rex will have to be coexpressed. Alternatively, the exact location of the *cis*-acting negative regulatory sequences will have to be identified and mutated or eliminated to permit expression of the *gag* proteins from the precursor protein or expression of p19 by itself.

6.5. Polyvalent Vaccine

A polyvalent vaccine is desirable to halt infection by a human retrovirus because there may be insufficient immunogenic T-cell epitopes associated with a single immunogen to stimulate T-cell responses across the full range of MHC types occurring in humans. In fact there are a number of HTLV-I-infected asymptomatic individuals who do not seroconvert uniformly to all HTLV-I proteins, particularly the HTLV-I gp46 envelope protein.[166] There is also evidence that various strains of mice and maybe rabbits possessing different MHC backgrounds have distinct antibody responses to recombinant HTLV-I envelope protein.[50,106] More recent studies suggest that this may also be true among different rat strains.[92,93] Humans have also been described as high or low responders to HTLV-I as defined by *in vitro* proliferation assays, with a correlation to particular MHC backgrounds.[167] Thus, consideration should be given to using *gag* and envelope proteins, not only as single-subunit vaccines, but also combined in a multiple-component subunit vaccine. Such consideration has already been given to hepatitis B and HIV-1 vaccines.[168–172]

7. HTLV-I ANIMAL VACCINE CHALLENGE STUDIES

The three most useful animal models for HTLV-I vaccine challenge studies are rabbit, rat, and monkey. Table II summarizes the challenge studies that have been conducted and the results obtained. In most cases at least partial protection and sometimes full protection has been achieved.

7.1. Rat Vaccine Studies

Not until recently has the rat model been characterized for use as a vaccine challenge model, and only one laboratory has reported data.[93] The results are not very encouraging since only partial protection was achieved. In this single study,[93] adenovirus, plasmid DNA immunization, and vaccinia virus vaccine strategies were employed in both WKY and F344 rats. The adenovirus and DNA immunization vectors used in prime–boost regimens in WKY rats did not induce antienvelope antibody responses, but did induce

TABLE II
Animal Models Relevant to HTLV-I Vaccine Design

Animal model	Vaccine approach	Results reported
WKY and F344 rat	Adenovirus,[93] DNA immunization,[93,117] vaccinia virus[93]	Partial protection
Outbred rabbit	Passive immunization with rabbit[115,116] or human immunoglobulin[114]	Protection
	Envelope-based vaccinia virus[114,134]	Protection[134]; however, no protection with mildly attenuated strain[114]
	Envelope-based NYVAC or ALVAC prime with same, or recombinant envelope protein boost[126]	Protection with poxvirus, only prime–boost
	DNA immunization[130] or synthetic envelope peptide[57,113]	Partial protection,[113,130] protection[57]
Cynomolgus monkey	Recombinant envelope protein subunit[54]	Protection (some transient infection)
Pig-tailed macaques	Inactivated whole virus[118,119]	Protection

CTL responses. Protection was observed in only a very low percentage of rats. It should be noted that the envelope plasmid DNA expression vector was not optimized for expression in this study. Somewhat better results were obtained in HTLV-I envelope recombinant vaccinia virus-immunized rats.[93] In this scenario, antibody responses, although not neutralizing, were obtained in the four animals, two of which were protected after challenge. In this study, viral exposure was via intraperitoneal inoculation, which is not a normal route of HTLV-I transmission. While mucosal challenge has not been reported for the rat, intravenous infection has been described[62] (G. A. Dekaban and J. Arp, manuscript in preparation). A recent, but limited DNA immunization study in the rat reported protection, and suggested that protection correlated with the presence of neutralizing antibodies.[117] However, the paucity of information on the number of rats in the study and the fact that there was no direct mention of whether complete or partial protection was obtained does not allow one to draw any real conclusions about these results. A caveat of the aforementioned rat studies is that they involved xenogeneic challenge, employing the MT-2 HTLV-I-infected human T-cell line, and that HTLV-I infection appears indolent in this animal model. In fact, even when rats were exposed to allogeneic or syngeneic HTLV-I-infected T-cells, only a transient infection occurred.[101] These data indicate that rats are not very susceptible to permanent HTLV-I infection. In the above-described rat vaccine studies,[93,101] the anti-HTLV-I immune response, as assessed by antibody responses, to the allogeneic and xenogeneic challenge infection was observed to be more vigorous than to the syngeneic challenge. The suggested explanation (by the authors) for this observation is that the heightened immune response to the infecting allogeneic and xenogeneic cells resulted in greater numbers of proliferating T-cells in the challenged rat and that this, in turn, provided a greater number of target cells for HTLV-I infection, thus contributing to a higher virus load.[101] Thus, the requirement for a xenogeneic challenge renders this system less likely to mimic natural human HTLV-I transmission. However, as long as the limitations of the rat challenge model are recognized, useful information may still be gleaned from its use.

7.2. Rabbit Vaccine Studies

Vaccine studies employing the outbred rabbit challenge model have also been successfully carried out, and is the most common model.[57,114,116,134] This model, like the rat challenge model, is a model for HTLV-I infection only, as infected rabbits do not develop HTLV-I-associated diseases under normal circumstances. Similarly, most challenge stocks for rabbit challenge studies are xenogeneic and are usually of human origin. In some studies, whole blood from an HTLV-I-infected rabbit was used to challenge rabbits passively immunized with anti-HTLV-I antibody of rabbit or human sources.[114,116] No systematic attempt has been made to use a rabbit T-cell line productively infected with HTLV-I as the source of challenge virus. Since rabbits have been shown to be successfully infected with whole blood from allogeneic HTLV-I-infected rabbits, it does not appear that the problems encountered with allogeneic challenge in the rat model will be seen in the rabbit model. Studies that determined the relative infectivity and minimum infectious dose of different strains of HTLV-I for rabbits have been completed.[50,57,126,171] Acknowledging the important advantages of the rabbit challenge model, it appears that the development of reasonably designed rabbit vaccine trials for HTLV-I has a great deal of potential.

Previous vaccine protection studies in rabbits have demonstrated some degree of efficacy of several envelope-based vaccines, including poxvirus vectors,[114,126,134] DNA immunization vectors,[130] and a synthetic peptide.[57,113] The use of weakly attenuated recombinant vaccinia viruses required only a single immunization to obtain protection in one study,[134] but failed in another.[114] However, the use of vaccine strains of vaccinia virus is accompanied by potential adverse side effects. These would likely make them unsuitable for human use, especially in developing countries, where immunosuppression and dermatological conditions preclude the use of vaccinia virus immunization. Thus, highly mutated vaccinia viruses, such as NYVAC or the canary pox (ALVAC) vectors, are being studied as effective alternatives that possess many of the beneficial traits of poxvirus vectors while greatly improving the safety of their use. They have already been tested successfully in rabbits as HTLV-I envelope vaccine vectors; however, in this study a prime–boost approach was required.[126] In this challenge study, an HTLV-I envelope recombinant poxvirus served as the priming injection and was followed by boosting with either recombinant baculovirus-derived HTLV-I precursor envelope protein gp63 or another injection of poxvirus. Interestingly, the rabbits immunized with only the poxvirus envelope vector were the only animals protected from a xenogeneic challenge with human HTLV-I-infected T-cells. This suggests that the immunity induced by an HTLV-I envelope recombinant pox vector immunization regimen was better than one involving the use of a nonnative recombinant HTLV-I envelope protein with an alum adjuvant as the boost. The protein and adjuvant boost may have suppressed the cell-mediated immune response in such a way that the poxvirus-primed immune responses directed to the HTLV-I envelope protein were no longer effective. Because poxviruses are known to induce a strong cell-mediated or a Th1-type of immune response, the success of these HTLV-I envelope-based poxvirus vector vaccine studies suggests that cell-mediated immunity may play a critical role in the protective immunity against HTLV-I infection. Unfortunately, except for limited evaluations of antibody responses, there has been no detailed examination of the cellular immune responses in the vaccine studies carried out so far, and thus the correlates of protection have not been addressed. In the early recombinant vaccinia virus studies, neutralizing antibody responses were not assessed.[134] The more recent vaccinia, NYVAC, and ALVAC studies did indeed test for neutralizing antibody, but did not observe any protection.[114,126] Thus, the mechanisms of protection remain unresolved.

The importance of neutralizing antibody responses is apparent from passive antibody immunization[72,115,116] and synthetic peptide vaccine studies[57,113] in the rabbit model and monkeys.[117] Several passive antibody immunization studies have demonstrated that neutralizing antibody directed against HTLV-I proteins can be protective, although the permissible time between the administration of antibody and virus challenge is relatively short.[72,115] Serum antibodies from HTLV-I-infected humans or HTLV-I-infected rabbits were equally effective in conferring passive immunity. Importantly, these passive antibody transfer experiments also demonstrated that human antibody could prevent infection even when phylogenetically distinct HTLV-I isolates were tested.[72] In addition, rabbit vaccine studies involving the use of synthetic peptides encompassing regions of gp46 known to contain the virus-neutralizing domains have been shown to be capable of inducing the production of neutralizing antibodies that protected against HTLV-I infection upon challenge.[57] In those cases where neutralizing antibody was not induced, protection was not obtained.[113] The success of the synthetic peptide vaccine study was undoubtedly critical in demonstrating the importance of neutralizing antibodies in conferring protection against HTLV-I infection, providing significant evidence that synthetic peptides can elicit a protective response. However, there are certain caveats that must be considered with respect to the synthetic peptide vaccine experiments. First, this study must be analyzed with caution since large amounts of synthetic peptide conjugated to the carrier molecule, ovalbumin, were used. More clinically relevant and acceptable peptide carriers for human use, such as tetanus or diphtheria toxoid, will have to be assessed. Second, Freund's complete adjuvant was used in the priming injections of this study, followed by multiple boosts containing Freund's incomplete adjuvant, to induce the production of neutralizing antibodies. Currently, only alum is licensed as an adjuvant for use in humans, and unfortunately it is likely too weak to be effective in stimulating a strong immune response. Third, despite high antipeptide antibody titers, the neutralizing antibody titers were low. Thus, critical questions have yet to be answered as to whether other clinically acceptable adjuvants and peptide carriers will be able to induce sufficient quantities of neutralizing antibody and ultimately afford protection against HTLV-I infection for sustained periods of time.

7.3. Nonhuman Primate Vaccine Studies

A few HTLV-I vaccine studies have been conducted in different species of monkeys.[54,94,118,119] The number of monkeys employed in these studies by necessity is usually small, rendering the assessment of efficacy difficult. The challenge virus inoculum consists of either xenogeneic HTLV-I-infected T-cells or allogeneic STLV-I-infected T-cells. In one vaccine study,[54] cynomolgus monkeys were immunized with an *E. coli*-derived HTLV-I envelope–β-galactosidase fusion protein mixed with complete Freund's adjuvant. High doses (100–250 μg) of the purified fusion protein were administered intradermally twice, followed by further boosting either intravenously or intradermally. The monkeys were then challenged with the human HTLV-I-infected T-cell line MT-2. Two of the six vaccinated monkeys appeared to be transiently infected by the challenge virus. The remaining four monkeys that exhibited the highest levels of neutralizing antibody were protected from HTLV-I infection. In contrast, all four nonvaccinated control monkeys became infected within 2–3 weeks postchallenge. This study suggests that the recombinant, nonnative envelope protein is capable of protecting monkeys from HTLV-I infection. However, whether similar protection could be obtained with the clinically relevant adjuvant alum is unknown. Unfortunately, it is questionable whether or not these results, obtained with large amounts of recombinant envelope protein coadministered with Freund's complete adjuvant, can be extrapolated to a successful vaccine in humans.

In another HTLV-I vaccine study, an inactivated whole-virus vaccine was prepared and tested in pig-tailed macaques (*Macaca nemestrina*).[118,119] The authors suggested that protection was achieved based on the presence or absence of reverse transcriptase activity in cultures of mononuclear cells taken after challenge from vaccinated and control monkeys at different time points. Only control monkeys had positive reverse transcriptase activity in their corresponding cultures. This method of virus detection is not a very sensitive means of detecting HTLV-I infection and, unlike the cynomolgus study, direct virus isolation assays were not performed. Although encouraging, this limited primate study did not present definitive proof that the inactivated virus vaccine approach was protective. Undoubtedly, the experiments need to be repeated and the challenged animals more thoroughly evaluated by polymerase chain reaction and direct virus isolation for evidence of protection.

8. SUMMARY

While HTLV-I vaccine development is justifiable in its own right, the vaccine research has progressed slowly due to the large shadow cast by the urgent need for an HIV-1 vaccine. In several ways, HTLV-I may serve as a simplified model for human retrovirus vaccine development as it is genetically stable and there are reasonable small-animal models of HTLV-I infection. Importantly, these animal models allow testing for the prevention of infection by several different routes of transmission, including intravenous transmission, sexual transmission, and transmission via breast milk. With respect to HTLV-I vaccine development, it is clear that much experimental testing remains to be carried out. The importance of neutralizing antibody has been experimentally recognized. In the absence of such antibody, cell-mediated immunity may also play a very important role since it, too, can provide protection against HTLV-I infection. To determine more definitively the correlates of protection and permit optimization of vaccination regimens, it is of pivotal importance to improve the inbred rat model and further develop the rabbit model of HTLV-I infection. Up until now, only relatively poor immune responses have been elicited by DNA immunization in animals other than mice. This calls for the development of improved expression vectors by, perhaps, incorporating genes for costimulatory molecules and/or cytokines. In addition, developing safe and effective adjuvants and/or coadministered immune-enhancing agents would be critical for the success of an HTLV-I vaccine, as well as benefiting other vaccine initiatives.

It would also be interesting to determine whether HTLV-I vaccines are capable of providing cross-protection against HTLV-II infection. Both rabbits and rats have been shown to be susceptible to HTLV-II infection, allowing this question to be addressed.[101,174] The necessity of an HTLV-II vaccine is currently difficult to justify since there is insufficient evidence to clearly correlate HTLV-II infection with disease.[175]

If animal vaccine and challenge studies for HTLV-I are successful, the following question will then arise: which human population would be an appropriate target for carrying out human HTLV-I vaccine trials? Phase I and II clinical trials could be carried out anywhere. However, efficacy trials would need to be carried out in endemic areas such as Africa, the Caribbean, South America, and in the Amerindian communities of British Columbia. Alternatively, well-characterized populations of intravenous drug abusers and sexual workers might serve as target test populations. This would also provide an opportunity to assess whether or not there is an effect by HTLV-I vaccination on the prevalence of HTLV-II infection. Prenatal screening for HTLV-I-infected pregnant women would also identify a target population for a vaccine that would boost maternal immunity against the virus. Alternatively, the newborn could be passively immunized with human monoclonal neutralizing antibodies, followed by active immunization of the child at a later age. The issue of whether it is necessary to seek an HTLV-I vaccine that induces long-lasting, sterilizing immunity remains to be resolved. This would be the ultimate goal for an HTLV-I vaccine. Whether such a goal is realistic remains to be seen; however, the probability of it being realized is enhanced by the genetic stability of the HTLV-I genome.

REFERENCES

1. Poiesz, B. J., Ruscetti, F. W., Gazdar, A. F., Bunn, P. A., Minna, J. D., and Gallo, R. C., 1980, Detection and isolation of type C retrovirus particles from fresh and cultured lymphocytes of a patient with cutaneous T-cell lymphoma, *Proc. Natl. Acad. Sci. USA* **77**:7415–7419.
2. Yoshida, M., Miyoshi, I., and Hinuma, Y., 1992, Isolation and characterization of retrovirus from cell lines of human adult T-cell leukemia and its implication in the disease, *Proc. Natl. Acad. Sci. USA* **79**:2031–2035.
3. Manns, A., and Blattner, W. A., 1991, The epidemiology of the human T-cell lymphotrophic virus type I and type II: Etiologic role in human disease, *Transfusion* **31**:67–75.
4. Gessain, A., and de The, G., 1996, in: *Advances in Virus Research* (K. Maranorosch, F. A. Murphy, and A. J. Shatkin, eds.), Academic Press, New York.
5. Vernant, J.-C., Maurs, L., Gessain, A., Barin, F., Gout, O., Delaporte, K., Sanhadji, K., Buisson, G., and de The, G., 1987, Endemic tropical spastic paraparesis associated with human T-lymphotropic virus type I: A clinical and seroepidemiologic study of 25 cases, *Ann. Neurol.* **21**:123–130.
6. Osame, M., Matsumoto, M., Usuku, K., Izumo, S., Ijichi, N., Amitani, H., Tara, M., and Igata, A., 1987, Chronic progressive myelopathy associated with elevated antibodies to human T-lymphotropic virus type I and adult T-cell leukemialike cells, *Ann. Neurol.* **21**:117–122.
7. Osame, M., Usuku, K., Izumo, S., Ijuchi, N., Amatani, H., Igata, A., Matsumoto, M., and Tara, M., 1986, HTLV-1 associated myelopathy, a new clinical entity, *Lancet* **i**:1031–1032.
8. Mann, D., DeSantis, P., Mark, G., Pfeifer, A., Newman, M., Gibbs, N., Popovic, M., Sarngadharan, M., Gallo, R. C., Clark, J., and Blattner, W. A., 1987, HTLV-I-associated B-cell CLL: Indirect role for retrovirus in leukemogenesis, *Ann. N.Y. Acad. Sci.* **236**:1103–1106.
9. Nakada, K., Kohakura, M., Komoda, H., and Hinuma, Y., 1984, High incidence of HTLV antibody in carriers of *Strongyloides stercoralis, Lancet* **1**:663.
10. Katsuki, T., Katsuki, K., Imai, J., and Hinuma, Y., 1987, Immune suppression in healthy carriers of adult T-cell leukemia retrovirus (HTLV-1): Impairment of T-cell control of Epstein-Barr virus-infected B-cells, *Jpn. J. Cancer Res. (Gann)* **78**:639–642.
11. Matsuzaki, H., Asou, N., Kawaguchi, Y., Hata, H., Yoshinaga, T., Kinuwaki, E., Ishii, T., Yamaguchi, K., and Takasuki, K., 1990, Human T-cell leukemia virus type 1 associated with small cell lung cancer, *Cancer* **66**:1763–1768.
12. Asou, N., Kumagai, T., Uekihara, S., Ishii, M., Sato, M., Sakai, K., Nishimura, H., Yamaguchi, K., and Takatsuki, K., 1986, HTLV-I seroprevalence in patients with malignancy, *Cancer* **58**:903–907.
13. Taguchi, H., Kobayashi, M., and Miyoshi, I., 1991, Immunosuppression by HTLV-I infection, *Lancet* **337**:308–309.
14. Welles, S. L., Tachibana, N., Okayama, A., Shioiri, S., Ishihara, S., Murai, K., and Mueller, N. E., 1994, Decreased reactivity to PPD among HTLV-I carriers in relation to virus and hematological status, *Int. J. Cancer* **56**:337–340.
15. Legrande, L., Hanchard, B., Fletcher, V., Cranston, B., and Blattner, W., 1990, Infective dermatitis of Jamaican children: A marker for HTLV-I infection, *Lancet* **336**:1345–1347.

16. Ijichi, S., Matsuda, T., Maruyama, I., Izumihara, T., Kojima, K., Niimura, T., Maruyama, Y., Sonoda, S., Yoshida, A., and Osame, M., 1990, Arthritis in a human T lymphotropic virus type I (HTLV-I) carrier, *Ann Rheum. Dis.* **49:**718–721.

17. Mochizuki, M., Watanabe, T., Yamaguchi, K., Yoshimura, K., Nakashima, S., Shirao, M., Araki, S., Takatsuki, K., Mori, S., and Miyata, N., 1992, Uveitis associated with human T-cell lymphotropic virus type I, *Am. J. Ophthalmol.* **114:**123–129.

18. Mochizuki, M., Watanabe, T., Yamaguchi, K., Takatsuki, K., Yoshimura, K., Shirao, M., Nakashima, S., Mori, S., Araki, S., and Miyata, N., 1991, HTLV-I uveitis: A distinct clinical entity caused by HTLV-I, *Jpn. J. Cancer Res.* **83:**236–239.

19. Mochizuki, M., Tajima, K., Watanabe, T., and Yamaguchi, K., 1994, Human T lympho- tropic virus type 1 uveitis, *Br. J. Ophthalmol.* **78:**149–154.

20. Gessain, A., and Gout, O., 1992, Chronic myelopathy associated with human T-lympho- tropic virus type I (HTLV-I), *Ann Intern. Med.* **117:**933–946.

21. Kaplan, J. E., Osame, M., Kubota, H., Nishitani, H., Maeda, Y., Khabbaz, R. F., and Janssen, R., 1990, The risk of development of HTLV-I-associated myelopathy/tropical spastic paraparesis among persons infected with HTLV-I, *J. Acquired Immune Defic. Syndr.* **3:**1096– 1101.

22. Mueller, N., 1991, The epidemiology of HTLV-I infection, *Cancer Causes Control Pap. Symp.* **2:**37–52.

23. Tajima, K., and The T- and B-Cell Malignancy Study Group and Coauthors, 1990, The 4th nation-wide study of adult T-cell leukemia/lymphoma (ATL) in Japan: Estimates of risk of ATL and its geographical and clinical features, *Int. J. Cancer* **45:**237–243.

24. Bomford, R., Kazanji, M., and de The, G., 1997, Vaccine against human T cell leukemia- lymphoma virus type I: Progress and prospects, *AIDS Res. Hum. Retrovir.* **12:**403–405.

25. Kinoshita, K., Amagasaki, T., Hino, S., Doi, H., Yamanouchi, K., Ban, N., Momita, S., Ikeda, S., Kamihira, S., Ichimaru, M., Katamine, S., Miyamoto, T., Tsuji, Y., Ishimaru, T., Yamabe, T., Ito, M., Kamura, S., and Tsuda, T., 1987, Milk-borne transmission of HTLV-I from carrier mothers to their children, *Jpn. J. Cancer Res. (Gann)* **78:**674–680.

26. Narita, M., Shibata, M., Togashi, T., and Koga, Y., 1991, Vertical transmission of human T cell leukemia virus type I, *J. Infect. Dis.* **163:**204.

27. Stuver, S. O., Tachibana, N., Okayama, A., and Mueller, N. E., 1996, Evaluation of morbidity among human T lymphotropic virus type 1 carriers in Miyazaki, Japan, *J. Infect. Dis.* **173:**584–591.

28. Williams, A. E., Fang, C. T., Slamon, D., Poiesz, B. J., Sandler, S., Darr, W., Shulman, G., McGowan, E., Douglas, D., Bowman, R., Peetoom, F., Kleinman, S., Lenes, B., and Dodd, R., 1988, Seroprevalence and epidemiological correlates of HTLV-I infection in U.S. blood donors, *Ann. N.Y. Acad. Sci.* **240:**643–649.

29. Lee, H. H., Swanson, P., Rosenblatt, J. D., Chen, I. S., Sherwood, W. C., Smith, D. E., Tegtmeier, G. E., Fernando, L. P., Fang, C. T., and Osame, M., 1991, Relative prevalence and risk factors of HTLV-I and HTLV-II infection in US blood donors, *Lancet* **337:**1435– 1439.

30. Lee, H. H., Weiss, S. H., Brown, L. S., Mildvan, D., Shorty, V., Saravolatz, L., Chu, A., Ginzburg, H. M., Markowitz, N., and Des, J. D., 1990, Patterns of HIV-1 and HTLV-I/II in intravenous drug abusers from the middle atlantic and central regions of the USA, *J. Infect. Dis.* **162:**347–352.

31. Lee, H., Swanson, P., Shorty, V., Zack, J., Rosenblatt, J. D., and Chen, I. S. Y., 1989, High rate of HTLV-II infection in seropositive IV drug abusers in New Orleans, *Ann. N.Y. Acad. Sci.* **244:**471–476.

32. Stuver, S., Tachibana, S., Okayama, A., Shioiri, S., Tsunetoshi, Y., Tsuda, K., and Mueller,

N., 1993, Heterosexual transmission of human T cell leukemia/lymphoma virus type I among married couples in southwestern Japan: An initial report from the Miyazaki Cohort Study, *J. Infect. Dis.* **167**:57–65.

33. Dekaban, G. A., King, E., Waters, D., and Rice, G., 1992, Nucleotide sequence analysis of an HTLV-I isolate from a Chilean patient with HAM/TSP, *AIDS Res. Hum. Retrovir.* **8**:1201–1207.

34. Power, C., Weinshenker, B. G., Dekaban, G. A., Kaufman, J. C. E., and Rice, G. P. A., 1991, Pathological and molecular biological features of a myelopathy associated with HTLV-I infection, *Can. J. Neurol. Sci.* **18**:352–355.

35. Dekaban, G. A., Inwood, M., Waters, D., Drouin, J., and Teitel, J., 1992, Absence of human T-lymphotropic virus types I and II infection in an Ontario hemophilia population, *Transfusion* **32**:513–516.

36. Picard, F. J., Coulthart, M. B., Oger, J., King, E. E., Kim, S., Arp, J., Rice, G. P. A., and Dekaban, G. A., 1995, Human T-lymphotropic virus type I (HTLV-I) in coastal natives of British Columbia: Phylogenetic affinities and possible origins, *J. Virol.* **69**:7248–7256.

37. The HTLV European Research Network, 1996, Seroepidemiology of the human T-cell leukaemia/lymphoma viruses in Europe, *J. Acquired Immune Defic. Syndr.* **13**:68–77.

38. Schechter, M., Moulton, L. H., and Harrison, L. H., 1997, HIV viral load and CD4 lymphocyte counts in subjects coinfected with HTLV-1 and HIV-1, *J. Acquired Immune Defic. Syndr.* **15**:308–311.

39. Schechter, M., Harrison, L. H., Halsey, N. A., Trade, G., Santino, M., Moulton, L. II., and Quinn, T. C., 1994, Coinfection with human T-cell lymphotropic virus type I and HIV in Brazil. Impact on markers in HIV disease progression, *JAMA* **271**:353–357.

40. Yamamoto, N., Okada, M., Koyanagi, Y., Kannagi, M., and Hinuma, Y., 1982, Transformation of human leukocytes by cocultivation with an adult T cell leukemia producer cell line, *Ann. N.Y. Acad. Sci.* **217**:737–739.

41. Okochi, K., and Sato, H., 1986, Transmission of adult T-cell leukemia virus (HTLV-I) through blood transfusion and its prevention, *AIDS Res.* **2**:S157–S161.

42. Fan, N., Gavalchin, J., Paul, B., Wells, K., Lane, M. J., and Poiesz, B. J., 1992, Infection of peripheral blood mononuclear cells and cell lines by cell-free human T-cell lymphoma/leukemia virus type I, *J. Clin. Microbiol.* **30**:905–910.

43. Clapham, P., Nagy, K., Cheingsong-Popov, R., Exley, M., and Weiss, R., 1983, Productive infection and cell-free transmission of human T-cell leukemia virus in a nonlymphoid cell line, *Ann. N.Y. Acad. Sci.* **222**:1125–1127.

44. Macchi, B., Popovic, B., Allavena, P., Ortaldo, J., Gallo, R. C., and Bonmassar, E., 1987, *In vitro* susceptibility of different human T-cell subpopulations and resistance of large granular lymphocytes to HTLV-I infection, *Int. J. Cancer* **40**:1–6.

45. Macchi, B., Popovic, M., Allavena, P., Rossi, P., Gallo, R. C., and Bonmassar, E., 1987, Differential susceptibility of human mononuclear cells to infection with HTLV-I, *Int. J. Tiss. React.* **9**:195–198.

46. Koyanagi, Y., Itoyama, Y., Nakamura, N., Takamatsu, K., Kira, J., Iwamasa, T., Goto, I., and Yamamoto, N., 1993, *In vivo* infection of human T-cell leukemia virus type I in non-T cells, *Virology* **196**:25–33.

47. Ho, D. D., Rota, T. R., and Hirsch, M. S., 1984, Infection of human endothelial cells by human T-lymphotropic virus type I, *Proc. Natl. Acad. Sci. USA* **81**:7588.

48. Hoffman, P. M., Dhib-Jalbut, S., Mikovits, J. A., Robbins, D. S., Wolf, A. L., Bergey, G. K., Lohrey, N. C., Weislow, O. S., and Ruscetti, F. W., 1992, Human T-cell leukemia virus type I infection of monocytes and microglial cells in primary human cultures, *Proc. Natl. Acad. Sci. USA* **89**:11784–11788.

49. Suga, T., Kameyama, T., Shimotohno, K., Matsumura, M., Tanaka, H., Kushida, S., Ami, Y., Uchida, M., Uchida, K., and Miwa, M., 1991, Infection of rats with HTLV-I: A small-animal model for HTLV-I carriers, *Int. J. Cancer* **49:**764–769.

50. Cockerell, G. L., Lairmore, M. D., De, B., Rovnak, J., Hartley, T. M., and Miyoshi, I., 1990, Persistent infection of rabbits with HTLV-I: Patterns of anti-viral antibody reactivity and detection of virus by gene amplification, *Int. J. Cancer* **45:**127–130.

51. Eguchi, T., Kubonishi, I., Daibata, M., Yano, S., Imamura, J., Ohtsuki, Y., and Miyoshi, I., 1988, Serial transplantation of an HTLV-I-transformed hamster lymphoid cell line into hamsters, *Int. J. Cancer* **41:**868–872.

52. Seto, A., Kawanishi, M., Matsuda, S., Ogawa, K., Eguchi, T., and Miyoshi, I., 1987, Induction of preleukemic stage of adult T cell leukemia-like disease in rabbits, *Jpn. J. Cancer Res. (Gann)* **78:**1150–1155.

53. Yoshiki, T., Kondo, N., Chubachi, T., Tateno, M., Togashi, T., and Itoh, T., 1987, Rat lymphoid cell lines with HTLV-I production. III. Transmission of HTLV-I into rats and analysis of cell surface antigens associated with HTLV-I, *Arch. Virol.* **97:**181–196.

54. Nakamura, H., Hayami, M., Ohta, Y., Ishikawa, K., Tsujimoto, H., Kiyokawa, T., Yoshida, M., Sasagawa, A., and Honjo, S., 1987, Protection of cynomolgus monkeys against infection by human T-cell leukemia virus type I by immunization with viral *env* gene products produced in *Escherichia coli*, *Int. J. Cancer* **40:**403–407.

55. Truckenmiller, M. E., Kulaga, H., Gugel, E., Dickerson, D., and Kindt, T. J., 1989, Evidence for dual infection of rabbits with the human retroviruses HTLV-I and HIV-I, *Res. Immunol.* **140:**527–544.

56. Ishiguro, N., Abe, M., Seto, K., Sakurai, H., Ikeda, H., Wakisaka, A., Togashi, T., Tateno, M., and Yoshiki, T., 1992, A rat model of human T lymphocyte virus type 1 (HTLV-I) infection. 1. Humoral antibody response, provirus integration, and HTLV-I-associated myelopathy/tropical spastic paraparesis-like myelopathy in seronegative HTLV-I carrier rats, *J. Exp. Med.* **176:**981–989.

57. Tanaka, Y., Tanaka, R., Terada, E., Koyanagi, Y., Miyano-Kurosaki, N., Yamamoto, N., Baba, E., Nakamura, M., and Shida, H., 1994, Induction of antibody responses that neutralize human T-cell leukemia virus type I infection *in vitro* and *in vivo* by peptide immunization, *J. Virol.* **68:**6323–6331.

58. Kalyanaraman, V. S., Sarngadharan, M. G., Bunn, P. A., Minna, J. D., and Gallo, R. C., 1981, Antibodies in human sera reactive against an internal structural protein of human T-cell lymphoma virus, *Nature* **294:**271–273.

59. Manns, A., Murphy, E. L., Wilks, R., Haynes, G., Figueroa, J., Hanchard, B., Barnett, M., Drummond, J., Waters, D., Cerney, M., Seals, J., Alexander, S. S., Lee, H., and Blattner, W. A., 1991, Detection of early human T-cell lymphotropic virus type I antibody patterns during seroconversion among transfusion recipients, *Blood* **77:**896–905.

60. Sinclair, A., Habeshaw, J., Muir, L., Chandler, P., Forster, S., Cruickshank, K., and Dalgleish, A. G., 1988, Antibody-dependent cell-mediated cytotoxicity: Comparison between HTLV-I and HIV-I assays, *AIDS* **2:**465–472.

61. Clapham, P., Nagy, K., and Weiss, R., 1984, Pseudotypes of human T-cell leukemia virus types 1 and 2: Neutralization by patient sera, *Proc. Natl. Acad. Sci. USA* **81:**2886–2889.

62. Nagy, K., Clapham, P., Cheingsong-Popov, R., and Weiss, R., 1983, Human T-cell leukemia virus type I: Induction of syncytia and inhibition by patients' sera, *Int. J. Cancer* **32:**321–328.

63. Jacobson, S., Shida, H., McFarlin, D. E., Fauci, A., and Koenig, S., 1990, Circulating CD8[+] cytotoxic T lymphocytes specific for HTLV-I pX in patients with HTLV-I associated neurological disease, *Nature* **348:**245–248.

64. Parker, C. E., Daenke, S., Nightingale, S., and Bangham, C. R., 1992, Activated, HTLV-I-

specific cytotoxic T-lymphocytes are found in healthy seropositives as well as in patients with tropical spastic paraparesis, *Virology* **188:**628–636.

65. Kunitomi, T., Takigawa, H., Sugita, M., Nouno, S., Suemune, M., Inoue, M., Kodani, N., Oda, M., and Ikeda, M., 1990, Antibody-dependent cellular cytotoxicity and natural killer activity against HTLV-I infected cells, *Acta Paediatr. Jpn.* **32:**16–19.

66. Zhang, X., Yang, L., Ho, D., Kuritzkes, D., Chen, I. S. Y., Ching, W., Chen, Y. M., and Schooley, R., 1992, Human T lymphotropic virus types I- and II-specific antibody-dependent cellular cytotoxicity: Strain specificity and epitope mapping, *J. Infect. Dis.* **165:**805–812.

67. Komurian, F., Pelloquin, F., and de The, G., 1991, *In vivo* genomic variability of human T-cell leukemia virus type I depends more upon geography than upon pathologies, *J. Virol.* **65:**3770–3778.

68. Hoshino, H., Nakamura, T., Tanaka, Y., Miyoshi, I., and Yanagihara, R., 1993, Functional conservation of the neutralizing domains on the external envelope glycoprotein of cosmopolitan and Melanesian strains of human T cell leukemia/lymphoma virus type I, *J. Infect. Dis.* **168:**1368–1378.

69. Benson, J., Tschachler, E., Gessain, A., Yanagihara, R., Gallo, R. C., and Franchini, G., 1994, Cross-neutralizing antibodies against cosmopolitan and Melanesian strains of human T cell leukemia/lymphotropic virus type I in sera from inhabitants of Africa and the Solomon Islands, *AIDS Res. Hum. Retrovir.* **10:**91–96.

70. Levy, J. A., 1998, *HIV and the Pathogenesis of AIDS*, 2nd ed., ASM Press, Chapter 7, pp 137 171.

71. Hoshino, H., Nakamura, T., Tanaka, Y., Miyoshi, I., and Yanagihara, I., 1993, Functional conservation of the neutralizing domains on the external envelope glycoprotein of cosmopolitan and Melanesian strains of human T cell leukemia/lymphoma virus type I, *J. Infect. Dis.* **168:**1368–1373.

72. Tanaka, Y., Ishii, K., Sawada, T., Ohtsuki, Y., Hoshino, H., Yanagihara, R., and Miyoshi, I., 1993, Prophylaxis against a Melanesian variant of human T-lymphotropic virus type-1 (HTLV-I) in rabbits using HTLV-I immune globulin from asymptomatically infected Japanese carriers, *Blood* **82:**3664–3667.

73. Jarrett, O., 1996, Efficacy of recombinant feline leukemia virus vaccines, *AIDS Res. Hum. Retrovir.* **12:**435–436.

74. Hoover, E. A., Mullins, J. I., Chu, H. J., and Wasmoen, T. L., 1997, Development and testing of an inactivated feline leukemia virus vaccine, *Semin. Vet. Med. Surg. (Small Anim.)* **10:**238–243.

75. Earl, P., Moss, B., Morrison, B., Wehrly, K., Nishio, J., and Chesebro, B., 1986, T-Lymphotropic priming and protection against Friend leukemia by vaccinia-retrovirus *env* gene recombinant, *Ann. N.Y. Acad. Sci.* **234:**728–731.

76. Miyazawa, M., Nishio, J., and Chesebro, B., 1992, Protection against Friend retrovirus-induced leukemia by recombinant vaccinia viruses expressing the *gag* gene, *J. Virol.* **66:**4497–4507.

77. Marx, P. A., Pederson, N. C., Lerche, N. W., Osborn, K. G., Lowenstine, L. J., Lacker, A. A., Maul, D. H., Kwang, H.-S., Kluge, J. D., Zaiss, C. P., Sharpe, V., Spinner, A. P., Allison, A. C., and Gardner, M. B., 1986, Isolation of a new serotype of simian acquired immune deficiency syndrome type D retrovirus from Celebes black macaques (*Macaca nigra*) with immune deficiency and retroperitoneal fibromatosis, *J. Virol.* **60:**431–435.

78. Brody, B. A., Hunter, E., Kluge, J. D., Lasarow, R., Gardner, M. B., and Marx, P. A., 1992, Protection of macaques against infection with simian type D retrovirus (SRV-1) by immunization with recombinant vaccine virus expressing the envelope glycoproteins of either SRV-1 or Mason–Pfizer monkey virus (SRV-3), *J. Virol.* **66:**3950–3954.

79. Sarin, P., and Gallo, R. C., 1998, Lymphotropic retroviruses of animals and man, *Adv. Vet. Sci. Comp. Med.* **32:**227–250.

80. Hopkins, S. G., and DiGiacomo, R. F., 1997, Natural transmission of bovine leukemia virus in dairy and beef cattle, *Vet. Clin. North Am. Food Anim. Pract.* **13:**107–128.

81. Ohishi, K., Suzuki, H., Yamamoto, T., Maruyama, T., Miki, K., Ikawa, Y., Numakunai, S., Okada, K., Ohshima, K., and Sugimoto, M., 1991, Protective immunity against bovine leukaemia virus (BLV) induced in carrier sheep by inoculation with a vaccinia virus-BLV *env* recombinant: Association with cell-mediated immunity, *J. Gen. Virol.* **72:**1887–1892.

82. Callebaut, I., Voneche, V., Mager, A., Fumiere, O., Krchnak, V., Merza, M., Zavada, J., Mammerickx, M., Burny, A., and Portetelle, D., 1993, Mapping of B-neutralizing and T-helper cell epitopes on the bovine leukemia virus external glycoprotein gp51, *J. Virol.* **67:**5321–5327.

83. Gatei, M. H., Naif, H. M., Kumar, S., Boyle, D. B., Daniel, R. C. W., Good, M. F., and Lavin, M. F., 1993, Protection of sheep against bovine leukemia virus (BLV) infection by vaccination with recombinant vaccinia viruses expressing BLV envelope glycoproteins: Correlation of protection with CD4$^+$ T-cell response to gp51 peptide 51–70, *J. Virol.* **67:**1803–1810.

84. Gatei, M. H., Good, M. F., Daniel, R. C. W., and Lavin, M. F., 1993, T-cell responses to highly conserved CD4 and CD8 epitopes on the outer membrane protein of bovine leukemia virus: Relevance to vaccine development, *J. Virol.* **67:**1796–1802.

85. Ohishi, K., and Ikawa, Y., 1996, T cell-mediated destruction of bovine leukemia virus-infected peripheral lymphocytes by bovine leukemia virus *env*–vaccinia recombinant vaccine, *AIDS Res. Hum. Retrovir.* **12:**393–398.

86. Tanaka, Y., Tozawa, H., Koyanagi, Y., and Shida, H., 1990, Recognition of human T cell leukemia virus type I (HTLV-I) *gag* and *pX* gene products by MHC-restricted cytotoxic T lymphocytes induced in rats against syngeneic HTLV-I-infected cells, *J. Immunol.* **144:**4202–4211.

87. Noguchi, Y., Tateno, M., Kondo, N., Yoshiki, T., Shida, H., Nakayama, E., and Shiku, H., 1989, Rat cytotoxic T lymphocytes against human T-lymphotrophic virus type I-infected cells recognize *gag* gene and *env* gene encoded antigens, *J. Immunol.* **143:**3737–3742.

88. Hori, M., Ami, Y., Kushida, S., Kobayashi, M., Uchida, K., Abe, M., and Miwa, M., 1995, Intrauterine transmission of human T-cell leukemia virus type I in rats, *J. Virol.* **69:**1302–1305.

89. Hirose, S., Kotani, S., Uemura, Y., Fujishita, M., Taguchi, H., Ohtsuki, Y., and Miyoshi, I., 1988, Milk-borne transmission of human T-cell leukemia virus type-I in rabbits, *Virology* **162:**487–489.

90. Yoshiki, T., Tateno, M., Sakurai, H., Ishiguro, N., Ikeda, H., and Wakisaka, A., 1992, A rat model of HTLV-I infection, *Gann Monogr. Cancer Res.* **39:**205–216.

91. Lairmore, M. D., Roberts, B. D., Frank, D., Rovnak, J., Weiser, M. G., and Cockerell, G. L., 1991, Comparative biological responses of rabbits infected with human T-lymphotropic virus type I isolates from patients with lymphoproliferative and neurodegenerative disease, *Int. J. Cancer* **49:**1–7.

92. Ibrahim, F., Fiette, L., Gessain, A., Buissan, N., de The, G., and Bonford, R., 1994, Infection of rats with human T-cell leukemia virus type-I susceptibility of inbred strains, antibody response and provirus location, *Int. J. Cancer* **58:**446–451.

93. Kazanji, M., Bomford, R., Bessereau, J.-L., and de The, G., 1997, Expression and immunogeneity in rats of recombinant adenovirus 5 DNA plasmids and vaccinia virus containing the HTLV-I *env* gene, *Int. J. Cancer* **71:**300–307.

94. Kazanji, M., Moreau, J. P., Mahieux, R., Bonnemains, B., Bomford, R., Gessain, A., and de The, G., 1997, HTLV-I infection in squirrel monkeys (*Saimiri sciureus*) using autologous, homologous, or heterologous HTLV-I-transformed cell lines, *Virology* **231:**258–268.

95. Neurath, A. R., Adamowicz, P., Kent, S. B. H., Riottot, M. M., Strick, N., Parker, K., Offensperger, W., Petit, M. A., Wahl, S., Budkowska, A., Girard, M., and Pillot, J., 1986, Characterization of monoclonal antibodies specific for the pre-S2 region of the hepatitis B virus envelope protein, *Mol. Immunol.* **23:**991–997.

96. Petit, M. A., Maillard, P., Capel, F., and Pillot, J., 1986, Immunochemical structure of the hepatitis B surface antigen vaccine—II. Analysis of antibody responses in human sera against the envelope proteins, *Mol. Immunol.* **23:**511–523.

97. Waters, J., Pignatelli, S., Galpin, K., Ishihara, K., and Thomas, H. C., 1986, Virus-neutralizing antibodies to hepatitis B virus: The nature of an immunogenic epitope on the S gene peptide, *J. Gen. Virol.* **67:**2467–2473.

98. Ogawa, K., Matsuda, S., and Seto, A., 1989, Induction of leukemic infiltration by allogeneic transfer of HTLV-I-transformed T cells in rabbits, *Leuk. Res.* **13:**399–406.

99. Seto, A., Kawanishi, M., Matsuda, S., Ogawa, K., and Miyoshi, I., 1988, Adult T cell leukemia-like disease experimentally induced in rabbits, *Jpn. J. Cancer Res. (Gann)* **79:** 335–341.

100. Kushida, S., Mizusawa, H., Matsumura, M., Tanaka, H., Ami, Y., Hori, M., Yagami, K., Kameyama, T., Tanaka, Y., Yoshida, A., Nyunoya, H., Shimotohno, K., Iwasaki, Y., Uchida, K., and Miwa, M., 1994, High incidence of HAM/TSP-like symptoms in WKA rats after administration of human T-cell leukemia virus type I-producing cells, *J. Virol.* **68:**7221–7226.

101. Kazanji, M., Ibrahim, F., Fiette, L., Bomford, R., and de The, G., 1997, Role of the genetic background of rats in infection by HTLV-I and HTLV-II and the development of associated diseases, *Int. J. Cancer* **73:**131–136.

102. Ishihara, S., Tachibana, N., Okayama, A., Murai, K., Tsuda, K., and Mueller, N., 1992, Successful graft of HTLV-I-transformed human T-cells (MT-2) in severe combined immunodeficiency mice treated with anti-asialo GM-1 antibody, *Jpn. J. Cancer Res.* **83:**320–328.

103. Feuer, G., Zack, J. A., Harrington, W., Valderama, R., Rosenblatt, J. D., Wachsman, W., Baird, S. M., and Chen, I. S. Y., 1993, Establishment of human T-cell leukemia virus type I T-cell lymphomas in severe combined immunodeficient mice, *Blood* **82:**722–731.

104. Schiedner, G., Morral, N., Parks, R. J., Wu, Y., Koopsmans, S. C., Langston, C., Graham, F. L., Beaudet, A. L., and Kochanek, S., 1998, Genomic DNA transfer with a high-capacity adenovirus vector results in improved *in vivo* gene expression and decreased toxicity, *Nature Genet.* **18:**180–183.

105. Sommerfelt, M., Williams, B., Clapham, P., Solomon, E., Goodfellow, P., and Weiss, R., 1988, Human T cell leukemia viruses use a receptor determined by human chromosome 17, *Ann N.Y. Acad. Sci.* **242:**1557–1560.

106. Ford, C. M., Arp, J., Palker, T. J., King, E., and Dekaban, G. A., 1992, Characterization of the antibody response to three different versions of the HTLV-I envelope protein expressed by recombinant vaccinia viruses: Induction of neutralizing antibody, *Virology* **191:**448–455.

107. Arp, J., Ford, C. M., Palker, T. J., King, E., and Dekaban, G. A., 1993, Expression and immunogenicity of the entire human T cell leukaemia virus type I envelope protein produced in a baculovirus system, *J. Gen. Virol.* **74:**211–222.

108. Blakeslee, J. R., McClure, M. H., Anderson, D. C., Bauer, R. M., Huff, L. Y., and Olsen, R. G., 1986, Chronic fatal disease in gorillas seropositive for simian T-lymphotropic virus I antibodies, *Cancer Lett.* **37:**1–6.

109. Homma, T., Kanki, P. J., King, N. M., Hunt, R. D., O'Connell, M. J., Letvin, N. L., Daniel, M. D., Desrosiers, R. C., Yang, C. S., and Essex, M., 1984, Lymphoma in macaques: Association with virus of human T-lymphotropic family, *Ann. N.Y. Acad. Sci.* **225:**716–718.

110. Tsujimoto, H., Noda, Y., Ishibishi, K., Nakamura, H., Fukasawa, M., Sakakibara, I., Sasa-

gawa, M., Honjo, S., and Hayami, M., 1987, Development of adult T-cell leukemia-like disease in African green monkey associated with clonal integration of simian T-cell leukemia virus type I, *Cancer Res.* **47**:269–274.

111. Schatzl, H., Tschikobava, M., Rose, D., Voevodin, A., Nitschko, H., Sieger, E., Busch, U., von der Helm, K., and Lapin, B., 1993, The Sukhumi primate monkey model for viral lymphomogenesis: High incidence of lymphomas with presence of STLV-I and EBV-like virus, *Leukemia* **7**(Suppl. 2):S86–S92.

112. Shida, H., Tochikura, T., Sato, T., Konno, T., Hirayoshi, K., Seki, M., Ito, Y., Hatanaka, M., Hinuma, Y., Sugimoto, M., Takahashi-Nishimaki, F., Maruyama, T., Miki, K., Suzuki, K., Morita, M., Sashiyama, H., and Hayami, M., 1987, Effect of recombinant vaccinia viruses, *EMBO J.* **6**:3379–3384.

113. Takehara, N., Iwahara, Y., Uemura, Y., Sawada, T., Ohtsuki, Y., Hoshino, H., and Miyoshi, I., 1989, Effect of immunization on HTLV-I infection in rabbits, *Int. J. Cancer* **44**:332–336.

114. Hakoda, E., Machida, H., Tanaka, Y., Morishita, N., Sawada, T., Shida, H., Hoshino, H., and Miyoshi, I., 1995, Vaccination for rabbits with recombinant vaccinia virus carrying the envelope gene of human T-cell lymphotropic virus type I, *Int. J. Cancer* **60**:567–570.

115. Kataoka, R., Takehara, N., Iwahara, Y., Sawada, T., Ohtsuki, Y., Dawei, Y., Hoshino, H., and Miyoshi, I., 1990, Transmission of HTLV-I by blood transfusion and its prevention by passive immunization in rabbits, *Blood* **76**:1657–1661.

116. Miyoshi, I., Takehara, N., Sawada, T., Iwahara, Y., Kataoka, R., Yang, D., and Hoshino, H., 1992, Immunoglobulin prophylaxis against HTLV-I in a rabbit model, *Leukemia* **6**:24–26.

117. Murata, N., Hakado, E., Machida, H., Ikezoe, T., Sawada, T., Hoshino, H., and Miyoshi, I., 1996, Prevention of human T-lymphotropic virus type I infection in Japanese macaques by passive immunization, *Leukemia* **10**:1971–1974.

118. Dezzutti, C. S., Frazier, D., Lafrado, L., and Olsen, R., 1990, Evaluation of a HTLV-I subunit vaccine in prevention of experimental STLV-I infection in *Macaca nemestrina, J. Med. Primatol.* **19**:305–316.

119. Dezzutti, C. S., Frazier, D., Huff, L., Stromberg, P., and Olsen, R., 1990, Subunit vaccine protects *Macaca nemestrina* (Pig-tailed Macaque) against simian T-cell lymphotropic virus type I challenge, *Cancer Res.* **50**:5687s–5691s.

120. Grassmann, R., Dengler, C., Muller-Fleckenstein, I., Fleckenstein, B., McGuire, K., Dokhelar, M. C., Sodroski, J., and Haseltine, W., 1989, Transformation to continuous growth of primary human T lymphocytes by human T-cell leukemia virus type I X-region genes transduced by a *Herpesvirus saimiri* vector, *Proc. Natl. Acad. Sci. USA* **86**:3351–3355.

121. Franchini, G., Mulloy, J. C., Koralnik, I. J., Lo Monico, A., Sparkowski, J. J., Andresson, T., Goldstein, D. J., and Schlegel, R., 1993, The human T-cell leukemia/lymphotropic virus type I p12I protein cooperates with the E5 oncoprotein of bovine papillomavirus in cell transformation and binds the 16-kilodalton subunit of the vacuolar H$^+$ ATPase, *J. Virol.* **67**:7701–7704.

122. Vile, R. G., Schulz, T. F., Danos, O. F., Collins, M. K., and Weiss, R., 1991, A murine cell line producing HTLV-I pseudotype virions carrying a selectable marker gene, *Virology* **180**:420–424.

123. Wilson, C., Reitz, M., Okayama, H., and Eiden, M., 1989, Formation of infectious hybrid virions with Gibbon ape leukemia virus and human T-cell leukemia virus retroviral envelope glycoproteins and the *gag* and *pol* proteins of Moloney murine leukemia virus, *J. Virol.* **63**:2374–2378.

124. Denesvre, C., Carrington, C., Corbin, A., Takeuchi, Y., Cosset, F.-C., Schulz, T., Sitbon, M., and Sonigo, P., 1996, TM domain swapping of murine leukemia virus and human T-cell leukemia virus envelopes confers different infectious abilities despite similar incorporation into virions, *J. Virol.* **70**:4380–4386.

125. Hu, S.-L., Abrams, K., Barber, G., Moran, P., Zarling, J., Langlois, A., Kuller, L., Morton, W., and Benveniste, R., 1992, Protection of macaques against SIV infection by subunit vaccines of SIV envelope glycoprotein gp160, *Ann. N.Y. Acad. Sci.* **255:**456–459.

126. Franchini, G., Tartaglia, J., Markham, P., Benson, J., Fullen, J., Wills, J., Arp, J., Dekaban, G. A., Paoletti, E., and Gallo, R., 1995, Highly attenuated HTLV type I$_{env}$ poxvirus vaccines induce protection against a cell associated HTLV-I challenge in rabbits, *AIDS Res. Hum. Retrovir.* **11:**307–313.

127. Lubeck, M. D., Natuk, R., Myagkikh, M., Kalyan, N., Aldrich, K., Sinangil, F., Alipanah, S., Murthy, S. C. S., Chanda, P., Nigida, S. M., Markham, P., Zolla-Pazner, S., Steimer, K., Wade, M., Reitz, M. S., Arthor, L. O., Mizutani, S., Davis, A., Hung, P. P., Gallo, R. C., Eichberg, J , and Rober-Guroff, M., 1997, Long-term protection of chimpanzees against high-dose HIV-I challenge induced by immunization, *Nature Med.* **3:**651–658.

128. Kim, J. J., Begarazzi, M. L., Trivedi, N., Hu, Y., Kazahaya, K., Wilson, D. M., Ciccarelli, R., Chaattergoon, M. A., Dang, K., Mahalingam, S., Chalin, A. A., Agagjanyan, M. G., Boyer, J. D., Wang, B., and Weiner, D. B., 1997, Engineering of *in vivo* immune responses to DNA immunization via codelivery of costimulatory molecules genes, *Nature Biotechnol.* **15:**651–646.

129. Fynan, E., Webster, R., Fuller, D., Haynes, J. R., Santoro, J., and Robinson, H., 1993, DNA vaccines: Protective immunizations by parenteral, mucosal, and gene-gun inoculations, *Proc. Natl. Acad. Sci. USA* **90:**11478–11482.

130. Ugen, K. E., Agadjanyan, M. G., Wang, B., Merva, M., Williams, W. V., and Weiner, D. B., 1994, Genetic immunization against HTLV-I, *AIDS Res. Hum. Retrovir.* **10:**445.

131. Agadjanyan, M. G., Wang, B., Nyland, S. B., Weiner, D. B., and Ugen, K. E., 1998, DNA plasmid based vaccination against the oncogenic human T cell leukemia virus type I, *Curr. Top. Microbiol. Immunol.* **226:**175–192.

132. Boyle, J. S., Koniaras, C., and Lew, A. M., 1997, Influence of cellular location of expressed antigen on the efficacy of DNA vaccination: Cytotoxic T lymphocyte and antibody re sponses are suboptimal when antigen is cytoplasmic after intramuscular DNA immunization, *Int. Immunol.* **9:**1897–1906.

133. Ruegg, C., Monell, C., and Strand, M., 1989, Identification, using synthetic peptides, of the minimum amino acid sequence from the retroviral transmembrane protein p15E required for inhibition of lymphoproliferation and its similarity to gp21 of human T-lymphotropic virus types I and II, *J. Virol.* **63:**3250–3256.

134. Lafrado, L., Lewis, M., Mathes, I., and Olsen, R., 1987, Suppression of *in vitro* neutrophil function by feline leukaemia virus (FeLV) and purified FeLV-p15E, *J. Gen. Virol.* **68:**507–513.

135. Shida, H., Tochikura, T., Sato, T., Konno, T., Hirayoshi, K., Seki, M., Ito, Y., Hatanaka, M., Hinuma, Y., Sugimoto, M., Takahashi-Nishimaki, F., Maruyama, T., Miki, K., Suzuki, K., Morita, M., Sashiyama, H., and Hayami, M., 1987, Effect of the recombinant vaccinia viruses that express HTLV-I envelope gene on HTLV-I infection, *EMBO J.* **6:**3379–3384.

136. Kageyama, S., Katsumoto, T., Taniguchi, K., Ismail, S. I., Shimmen, T., Sasao, F., Gao, M., Owatari, S., Wakamiya, N., Tsuchie, H., Ueda, S., and Kurimura, T., 1996, Neutralization of human immunodeficiency virus type I (HIV-I) with antibody from carriers' plasma against HIV-1 protein p17, *Acta Virol.* **40:**195–200.

137. Papsidero, L. D., Sheu, M., and Ruscetti, F. W., 1989, Human immunodeficiency virus type-I-neutralizing monoclonal antibodies which react with p17 core protein, *J. Virol.* **63:**267–279.

138. Emini, E. A., Schleif, W. A., Quintero, J. C., Conard, P. G., Eichberg, J., Vlasuk, G. P., Lehman, E. D., Polokoff, M. A. Schaeffer, T. F., Schultz, L. D., Hofmann, K. J., Lewis, J. A., and Larson, V. M., 1990, Yeast-expressed p55 precursor core protein of human immuno-

deficiency virus type 1 does not elicit protective immunity in chimpanzees, *AIDS Res. Hum. Retrovir.* **6:**1247–1250.

139. Buseyne, F., McChesney, M., Porrot, F., Kovarik, S., Guy, B., and Riviere, Y., 1993, Gag-specific cytotoxic T lymphocytes from human immunodeficiency virus type 1-infected individuals: Gag epitopes are clustered in three regions of the p24^{gag} protein, *J. Virol.* **67:**694–702.

140. Wiktor, S. Z., Pate, E. J., Murphy, E. L., Palker T. J., Champegnie, E., Ramlal, A., Cranston, B., Hanchard, B., and Blattner, W. A., 1993, Mother-to-child tranmission of human T-cell lymphotropic virus type I (HTLV-I) in Jamaica: Association with antibodies to envelope glycoprotein (gp46) epitopes. *J. Acquired Immune Defic. Syndr.* **6:**1162–1167.

141. Green, J., Hinrichs, S. H., Vogel, J., and Jay, G., 1989, Exocrinopathy resembling Sjogren's syndrome in HTLV-I *tax* transgenic mice, *Nature* **341:**72–74.

142. Parker, C. E., Nightingale, S., Taylor, G. P., Weber, J., and Bangham, C. R., 1994, Circulating anti-Tax cytotoxic T lymphocytes from human T-cell leukemia virus type I-infected people, with and without tropical spastic paraparesis, recognize multiple epitopes simultaneously, *J. Virol.* **68:**2860–2868.

143. Greene, W. C., Bohnlein, E., and Ballard, D. W., 1989, HIV-1, HTLV-I and normal T-cell growth: Transcriptional strategies and surprises, *Immunol. Today* **10:**272–278.

144. Kelly, K., Davis, P., Mitsuya, H., Irving, S., Wright, J., Grassmann, R., Fleckenstein, B., Wano, Y., Greene, W. C., and Siebenlist, U., 1992, A high proportion of early response genes are constitutively activated in T cells by HTLV-I, *Oncogene* **7:**1463–1470.

145. Gartenhaus, R., Lunardi-Iskandar, Y., Berneman, Z., Reitz, M., Gallo, R. C., and Klotman, M., 1994, Soluble recombinant HTLV-I tax protein has proliferative effects on cells of lymphoid origin, *J. Clin. Invest.* **:**1–34.

146. Schonbach, C., Nokihara, K., Bangham, C. R., Kariyone, A., Karaki, S., Shida, H., Takatsu, K., Egawa, K., Weismuller, K., and Takiguchi, M., 1996, Identification of HTLV-I-specific CTL directed against synthetic and naturally processed peptides in HLA-B*3501 transgenic mice, *Virology* **226:**102–112.

147. Lal, R. B., Giam, C. Z., Coligan, J. E., and Rudolph, D. L., 1994, Differential immune responsiveness to the immunodominant epitopes of regulatory proteins (Tax and Rex) in human T cell lymphotropic virus type I-associated myelopathy, *J. Infect. Dis.* **169:**496–503.

148. Kuga, T., Hattori, S., Yoshida, M., and Taniguchi, T., 1986, Expression of human T-cell leukemia virus type I envelope protein in *Saccharomyces cerevisiae*, *Gene* **44:**337–340.

149. Arp, J., LeVatte, M., Rowe, J., Perkins, S., King, E., Leystra- Lantz, C., Foung, S. K., and Dekaban, G. A., 1996, A source of glycosylated human T-cell lymphotropic virus type 1 envelope protein: Expression of gp46 by the vaccinia virus/T7 polymerase system, *J. Virol.* **70:**7349–7359.

150. Hadlock, K. G., Rowe, J., Perkins, S., Bradshaw, P., Song, G. Y., Cheng, C., Yang. J., Gascon, R., Halmos, J., Rehman, S. M. M., McGrath, M. S., and Foung, S. K., 1997, Neutralizing human monoclonal antibodies to conformational epitopes of human T-cell lymphotropic virus type 1 and 2 gp46, *J. Virol.* **71:**5828–5840.

151. Palker, T. J., Tanner, M., Scearce, R., Streilein, R., Clark, M., and Haynes, B., 1989, Mapping of immunogenic regions of human T-cell leukemia virus type I (HTLV-I) gp46 and gp21 envelope glycoproteins with *env*-encoded synthetic peptides and a monoclonal antibody to gp46, *J. Immunol.* **142:**971–978.

152. Tanaka, Y., Tanaka, R., and Hoshino, H., 1994, Identification of a novel neutralization epitope on envelope gp46 antigen of human T-cell-leukemia virus-type-II (HTLV-II), *Int. J. Cancer* **59:**655–660.

153. Baba, E., Nakamura, M., Tanaka, Y., Kuroki, M., Itoyama, Y., Nakano, S., and Niho, Y., 1993, Multiple neutralizing B-cell epitopes of human T-cell leukemia virus type I (HTLV-I) identified by human monoclonal antibodies, *J. Immunol.* **151:**1013–1024.

154. Palker, T. J., Riggs, E., Spragion, D., Muir, A., Scearce, R., Randall, R., McAdams, M., McKnight, A., Clapham, P., Weiss, R., and Haynes, B., 1992, Mapping of homologous, amino-terminal neutralizing regions of human T-cell lymphotropic virus type I and II gp46 envelope glycoproteins, *J. Virol.* **66:**5879–5889.

155. Manca, F., Li Pira, G., Fenoglio, D., Valle, M., Kunkl, A., Ferraris, A., Lancia, F., Saverino, D., Mortara, L., Balderas, R., Arp, J., Dekaban, G., Dalgleish, A., Lozzi, L., and Theofilopoulos, A., 1995, Recognition of human T-leukemia virus (HTLV-I) envelope by human CD4$^+$ T-cell lines from HTLV-I seronegative individuals: Specificity and clonal heterogeneity, *Blood* **85:**1547–1554.

156. Bab, E., Nakamura, M., Ohkuma, K., Kira, J. I., Tanaka, Y., Nakano, S., and Niho, Y., 1995, A peptide based human T-cell leukemia virus type I vaccine containing T and B cell epitopes that induces high titers of neutralizing antibodies, *J. Immunol.* **154:**399–412.

157. Ralston, S., Hoeprich, P., and Akita, R., 1989, Identification and synthesis of the epitope for a human monoclonal antibody which can neutralize human T-cell leukemia/lymphotropic virus type I, *J. Biol. Chem.* **264:**16343–16346.

158. Desgranges, C., Souche, S., Vernant, J. C., Smadja, D., Vahlne, A, and Horal, P., 1994, Identification of novel neutralization-inducing regions of the human T cell lymphotropic virus type I envelope glycoproteins with human HTLV I seropositive sera, *AIDS Res. Hum. Retrovir.* **10:**163–173.

159. Lafrado, L., Lewis, M., Mathes, L., and Olsen, R., 1987, Suppression of *in vitro* neutrophil function by feline leukemia virus (FeLV) and purified FeLV p15E, *J. Gen. Virol.* **68:**507–513.

160. Tanaka, Y., Zeng, L., Shiraki, H., Shida, H., and Tozawa, H., 1991, Identification of a neutralization epitope on the envelope gp46 antigen of human T-cell leukemia virus type I and induction of neutralizing antibody by peptide immunization, *J. Immunol.* **147:**354–360.

161. Jacobson, S., Reuben, J., Streilein, R., and Palker, T. J., 1991, Induction of CD4$^+$ human T lymphotropic virus type-I-specific cytotoxic T lymphocytes from patients with HAM/TSP, *J. Immunol.* **146:**1155–1162.

162. Palker, T. J., Scearce, R., Copeland, T. D., Oroszlan, S., and Haynes, B., 1986, C-terminal region of human T cell lymphotropic virus type I (HTLV-I) p19 core protein is immunogenic in humans and contains an HTLV-I-specific epitope, *J. Immunol.* **136:**2393–2397.

163. Noguchi, Y., Noguchi, T., Sato, T., Yokoo, Y., Itoh, S., Yoshida, M., Yoshiki, T., Akiyoshi, K., Sunamoto, J., Nakayama, E., and Shiku, H., 1991, Priming for *in vitro* and *in vivo* anti-human T lymphotropic virus type I cellular immunity by virus-related protein reconstituted into liposome, *J. Immunol.* **146:**3599–3603.

164. Dokhelar, M. C., Pickford, H., Sodroski, J., and Haseltine, W., 1989, HTLV-I p27rex regulates *gag* and *env* protein expression, *J. Acquired Immune Defic. Syndr.* **2:**431–440.

165. Schwartz, S., Felber, B., and Pavlakis, G., 1992, Distinct RNA sequences in the gag region of human immunodeficiency virus type 1 decrease RNA stability and inhibit expression in the absence of Rev protein, *J. Virol.* **66:**150–159.

166. Lal, R. B., Owen, S. M., Rudolph, D., and Levine, P. H., 1994, Sequence variation within the immunodominant epitope-coding region from the external glycoprotein of human T lymphotropic virus type II in isolates from Seminole indians, *J. Infect. Dis.* **169:**407–411.

167. Sonoda, S., Yashiki, S., Takahashi, K., Arhna, N., Daitoku, Y., Matsumoto, M., Matsumoto, T., Tara, M., Shinmyoszu, K., Sato, K., Inoko, H., Ando, A., and Tsuji, K., 1987, Altered HLA antigens expressed on T and B lymphocytes of adult T-cell leukemia/lymphoma patients and their relatives, *Int. J. Cancer* **40:**629–634.

168. Haynes, B., 1993, Scientific and social issues of immunodeficiency virus vaccine development, *Ann. N.Y. Acad. Sci.* **260:**1279–1286.

169. Milich, D. R., and McLachlan, A., 1986, The nucleocapsid of hepatitis B virus is both a

T-cell-independent antigen and a T-cell-dependent antigen, *Ann. N.Y. Acad. Sci.* **234:** 1398–1401.

170. Haynes, J. R., Cao, S. X., Rovinski, B., Sia, C., James, O., Dekaban, G. A., and Klein, M. H., 1991, Production of immunogenic HIV-I viruslike particles in stably engineered monkey cell lines, *AIDS Res. Hum. Retrovir.* **1:**17–27.

171. Weiss, S. H., 1994, The evolving epidemiology of human T lymphotropic virus type II [editorial], *J. Infect. Dis.* **169:**1080–1083.

172. Calender, A., Gessain, A., Essex, M., and de Thé, G., 1984, Seroepidemiology of human T-cell leukemia virus in the French West Indies: Antibodies in blood donors and patients with lymphoproliferative diseases who do not have AIDS, *Ann. N.Y. Acad. Sci.* **437:**175–176.

173. Lairmore, M. D., Roberts, B. D., Frank, D., Rovnak, J., Weiser, M. G., and Cockerell, G. L., 1992, Comparative biological responses of rabbits infected with human T-lymphotropic virus type I isolates from patients with lymphoproliferative and neurodegenerative disease, *Int. J. Cancer* **50:**124–130.

174. Cockerell, G. L., Weiser, M. G., Rovnak, J., Wicks-Beard, B., Roberts, B. D., Post, A., Chen, I. S. Y., and Lairmore, M. D., 1991, Infectious transmission of human T-cell lymphotropic virus type II in rabbits, *Blood* **78:**1532–1537.

175. Abiad, H., and Hershow, R., 1997, Current understanding of HTLV-II and its role in disease, *Infect. Med.* **14:**815–820.

7

Immune Responses against HIV-2

EWA BJÖRLING

1. INTRODUCTION

Human immunodeficiency virus type 2 (HIV-2), the second type of lenti-virus discovered to cause immunodeficiency in humans, was first isolated from a West African patient.[1-3] HIV-2 infection has been documented in Africa, Europe, the Americas, and Asia, but is still mostly confined to West Africa and Portugal. Since HIV-2 has a more restricted geographic distribution and a somewhat reduced disease-causing potential, research involving HIV-2 has been less extensive when compared with the globally occurring HIV-1.

The primate lentiviruses are classified according to their phylogenetic relationships, which have been determined by comparing their nucleotide sequences. The most common approach is to assess how often particular clusters of sequences occur when the data are randomly resampled.[4] The classification of a phylogenetic tree of primate lentiviruses leads to a first level of different lineages, and the second level of classification is to identify subgroups within each of these clusters.[5-9]

HIV-2 has been divided into several different subtypes, and nearly all of them have been identified in West Africa. More recently, epidemic spreading of HIV-2 subtype A isolates was demonstrated in India. The most common subtypes today are A and B, which are both prevalent throughout West Africa, while the other subtypes are only represented by a few isolates.

HIV-2, like other lentiviruses, has the capacity to direct the synthesis of a large number of different proteins, of which some are important for the

EWA BJÖRLING   •   Microbiology and Tumorbiology Center, Karolinska Institute, S-171 77 Stockholm, Sweden.

Human Retroviral Infections, edited by Kenneth E. Ugen *et al.*
Kluwer Academic / Plenum Publishers, New York, 2000.

143

formation of the virus particles, whereas other proteins of a nonstructural nature appear in infected cells and may be released through secretion by the cell. Each of these proteins has immunogenic sites which may elicit cell-mediated and humoral immune responses. For an extended knowledge of HIV immunobiology and development of treatment against acquired immune deficiency syndrome (AIDS) it is therefore necessary to obtain a comprehensive understanding of the dynamics and specificities of these immune reactions.

2. PROPERTIES OF DIFFERENT HIV-2 PROTEINS

Two proteins are cleaved from the *env* polyprecursor: the glycosylated outer protein gp125 and the transmembraneous protein gp36. The envelope proteins, as with other enveloped viruses, play an important role in interaction with the receptor, fusion, induction of cytopathic effects and development of humoral and cellular immune responses. The two glycoproteins are known to be highly antigenic and immunogenic.

The *env* gene accumulates most of the variation and the gp125 protein exhibits more variability than the gp36 protein of HIV-2.[5,6,10] Many of the cysteine residues are conserved among all the HIV strains sequenced, suggesting a common three-dimensional structure.

The large glycoprotein has diversity in the number and relative positions of the glycosylation sites in different HIV strains. This could be the result of extensive base changes in the hypervariable regions which comprise a large part of the potential N-linked glycosylation sites of the highly glycosylated gp125 protein. The hypervariable regions, consisting of 20–80 amino acid residues, are surrounded by cysteine residues, suggesting that they form loops. These variations may influence the specific immunogenicity of the protein.[11] Mapping of structures on the HIV glycoproteins indicates that by way of contrast many conserved regions may be inaccessible for antibodies in the natural configuration. The *gag* gene encodes for the matrix and capsid proteins. p26 elicits a strong antibody response during infection, and the presence of anti-p26 antibodies is an important tool in diagnostic tests.

The enzymatic proteins encoded by the *pol* gene are protease, reverse transcriptase including RNAse H activity, and endonuclease (integrase).

3. IMMUNE RESPONSES AGAINST HIV-2

Many viral infections are controlled by the host immune system and can thereby also be eliminated. Both cellular and humoral arms of the immune system are involved in this process. Cell-mediated immunity can reduce viral load after infection and in some cases completely eliminates the virus. Neutralizing antibodies can conversely prevent the infection from occurring.

For HIV there are several problems for the immune system to overcome. HIV in its latent proviral form can infect a small number of host cells while remaining invisible for both arms of the immune system until the infected cells are activated. In those cells in which the virus replicates with a high rate there are many mutations. Thus, there are many opportunities for the virus to escape the immune system. Furthermore, HIV infects the CD4+ T cells of the immune system such as T helper cells, dendritic cells, and macrophages.

Since the majority of the functional immune response is directed against the envelope glycoproteins and *gag* proteins, a more detailed description of the different epitopes, B cell (Table I), antibody-dependent cellular cytotoxicity (Table II), and cytotoxic T be mostly focused on the structural proteins. Both linear sites defined by peptide reagents and discontinuous sites have been identified. This type of site mapping data primarily gives information about antigenic linear sites. A single site may contain one or many epitopes and the latter term for B cell epitopes will only be employed when it has been identified with a clonal immune reagent.

3.1. Humoral Immunity

HIV-2 infection gives rise to a strong humoral immune response directed against most viral structural proteins. Most of these antibodies do not interfere with biological activity of the virus, have no control of virus replication or virus inhibition, and only bind the specific virus proteins. These antibodies can still be important in the context of serodiagnosis and epidemiology, and also as tools for analysis of the structure of viral proteins. Biologically active antibodies include those that can inhibit virus infectivity, neutralizing antibodies, and those that can mediate killing of virus-infected cells through the mechanism of antibody-dependent cellular cytotoxicity (ADCC).

Antibodies responsible for neutralizing activity have been extensively studied for HIV-1. They have been characterized to predominantly identify discontinuous sites, and of particular importance in this context may be such discontinuous structures being responsible for CD4 binding of native gp120.[12–16] Anti-CD4 antibodies seem to be associated with long-term controlled infection and are lost in individuals with disease progression.[17]

HIV-2 infections, like HIV-1 infections, result in the production of neutralizing antibodies predominantly directed against regions in the envelope glycoproteins,[18,19] and there is also *in vitro* evidence of cross-neutralizing antibodies.[20–22]

TABLE I
HIV-2 B-Cell Epitopes Located in the Envelope Glycoproteins

Isolate	Region	Amino acids	Sequence[a]	Specification	Reference
ISY	C1	30–44, 69–83	GVPVW..., NVTEA	Antigenic site, human sera	27
LAV-2/ROD	C1	43–54	DDYQE...	Rat mAb	25
SBL-6669	C1	44–56	ATKNR...	Antigenic site, human sera	26
ROD	V1	118–125, 125–141	GNNTT..., STSTT	Antigenic site, human sera	28
SBL-6669	V1	119–137	AVATT...	Guinea-pig antisera	29
LAV-2/ROD	V1	125–133	SEDTP...	Rat mAb	25
LAV-2/ROD	V2	140–148	GEETI...	Rat mAb	25
LAV-2/ROD	V2	149–154	FNMTG...	Rat mAb	25
SBL-6669	V2	160–189	CQFNM...	Antigenic site, human sera; guinea-pig antisera	31
LAV-2/ROD	V2	167–175	YSKDV...	Rat mAb	25
ISY	C2	196–215	CNTSV	Antigenic site, human sera	27
LAV-2/ROD	C2	223–234	DTNYS...	Rat mAb	25
SBL-6669	C2	228–248	CNDTN	Antigenic site, human sera; guinea-pig antisera	31
ISY	c2	234–248	SGFEP	Antigenic site, human sera	27
ISY	V3	297–330	RRPEN...	Antigenic site, human sera	41
ISY	V3	301–315	NKTVV	Antigenic site, human sera	27
ROD	V3	303–324	LHCKR...	Antigenic site, human sera	28
LAV-2/ROD	V3	304–311	LMSGH...	Rat mAb	25
ISY	V3	305–320	VPITL	Antigenic site, human sera	27

LAV-2/ROS	V3	306–311	SGHVF...	Rat mAb	25
ROD	V3	310–335	NKIVK	Mouse mAb	44
SBL-6669	V3	311–330, 318–337	SGRRF..., QKIIN	Guinea-pig antisera	29
SBL-6669	V3	311–326, 322–337	SGRRF..., NKKPR	Mouse mAb	43
ISY	V3	318–334	QKIIN...	Antigenic site, human sera	27
ISY	V3	332–355	FKGEW...	Antigenic site, human sera	27
ROD	C3	366–392	TNDTR...	Mouse mAb	44
ROD	C3	340–358	CWFKG	Antigenic site, human sera	28
ISY	V4	398–416	VENKT...	Antigenic site, human sera	27
SBL-6669	V4	399–419	CNMTW	Antigenic site, human sera; guinea-pig antisera	31
ROD	C4	436–452	YLPPR...	Antigenic site, human sera	28
SBL-6669	C-terminal	472–493, 489–509	ELGDY..., TAEKR	Guinea-pig antisera	29
ISY	C-terminal	481–498	VTPIG...	Antigenic site, human sera	27
ROD	C-terminal	486–507	LVEIT...	Antigenic site, human sera	27
SBL-6669	gp36	573–595	AIEKY	Antigenic site, human sera	26
SBL-6669	gp36	595–614	CHTTV...	Antigenic site, human sera	26, 53
SBL-6669	gp36	634–649	EQAQI	Antigenic site, human sera	26
SBL-6669	gp36	644–658	MYELQ	Antigenic site, human sera; guinea-pig antisera	31
SBL-6669	gp36	714–729	HIHKD...	Guinea-pig antisera	29
Human primary			ConfDep	Human mAb	54
LAV-2/ROD		28.3e	ConfDep	Rat mAb	25
LAV-2/ROD		28.8e	ConfDep	Rat mAb	25
LAV-2/ROD		25.3f	ConfDep	Rat mAb	25

[a]ConfDep, conformation-dependent.

TABLE II
Antibody-Dependent Cellular Cytotoxicity (ADCC) Linear Sites
in the Envelope Glycoproteins of HIV-2

Region	Amino acids	Sequence	Comments	Reference
V3	291–311	NLTIL...	Guinea-pig antisera	29
V3	318–337	QKIIN...	Guinea-pig antisera	29
V3	291–337	NLTIL...	Human anti-HIV-2 ADCC positive sera reactivity against peptides	59
C4	446–461	IANID...	Guinea-pig antisera	29
C-terminal	472–509	ELGDY...	Human anti-HIV-2 ADCC positive sera reactivity against peptides	59

Understanding the specificity and effect of sites that are accessible to neutralizing antibodies is the most important aspect for construction of a nonreplicating immunogen based on envelope proteins for two reasons: the principal viral determinant binds to host receptors, and neutralizing antibodies are directed against the major antigenic determinant.

Envelope conformation should also be taken into consideration. Data for HIV-1 indicate that the mature envelope protein consists of a trimer on the surface, cleaved into gp120 and gp41, which associate into surface glycoprotein spikes of which the variable regions are exposed, shielding the conserved regions from the antibodies. Given the structural variability and flexibility of the virus envelope, one should remember that antigenic epitopes exposed on monomeric recombinant gp120 can be hidden by carbohydrates on the gp160 and are therefore no longer targets for functional antibodies.

Due to sequence variability, both intrasubtypically and between the different subtypes, the results of B cell site mapping with one HIV isolate may not be applicable to other isolates. The infrequency of cross-clade neutralizing antibodies has led some HIV researchers to try to define the structure of the epitopes to which these antibodies bind. The discovery and identification of a second host receptor for HIV entry gave new hope in the struggle to stop HIV infection. HIV infection is initiated by interaction of the envelope glycoproteins with at least two cellular receptors. For HIV-1 second receptors, it is now established that T cell-line-adapted isolates use the CXCR4 receptor, while macrophage-tropic isolates use the CCR5 receptor.[23] Unlike HIV-1, HIV-2 primary isolates seem to be more multitropic and promiscuous in their coreceptor usage in that apart from CCR5, they are able to use one or more of the CCR1, CCR2b, CCR3, or BOB receptors.[24]

3.1.1. HIV-2 B Cell Epitopes and Linear Sites (Table I)

3.1.1a. The Large Glycoprotein gp125. The large clycosylated protein gp125 consists of several conserved (C1–C5) and variable regions (V1–V5). McKnight and coworkers[25] described a rat monoclonal antibody (mAb) directed against the C1 region of the amino-terminal end of HIV-2. A linear antigenic site for human HIV-2-antibody-positive sera has also been indicated in this area.[26,27]

The V1 region of HIV-2 has been described to harbor one antigenic site for human anti-HIV-2 sera,[28] and within this region a neutralization target was defined using both a rat monoclonal antibody[25] and hyperimmune antipeptide sera from guinea-pigs.[29]

Accumulated evidence demonstrates that the V2 region is an important target for neutralizing antibodies of both subtypes of HIV. Most of these are strain-specific, however, and recognize conformationally sensitive structures located at the central portion of V2. In the case of HIV-2 there are data indicating that the V2 region is important as a target for antibody binding. By producing rat mAbs capable of neutralizing both HIV-2 isolates and SIVsm,[25] linear epitopes in the V2 region have been shown to harbor neutralization targets for HIV-2. Nonneutralizing mouse mAbs have also been generated to this region.[30] Recent results report that the majority of a panel of human HIV-2 sera recognize overlapping peptides corresponding to the central and C-terminal parts of the HIV-2 V2 region.[31] Babas and coworkers[32] immunized rabbits with peptides corresponding to the V2 region of HIV-2, but these sera were unable to inhibit syncytia formation induced by HIV-2 *in vitro.*

In the C2 region of HIV-2, a linear antigenic site reacting with human anti-HIV-2 antibodies has been demonstrated.[27,31] One of the more recently developed rat mAbs by McKnight and coworkers[25] was also directed against this region. Furthermore, animal anti-V2 peptide antisera with neutralizing capacity have recently been developed.[31]

Already in early studies of HIV-1 the V3 loop was identified as the principal neutralizing determinant.[33] Since then much attention has been focused on this region. In the original studies it was demonstrated that peptides corresponding to the V3 region could elicit neutralizing antibodies and that most of the neutralizing activity from experimentally infected or gp120-vaccinated animals was directed against the V3 domain.[33–40] This region, the principal neutralizing domain (PND), is associated with amino acids 296–343, depending on the HIV-1 isolate, and forms a loop due to disulfide bonding between the cysteines at each end. The amino acids at the tip of the loop are relatively conserved, and display only limited variation among virus isolates from different parts of the world. Important to remember in this context is that most of the studies in this area were performed with T-cell line-adapted HIV-1 and for primary isolates the situation appears to be strikingly different.

Using peptides representing regions of the glycoprotein gp125 of HIV-2 to screen HIV-2 human-antibody-positive sera, several groups have defined the HIV-2 V3 region as a linear antibody-binding site.[26-28,41,42]

Characterization of the potential neutralization activity of antibodies reacting with HIV-2 gp125 has produced conflicting results. There are reports which describe failure to induce neutralizing antibodies to the HIV-2 V3 region by peptide immunization.[32,41] Several groups[25,29,43,44] have identified the homologue of the HIV-1 V3 loop as a target for neutralization of HIV-2, through raising of neutralizing animal sera by peptide immunization, through blocking of the neutralizing capacity of anti-HIV-2 human sera by peptides corresponding to the V3 loop, and through development of neutralizing murine monoclonal antibodies directed against the V3 loop of HIV-2. Fine mapping of important individual amino acids for antibody binding in this region revealed two antigenic sites with conserved motifs within V3: Phe–His–Ser (amino acids 315–317) and Trp–Cys–Arg (amino acids 329–331). Potentially, these two sites can interact to represent a single discontinuous antigenic site. Traincard and coworkers[30] developed another monoclonal antibody against the V3 region of HIV-2, but this antibody does not neutralize the virus.

Broad cross-neutralizing activity including many primary HIV-2 isolates from Guinea Bissau was demonstrated with guinea-pig hyperimmune sera against V3-loop peptides.[43] It should be emphasized that in this context the V3 amino acid sequences of HIV-2 isolates exhibit a more stable picture with only a 7% variation as compared to 16% in different HIV-1 strains.[39]

Already in 1991 deWolf and coworkers[28] identified one linear antigenic site in the C3 region that was important for binding of human anti-HIV-2 sera. Subsequently, one murine monoclonal antibody reacting with a linear C3 HIV-2 site has been described,[44] although this mAb only exhibited a weak neutralizing activity.

The C4 region was reported early on to be important for CD4 binding of HIV-1, since deletion of a stretch of 12 amino acids could abolish the ligand activity[45] and mAbs directed to this region could block CD4 binding.[13,14,46,47] These early findings have led to the conclusion that a short linear site in C4 is of importance for CD4 binding. Since then, other studies have demonstrated that amino acid substitutions outside the C4 domain could also abolish CD4 binding to gp120, and that CD4 binding must hence be dependent upon structures including stretches of amino acids from different regions of gp120 which fold to physical proximity in the native protein.[48] A single amino acid in C4 of HIV-1, Trp-427, appears to play a particularly critical role in this respect.[48,49] HIV-2 has a similar picture, with amino acid residue Trp-428 playing a corresponding role.[50]

The V4 region appears to be more antigenic in human HIV-2-antibody-positive sera,[27,31] and in concordance with this, two groups[26,28] have identified a linear antigenic site important for human HIV-2-antibody binding in the C4 region. Low neutralizing activity with guinea-pig antisera against the V4 region has also been reported.[31]

The carboxy-terminal part of gp125 harbors an immunodominant site that reacts with the majority of HIV-2-positive sera.[26–28] Two overlapping peptides corresponding to this region could induce neutralizing antibodies in animals and also mediate a certain blocking of neutralization of human anti-HIV-2 sera.[29,43]

3.1.1b. The Transmembrane Protein gp36. The transmembraneous protein gp 41 of HIV-1 is responsible for viral fusion to the target cell and for syncytia formation.[51,52] By analogy, the same properties pertain to the gp36 protein of HIV-2. A highly immunodominant region, the principal antigenic domain (PAD), in HIV-2 gp36 has been mapped to amino acids 595–614 by demonstration of strongly reactive peptides in this region.[26,53]

Sites in the central part of gp36[31] and somewhat unexpectedly in the intracytoplasmic domains have also been reported to induce *in vitro* neutralizing HIV-2 antibodies in guinea-pigs.[29] These sites are most probably not exposed at the surface of mature virions, but speculatively it may be recognizable by antibodies after CD4 binding when the virus and cell membrane are in close apposition, thus blocking the fusion process. More work is needed to clarify the *in vivo* relevance of the reported results, since this study on anti-gp36-neutralizing antibodies has only been performed in animals.

By immunization with recombinant baculovirus-derived gp105, rat mAbs directed against conformational epitopes on the HIV-2 envelope glycoprotein have been produced. These mAbs have also been reported to harbor some neutralizing capacity of HIV-2.[25] One human conformational mAb with neutralizing capacity against HIV-2 has also been developed.[54]

3.2. Antibody-Dependent Cellular Cytotoxicity

In addition to inhibition of viral infectivity, antibodies can also participate in a cytotoxic mechanism, ADCC (Table II), in which the contact between the target and the effector cell is mediated by an antibody. The antibody binds to the antigen expressed on the surface of the target cell and through its Fc region to the Fc receptor on the effector cell. The most important ADCC effector cells are natural killer (NK) cells. ADCC has been demonstrated to function *in vivo* in several systems. For HIV it has been demonstrated *in vitro* that the viral targets for this biological activity, similarly to neutralizing antibodies, are mainly the envelope glycoproteins. ADCC antibodies appear early in HIV infection and are broadly cross-reactive, compared to neutralizing antibodies, which appear later and are more type-specific in the beginning with subsequent broadening. Since cross-ADCC activity has been shown with sera from HIV-infected individuals, it has been suggested that ADCC activity is directed to more conserved sites in the envelope glycoproteins. The role of neutralization and ADCC in immunotherapy against HIV is unknown. ADCC antibodies and neutralizing antibodies are complementary to each other, since ADCC antibodies kill virus-producing HIV-infected cells and make the free virions accessible to neutralizing antibodies. More detailed analysis has revealed both overlapping and nonoverlapping sites for neutralization in parallel with ADCC in these proteins.

The significance of ADCC in the comprehension of viral spread in the infected host remains unclear, with ADCC suggested to function as a protective effect and also to be a part of the pathogenic process.[55,56] HIV-specific ADCC activity has been reported in the majority of sera from HIV-infected individuals[55,57] and can be both group- and strain-specific.[58] The presence of autologous ADCC activity has been demonstrated in both HIV-1 and HIV-2 individuals. It has also been reported that HIV-2-specific ADCC activity seems to be more broadly reactive compared to that of HIV-1.[59] The ADCC response appears early after HIV infection,[60,61] although a dysfunction in the NK cell activity is apparent in some patients.[62] NK and ADCC functions are conducted essentially by the same cell type, NK cells. The defective NK activity in HIV-positive individuals is due to defective lysis mechanisms, although these cells can still bind to the target cells.[63] The reason for NK cell ADCC activity from these HIV-infected individuals is unknown, but several suggestions have been proposed, such as a mechanism at the recognition or postrecognition level or reduced or defective receptors. Besides defects in ADCC effectors, humoral effectors have also been discussed in HIV-positive individuals. The humoral defects could involve circulating immune complexes, irrelevant antibodies, or soluble CD16, since all these would compete with the ADCC-mediating antibodies.

3.2.1. ADCC Epitopes in HIV-2

Because the expression of the target antigen on the surface of target cells is a prerequisite for ADCC, only epitopes located on the envelope glycoproteins of HIV-2 have been sought.

Linear target sites for ADCC in HIV-2 have been identified in the envelope glycoproteins by using antipeptide guinea-pig sera representing the V3 domain and the C4 region.[29] In another study with the purpose of delineation of ADCC targets, human anti-HIV-2 ADCC-positive sera were tested against peptides representing different regions of the HIV-2 envelope glycoprotein. The V3 region and the C-terminal end of gp125 were thus suggested to be involved in ADCC.[59]

3.3. Cell-Mediated Immunity

Cell-mediated immunity is thought to play a major role in host defense against HIV infection by elimination of virus-infected cells. Unlike antibodies, cytotoxic T lymphocytes (CTL) recognize processed viral fragments (peptides) presented on the surface of an infected cell in association with a major histocompatibility class (MHC) molecule (Table III). The self-peptides identified in the human leukocyte antigen (HLA) structures are important for the conformation, stability, and cell surface expression of the class I molecules. The use of short synthetic peptides to identify CTL epitopes in the lysis assay has proved invaluable in the localization of CTL epitopes. Once a peptide has been identified, the optimal length of the peptide can be determined and the fine specificity of the peptide mapped. It has been demonstrated that cells sharing a given MHC class I molecule produce identical peptides from a given protein. This explains why a single viral epitope is usually recognized by the majority of humans with a particular HA-2 or HLA type.[64] Epitope selection is also influenced by the T cell receptor (TCR). The TCR, a heterodimer consisting of variable alpha and beta chains, accounts for the specificity of CTL.[65,66] TCR diversity is generated by somatic rearrangements of noncontiguous variable (V), diversity (D), and joining (J) regions.[66] Structural crystallographic analyses indicate that the TCR combining site is relatively flat except for a deep hydrophobic

TABLE III
Human Cytotoxic T Lymphocyte (CTL)
Peptide Epitopes in HIV-2

Region	Amino acids	Sequence	Comments	Reference
gag	130–138	PPSGK...	HLA-B35 restricted	77
gag	173–181	TPYDI...	HLA-B53 restricted	82
gag	241–250	TSTVE...	HLA-B5801 restricted	79
gag	245–253	NPVPV...	HLA-B35 restricted	77
nef	75–82	VPLRP	HLA-B35 restricted	77
pol	386–393	NPDVILIQ	HLA-B35 restricted	77

cavity between the hypervariable CDR3s of the alpha and beta chains,[67] and this highly variable CDR3 region is thought to interact with the antigen–MHC complex.[68,69]

Identification of HIV peptide epitopes has been facilitated by advanced in the understanding of CTL recognition of viral antigens through analysis of crystal structures of human leukocyte antigen (HLA) class I molecules, identification of natural epitopes isolated from infected or uninfected cells leading to allele-specific motifs, and furthermore through understanding of the assembly of class I molecules and the importance of the peptide for structure and stabilization. CTL activity in HIV-1 infection has been extensively studied and determined to be directed against several viral epitopes. Virus-specific CTLs recognize epitopes within both the structural proteins encoded by *env*, *pol*, and *gag* and the regulatory and accessory proteins.[70]

Despite the dramatic decrease of HIV load following initial appearance of CTL early after primary infection, the association of HIV-specific CTL activity, and a controlled viral load during the asymptomatic stage, CTLs are not efficient in clearance of HIV.[71,72] In a recent elegant study by Ogg and coworkers,[73] an inverse correlation between HIV-specific CTLs and plasma viral load was demonstrated by the use of HLA–peptide tetrameric complexes, but no significant association was detected between the clearance rate of productively infected cells and the frequency of HIV-specific CTLs.

One possible explanation could be the phenomenon of immunodominance, in which CTL responses are restricted to one or a few of several epitopes within a viral antigen, which gives a feasibility for emergence of virus variants that can escape recognition and killing by virus-specific CTLs.[74] These are similar problems as for antibody recognition failure against virus escape mutants in the humoral immune response against HIV. This immunodominance also suggests that the immune response to the most antigenic stimulus could outcompete the immune responses to other T-cell clones, thereby leading to an increased virus load. Another reason could be that even if there are multiple epitopes, the CTL response may be unevenly distributed in different tissues. Only a minor defect of the CTL response may be sufficient to give virus at different sites more opportunity to replicate before killing of the host cell. One should also take into consideration the conformational changes in the envelope glycoproteins upon dissociation during binding to the CD4 ligand, which possibly alter earlier exposed CTL epitopes.

3.3.1. CTL Epitopes in HIV-2

The cellular immune response against HIV-1 has been extensively studied compared to HIV-2, but CTL activity has also been demonstrated for HIV-2. Nixon and coworkers tested HIV-1 CTL specific for the HLA-B27-restricted *gag* epitope 265–279 on a peptide corresponding to the same region in HIV-2 which differed by 5 amino acids out of 15, and recorded cross-reactivity.[75] However, this may not occur for all peptides and the specific amino acids which differ between the two viruses could be crucial. Buseyne and coworkers[76] reported cross-reactivity between PBMC of HIV-1-infected patients and a recombinant HIV-2 p56 *gag* protein. In another study of HIV-exposed but uninfected Gambian women, four different HIV-1 and HIV-2 cross-reactive peptide epitopes were identified with CTLs from infected individuals with HLA-B35. The majority of HIV-infected donors with HLA-B35 had B35-restricted CTL recognizing one or more peptides from *gag*, p17 and p24, *pol*, and *nef*.[77] An HIV-1 gp120 peptide, eight amino acids long (amino acids 117–124), represents a CTL epitope in simian immunodeficiency virus (SIV)-infected rhesus monkeys and also exhibits functional cross-reactivity against HIV-2.[78] Another elegant study recently reported that HIV-2-infected HLA-B5801-positive individuals exhibited broad cross-recognition of HIV-1 subtypes because of a mounting T-cell response that tolerated extensive amino acid substitutions within HAL-B5801-restricted HIV-1 and HIV-2 epitopes.[79] The majority of HIV-2-infected individuals had cross-reactive CTL with the capacity to lyse target cells presenting *gag* proteins from different HIV-1 clades. It has also been reported that the CTL response against HIV-2 is dominated against the *gag* protein, and in the same study investigation of the inverse relation between HIV-2-specific cytotoxic activity and proviral load was performed.[80] HIV-2 *nef*-specific HLA-B7-restricted CTLs are efficient in lysis of HIV-infected CD4+ T cells.[81] Epitopes from HIV-2 *gag* presented by HLA53 and B35 have also been identified.[82] If HIV-2-positive patients have broader T-cell repertoires, maybe this could be one explanation for the different clinical outcomes of HIV-1 and HIV-2.

In animal models it have been shown that HIV-2 infected macaques had detectable *gag* CTL responses to HIV-2 proteins.[83,84] Furthermore, the sequestration of HIV-2-specific CD8+ CTLs in both the lymph nodes and spleens of macaques, despite the lack or low level of circulating CTLs in peripheral blood, has been demonstrated.[85]

3.3.2. Anti-HIV Suppressing Activity of CD8+ Cells

The inhibition of viral replication by a CD8+ T-cell-dependent, nonlytic, and non-MHC-restricted suppression has been observed *in vitro*.[86–88] The suppression of HIV mediated by noncytolytic CD8s has been detected early after infection and decreases with progression to disease.[87] This nonlytic suppression of HIV is mediated by a soluble factor produced by CD8+ cells.[86,89] The phenomenon of CD8+ antiviral activity in a non-MHC-restricted fashion has also been demonstrated in HIV-2-infected baboons, in which the CD8+ cells develop a substantial anti-HIV-2 activity, and this was suggested to be one of the reasons for the low frequency of pathogenesis in HIV-2-infected baboons.[90,91] Recently, it was also reported that human CD8+ T cells from unexposed HIV-seronegative blood donors are able to control HIV-1 and HIV-2 replication in experimentally infected autologous CD4 cells.[92]

3.3.3. Cytotoxic Natural Killer Cells

Natural killer (NK) cells are thought to represent a first line of defense against virally infected cells and tumor growth. NK cells differ from T or B cells in that they do not express known TCR or immunoglobulin receptor for antigen, and kill cells in a non-MHC-restricted fashion. Many studies have reported on the characterization of surface receptors on NK cells that are implicated in the mechanism of NK killing. In HIV-1-infected individuals NK cytotoxic function is depressed, and this function declines as a function of disease progression. Low NK cell numbers and their decreased functioning may contribute to the susceptibility of HIV-infected individuals to opportunistic infections. It has been suggested that NK cells most probably play an important role in ADCC (discussed earlier in this chapter), but the role of NK cells in the functional immune response against HIV remains very unclear and requires further investigation.

3.3.4. T-Helper (TH) Cell Responses

Since HIV infects CD4+ T cells, this virus has the ability to partly inhibit these cells, which are responsible for many immunological reactions. Human CD4+ T cells can be separated into TH1 and TH2 subsets. TH1 cells secrete interleukin-2 (IL-2) and interferon-gamma, and TH2 cells secrete IL-4, IL-6, and IL-10. HIV does not induce a definite TH1 to TH2 switch, but favors a shift from the TH0 phenotype in response to recall antigens. It also appears that HIV-1 replicates preferentially in CD4+ T cell producing TH2-type cytokines.[93] A TH2 response would lead to B-cell activation, and the high secretion of IL-10 by TH2 cells would be suppressive for TH1 function. Even though this relationship between TH1 and TH2 cells including different cellular factors has been studied for HIV-1, this remains to be investigated for HIV-2.

4. CONCLUSION

This chapter summarizes the work from many laboratories on the humoral and cellular responses against HIV-2. The knowledge of antigenic and immunogenic regions in HIV-2, preferentially in the envelope glycoproteins, combined with epitopes important for CTL responses against HIV-2 could be applied in several ways: to broaden the understanding of the natural immune response to HIV-2, and thereby suppress or enhance selected responses in the matter of controlling the HIV infection, to design subunit prophylactic vaccines, or in development of tools for immunotherapy.

Several crucial questions remain: Why does the existing immune response fail to clear the virus? What is the protecting factor for the partners of HIV-infected individuals who remain negative despite repeated exposure to the virus? Why is the neutralizing antibody response so weak or absent in later stages of the infection? What is the mechanism behind the decline of CTL activity? How does HIV-2 pathogenesis differ from that of HIV-1, and why? What is the influence of host cell adaptation in viral immunogenic properties? What are the critical features in virus-load relationships and what causes the immune system to break down rapidly in some individuals and not in others? What effect will suppression of virus load at time of the primary infection have on long-term events? Will it be acceptable to tolerate a vaccine- induced infection-permissive immunity safeguarding against development of disease?

In this context it is also of interest to mention that it has been suggested that HIV-2 naturally protects against HIV-1 infection,[94] and this may be a result of cross-reactive immunity to epitopes conserved between HIV-1 and HIV-2. Further studies in this area should also assist in the design of vaccine candidates that are broadly protective against HIV. There is still a reason for optimism in this field since the host–virus interactions present remarkable challenges, and future breakthroughs in our understanding of HIV immunobiology may provide effective tools for immunoprotection against HIV.

REFERENCES

1. Clavel, F., Guetard, D., Brun-Vezinet, F., Chamaret, S., Rey, M. A., Santos-Ferreira, M. O., Laurent, A. G., Dauguet, C., Katlama, C., Rouzioux, C., Klatzmann, D., Champalimaud, J. L., and Montagnier, L., 1986, Isolation of a new human retrovirus from West African patients with AIDS, *Science* **233**:343–346.
2. Kanki, P. J., M'Boup, S., Ricard, D., Barin, F., Denis, F., Boye, C., Sangare, L., Travers, K., Albaum, M., Marlink, R., Romet-Lemonne, J.-L., and Essex, M., 1987, Human T-lymphotropic virus type 4 and the human immunodeficiency virus in West Africa, *Science* **236**:827–831.
3. Albert, J., Bredberg, U., Chiodi, F., Böttiger, B., Fenyö, E.-M., Norrby, E., and Biberfeld, G., 1987, A new human retrovirus isolate of West African origin and its relationship to HTLV-IV, LAV-II and HTLV-IIIB, *AIDS Res. Hum. Retrovir.* **3**:3–10.
4. Felsenstein, J., 1985, Confidence limits on phylogenies: An approach using the bootstrap, *Evolution* **39**:783–791.
5. Myers, G., Korber, B., Smith, R. F., Berzofsky, J. A., and Pavlakis, G. N., 1993, *Human*

Retroviruses and AIDS 1993, Theoretical Biology and Biophysics, Los Alamos National Laboratory, Los Alamos, New Mexico.

6. Myers, G., Korber, B., Jeang, K. T., Henderson, L., Wain-Hobson, S., and Pavlakis, G. N., 1994, *Human Retroviruses and AIDS 1994*, Theoretical Biology and Biophysics, Los Alamos National Laboratory, Los Alamos, New Mexico.

7. Louwagie, J., McCutchan, F. E., Peeters, M., Brennan, T. P., Sanders-Buell, E., Eddy, G. A., van der Groen, G., Fransen, K., Gershy-Damet, G.-M., Deleys, R., and Burke, D. S., 1993, Phylogenetic analysis of *gag* genes from 70 international HIV-1 isolates provides evidence for multiple genotypes, *AIDS* **7**:769–780.

8. Jin, M. J., Hui, H., Robertson, D. L., Muller, M. C., Barré-Sinoussi, F., Hirsch, V. M., Allan, J. S., Shaw, P. M., and Hahn, B. H., 1994, Mosaic genome structure of simian immunodeficiency virus from West African green monkeys, *EMBO J.* **13**:2935–2947.

9. Breuer, J., Douglas, N. W., Goldman, N., and Daniels, R. S., 1995, Human immunodeficiency virus type 2 (HIV-2) *env* gene analysis: Prediction of glycoprotein epitopes important for heterotypic neutralization and evidence for three genotype clusters within the HIV-2a subtype, *J. Gen. Virol.* **76**:333–345.

10. Myers, G., Hahn, B. H., Mellors, J. W., Henderson, L. E., Korber, B., Jeang, K.-T., McCutchan, F. E., and Pavlakis, G. N., 1995, *Human Retroviruses and AIDS 1995*, Theoretical Biology and Biophysics, Los Alamos National Laboratory, Los Alamos, New Mexico.

11. Alexander, S., and Elder, J. H., 1984, Carbohydrate dramatically influences immune reactivity of antisera to viral glycoprotein antigens, *Science* **226**:1328–1330.

12. Kang, C. Y., Nara, P., Chamat, S., Caralli, V., Ryskamp, T., Haigwood, N., Newman, R., and Kohler, H., 1991, Evidence for non V3 specific neutralizing antibodies that interfere with gp120/CD4 binding in human immunodeficiency infected humans, *Proc. Natl. Acad. Sci. USA* **88**:6171–6175.

13. McKeating, J., Thali, M., Furman, C., Karwowska, S., Gorny, M. K., and Cordell, J., 1992, Amino acid residues of the HIV-1 gp120 critical for the binding of rat and human neutralizing antibodies that block the gp120-sCD4 interaction, *Virology* **190**:134–142.

14. Nakamura, G. R., Byrn, R., Rosenthal, K., Porter, J. P., Hobbs, M. R., Riddle, L., Eastman, D. J., Dowbenko, D., Gregory, T., Fendly, B. M., and Berman, P. W., 1992, Monoclonal antibodies to the extracellular domain of HIV-1 IIIB gp160 that neutralize infectivity, block binding to CD4 and react with diverse isolates, *AIDS Res. Hum. Retrovir.* **8**:1875–1885.

15. Thali, M., Furman, C., Ho, D. D., Robinson, J., Tilley, S., Pinter, A., and Sodroski, J., 1992. Discontinuous conserved neutralization epitopes overlapping the CD4 binding region of the HIV-1 gp120 envelope glycoprotein, *J. Virol.* **66**:5635–5641.

16. Posner, M. R., Cavacini, L. A., Emes, C., Power, J., and Byrn, R., 1993, Neutralization of HIV-1 by F105, a human monoclonal antibody to the CD4 binding site of gp120, *J. Acquired Immune Defic. Syndr.* **6**:7–14.

17. Cavacini, L. A., Emes, C. L., Power, J., Underdahl, J., Goldstein, R., Mayer, K., and Posner, M. R., 1993, Loss of serum antibodies to a conformational epitope of HIV-1/gp120 identified by a human monoclonal antibody is associated with disease progression. *J. Acquired Immune Defic. Syndr.* **6**:1093–1102.

18. Weiss, R. A., Clapham, P. R., Weber, J., Cheingsong-Popov, R., Dalgleish, A., Carne, A., Weller, I., and Tedder, R. S., 1985, Neutralisation of human T-lymphotropic virus type III by sera of AIDS and AIDS-risk patients, *Nature* **316**:69–72.

19. Ranki, A., Weiss, S. H., Valle, S. L., Antonen, J., and Krohn, K. J. E., 1987, Neutralizing antibodies in HIV (HTLV-III) infection: Correlation with clinical outcome and antibody response against different viral proteins, *Clin. Exp. Immunol.* **69**:231–239.

20. Weiss, R. A., Clapham, P., Weber, J. N., Whitby, D., Tedder, R. S., O'Connor, T., Chamaret, S., and Montagnier, L., 1988, HIV-2 antisera cross-neutralize HIV-1, *AIDS* **2**:95–100.

21. Böttiger, B., Karlsson, A., Naucler, A., Andreasson, P. Å, Mendes Costa, C., and Biberfeld,

G., 1989, Cross-neutralizing antibodies against HIV-1 (HTLV-IIIB and HTLV-III RF) and HIV-2 (SBL-6669 and a new isolate SBL-K135). *AIDS Res. Hum. Retrovir.* **5**:511–519.

22. Böttiger, B., Karlson, A., Andreasson, P. Å., Naucler, A., Mendes Costa, C., Norrby, E., and Biberfeld, G., 1990, Envelope cross reactivity between human immunodeficiency virus type 1 and 2 detected by different serological methods: Correlation between cross neutralization and reactivity against the main neutralizing site, *J. Virol.* **64**:3492–3499.

23. Cairns, J. S., and D'Souza, M. P., 1998, Chemokines and HIV-1 second receptors: The therapeutic connection, *Nature Med.* **4**:563–568.

24. Mörner, A., Björndal, Å., Albert, J., KewalRamani, V., Littman, D. R., Thorstensson, R., Fenyö, E.-M., and Björling, E., 1999, Primary human immunodeficiency virus type 2 (HIV-2) isolates, like immunodeficiency virus type 1 (HIV-1), frequently use CCR5 but show promiscuity in coreceptor usage, *J. Virol.* **73**:2343–2349.

25. McKnight, A., Shotton, C., Cordell, J., Jones, I., Simmons, G., and Clapham, P. R., 1996, Location, exposure and conservation of neutralizing and nonneutralizing epitopes on human immunodeficiency virus type 2 SU glycoprotein, *J. Virol.* **70**:4598–4606.

26. Norrby, E., Putkonen, P., Böttiger, B., Utter, G., and Biberfeld, G., 1991, Comparison of linear antigenic sites in the envelope proteins of human immunodeficiency virus (HIV) type 2 and type 1, *AIDS Res. Hum. Retrovir.* **7**:279–285.

27. Mannervik, M., Putkonen, P., Ruden, U., Kent, K. A., Norrby, E., Wahren, B., and Broliden, P.-A., 1992, Identification of B-cell antigenic sites on HIV-2 gp125, *J. Acquired Immune Defic. Syndr.* **5**:177–187.

28. de Wolf, F., Meloen, R. H., Bakker, M., Barin, F., and Goudsmit, J., 1991, Characterization of human antibody-binding sites on the external envelope of human immunodeficiency virus type 2, *J. Gen. Virol.* **72**:1261–1267.

29. Björling, E., Broliden, K., Bernardi, D., Utter, G., Thorstensson, R., Chiodi, F., and Norrby, E., 1991, Hyperimmune antisera against synthetic peptides representing the glycoprotein of human immunodeficiency virus type 2 can mediate neutralization and antibody-dependent cytotoxic activity, *Proc. Natl. Acad. Sci. USA* **88**:6082–6086.

30. Traincard, F., Rey-Cuille, M. A., Huon, I., Dartevelle, S., Mazie, J. C., and Benichou, S., 1994, Characterization of monoclonal antibodies to human immunodeficiency virus type 2 envelope glycoproteins, *AIDS Res. Hum. Retrovir.* **10**:1659–1667.

31. Skott, P., Achour, A., Norin, M., Thorstensson, R., and Björling, E., 1999, Characterization of neutralizing sites in the second variable (V2), second constant (C2) and fourth variable (V4) region in the gp125 and a conserved region in the gp36 of human immunodeficiency virus type 2, *Vir. Immunol.* **1**:79–88.

32. Babas, T., Benichou, S., Guetard, D., Montagnier, L., and Bahraoui, E., 1994, Specificity of antipeptide antibodies produced against V2 and V3 regions of the external envelope of human immunodeficiency virus type 2, *Mol. Immunol.* **31**:361–369.

33. Javaherian, K., Langlois, A. J., McDanal, C., Ross, K. L., Eckler, L. I., Jellis, C. L., Profy, A. T., Rusche, J. R., Bolognesi, D. P., Putney, S. D., and Matthews, T. J., 1989, Principal neutralizing domain of the human immunodeficiency virus type 1 envelope protein, *Proc. Natl. Acad. Sci. USA* **86**:6768–6772.

34. Matthews, T. J., Langlois, A. J., Robey, W. G., Chang, N. T., Gallo, R. C., Fischinger, P. J., and Bolognesi, D. P., 1986, Restricted neutralization of divergent human T-lymphotropic virus type III isolates by antibodies to the major envelope glycoprotein, *Proc. Natl. Acad. Sci. USA* **83**:9709–9713.

35. Putney, S. D., Matthews, T. J., Robey, W. G., Lynn, D. L., Robert-Guroff, M., Mueller, W. T., Langlois, A., Ghrayeb, J., Petteway, S. R., Weinhold, K. J., Fischinger, P. J., Wong-Staal, F., Gallo, R. C., and Bolognesi, D. P., 1986, HTLV-III/LAV-neutralizing antibodies to an *E. coli*-produced fragment of the virus envelope, *Science* **234**:1392–1395.

36. Robey, W. G., Arthur, L. O., Matthews, T. J., Langlois, A., Copeland, T. D., Lerche, N. W.,

Oroszlan, S., Bolognesi, D. P., Gilden, A. V., and Fischinger, P. J., 1986, Prospect for prevention of human immunodeficiency virus infection: Purified 120kDa envelope glycoprotein induces neutralizing antibody, *Proc. Natl. Acad. Sci. USA* **83:**7023–7027.

37. Goudsmit, J., Boucher, C. A. B., Meloen, R. H., Epstein, L. G., Smit, L., v. d. Hoek, L., and Bakker, M., 1988, Human antibody response to a strain-specific HIV-1 gp120 epitope associated with cell fusion inhibition, *AIDS* **2:**157–164.

38. Rusche, J. R., Javaherian, K., McDanal, C., Petro, J., Lynn, D. L., Grimalia, R., Langlois, A., Gallo, R. C., Arthur, L. O., Fischinger, P. J., Bolognesi, D. P., Putney, S., and Matthews, T. J., 1988, Antibodies that inhibit fusion of HIV-1 infected cells bind to 24 amino acid sequence of the viral envelope, gp120. *Proc. Natl. Acad. Sci. USA* **85:**3198–3202.

39. LaRosa, G. J., Davide, J. P., Weinhold, K., Waterbury, J., Profy, A., Lewis, J., Langlois, A., Dreesman, G., Boswell, R., Shadduck, P., Holley, L., Karplus, M., Bolognesi, D., Matthews, T., Emini, E., and Putney, S., 1990, Conserved sequence and structural elements in the HIV-1 principal neutralizing determinant, *Science* **249:**932–935.

40. Javaherian, K., Langlois, A. J., LaRosa, G. J., Profy, A. T., Bolognesi, D. P., Herlihy, W. C., Putney, S. D., and Matthews, T. J., 1990, Broadly neutralizing antibodies elicited by the hypervariable neutralizing determinant of HIV-1, *Science* **250:**1590–1593.

41. Robert-Guroff, M., Aldrich, K., Muldoon, R., Stern, T. L., Bansal, G. P., Matthews, T. J., Markham, P. D., Gallo, R. C., and Franchini, G., 1992, Cross-neutralization of human immunodeficiency virus type 1 and 2 and simian immunodeficiency virus isolates, *J. Virol.* **66:**3602–3608.

42. Mörner, A., Achour, A., Norin, M., Thorstensson, R., and Björling, E., 1999, Fine characterization of a V3-region neutralizing epitope in human immunodeficiency virus type 2, *Virus Research* **59:**49–60.

43. Björling, E., Chiodi, F., Utter, G., and Norrby, E., 1994, Two neutralizing domains in the V3 region in the envelope glycoprotein gp125 of HIV type 2, *J. Immunol.* **152:**1952–1959.

44. Matsushita, S., Matsumi, S., Yoshimura, K., Morikita, T., Murakami, T., and Takatsuki, K., 1995, Neutralizing monoclonal antibodies against human immunodeficiency virus type2 gp120, *J. Virol.* **69:**3333–3340.

45. Lasky, L. A., Nakamura, G., Smith, D. H., Fennie, C., Shimasaki, C., Patzer, E., Berman, P., Gregory, T., and Capon, D. J., 1987, Delineation of a region of the human immunodeficiency virus type 1 gp120 glycoprotein critical for interaction with the CD4 receptor, *Cell* **50:**975–985.

46. Dowbenko, D., Nakamura, G., Fennie, C., Shimasaki, C., Riddle, L., Harris, R., Gregory, T., and Lasky, L., 1988, Epitope mapping of the human immunodeficiency virus type 1 gp120 with monoclonal antibodies, *J. Virol.* **62:**4703–4711.

47. Cordell, J., Moore, J. P., Dean, C. J., Klasse, P. J., Weiss, R. A., and McKeating, J. A., 1991, Rat monoclonal antibodies to non-overlapping epitopes of HIV-1 gp120 block CD4 binding *in vitro, Virology* **185:**72–79.

48. Olshevsky, U., Helseth, E., Furman, C., Li, J., Haseltine, W., and Sodroski, J., 1990, Identification of individual human immunodeficiency virus type 1 gp120 amino acids important for CD4 receptor binding, *J. Virol.* **64:**5701–5707.

49. Cordonnier, A., Montagnier, L., and Emerman, M., 1989, Single amino acid changes in HIV envelope affect viral tropism and receptor binding, *Nature* **340:**571–574.

50. Keller, R., Peden, K., Paulous, S., Montagnier, L., and Cordonnier, A., 1993, Amino acid changes in the fourth conserved region of human immunodeficiency virus type 2 strain HIV-2$_{ROD}$ envelope glycoprotein modulate fusion, *J. Virol.* **67:**6253–6258.

51. Sodroski, J., Goh, W. C., Rosen, K., Campbell, K., and Haseltine, W. A., 1986, Role of the HTLV-III/LAV envelope in syncytium formation and cytopathicity, *Nature* **322:**470–474.

52. Kowalski, M., Potz, J., Basiripour, L., Dorfman, T., Goh, W. C., Terwilliger, E., Dayton, A.,

Rosen, C., Haseltine, W., and Sodroski, J., 1987, Functional regions of the envelope glycoprotein of human immunodeficiency virus type 1, *Science* **237**:1351–1355.

53. Norrby, E., Biberfeld, G., Chiodi, F., von Gegerfelt, A., Naucler, A., Parks, E., and Lerner, R., 1987, Discrimination between antibodies to HIV and to related retroviruses using site-directed serology, *Nature* **329**:248–250.

54. Kent, K. A., Powell, C., Corcoran, T., Jenkins, A., Almond, N., Jones, I., Robinson, J., and Stott, E. J., 1993, Characterization of neutralizing epitopes of simian immunodeficiency virus envelope using monoclonal antibodies, in: *Proceedings of the Huitieme Colloque des Cent Gardes*, pp. 167–172.

55. Lyerly, H. K., Reed, D. L., Matthews, T. J., Langlois, A. J., Ahearne, P. A., Petteway, S., Jr., and Weinhold, K. J., 1987, Anti-gp120 antibodies from HIV seropositive individuals mediate broadly reactive anti-HIV ADCC, *AIDS Res. Hum. Retrovir.* **3**:409–422.

56. Ahmad, A., and Menezes, J., 1996, Antibody-dependent cellular cytotoxicity in HIV infections, *FASEB J.* **10**:2928–2966.

57. Ljunggren, K., Böttiger, B., Biberfeld, G., Karlsson, A., Fenyö, E.-M., and Jondal, M., 1987, Antibody-dependent cellular cytotoxicity inducing antibodies against human immunodeficiency virus: Presence at different clinical stages, *J. Immunol.* **139**:2263–2267.

58. Ljunggren, K., Biberfeld, G., Jondal, M., and Fenyö, F.-M , 1989, Antibody dependent cellular cytotoxicity detects group and strain specific antigens among HIV-1 and HIV-2 isolates, *J. Virol.* **63**:3376–3381.

59. von Gegerfelt, A., Diaz-Pohl, C., Fenyö, E.-M., and Broliden, K., 1993, HIV-2/SIV specific antibody-dependent cellular cytotoxicity in HIV-2 infected individuals, *AIDS Res. Hum. Retrovir.* **9**:883–889.

60. Sawyer, L. A., Katzenstein, D. A., Henry, R. M., Bone, E. J., Vujcic, L. K., and Williams, C. C., 1990, Possible beneficial effects of neutralizing antibodies and antibody-dependent, cell-mediated cytotoxicity in human immunodeficiency virus infection, *AIDS Res. Hum. Retrovir.* **6**:341–356.

61. Ojo-Amaize, E., Nishanian, P. G., Heitjan, D. F., Rezai, A., Korns, E., Detels, R., Fahey, J., and Giorgi, J., 1989, Serum and effector cell antibody dependent cellular cytotoxicity (ADCC) remains high during human immunodeficiency virus disease progression, *J. Clin. Immunol.* **9**:454–461.

62. Ljunggren, K., Karlsson, A., Jondal, M., and Fenyö, E.-M., 1989, Natural and antibody-dependent cytotoxicity (ADCC) in different clinical stages of HIV infections, *Clin. Exp. Immunol.* **75**:184–190.

63. Bonavida, B., Katz, J., and Gottlieb, M. S., 1986, Mechanisms of defective NK cell activity in patients with AIDS and AIDS related complex. Defective trigger of NK cells for NKCF production of target cells and partial restoration with IL-2, *J. Immunol.* **140**:2565–2568.

64. Falk, K., Rotzschke, O., and Ramensee, H.-G., 1990, Cellular peptide composition governed by major histocompatibility complex class I molecules, *Nature* **348**:248–251.

65. Marrack, P., and Kappler, J., 1987, The T cell receptor, *Science* **238**:1073–1079.

66. Davis, M. M., and Bjorkman, P. J., 1988, T-cell receptor genes and T-cell recognition, *Nature* **334**:395–402.

67. Garcia, K. C., Degano, M., Stanfield, R. L., Brunmark, A., Jackson, M. R., Peterson, P. A., Teyton, L., and Wilson, I. A., 1996, Structure of an alpha beta T-cell receptor at 2.5Å and its orientation in the TCR–MHC complex, *Science* **274**:209–219.

68. Matis, L. A., 1990, The molecular basis of T-cell specificity, *Annu. Rev. Immunol.* **8**:65–82.

69. Jorgensen, J., Reay, P., Ehrlich, E. W., and Davis, M. M., 1992, Molecular components of T cell recognition, *Annu. Rev. Immunol.* **10**:835–873.

70. McMichael, A. J., and Walker, B. D., 1994, Cytotoxic T lymphocyte epitopes: Implications for HIV vaccines, *AIDS* **8**:S155–S173.

71. Ho, D. D., Neumann, A. U., Perelson, A. S., Chen, W., Leonard, J. M., and Markowitz, M., 1995, Rapid turn over of plasma virions and CD4 lymphocytes in HIV infection, *Nature* **373**:123–126.

72. Perelson, A. S., Neumann, A. U., Markowitz, M., Leonard, J. M., and Ho, D. D., 1996, HIV-1 dynamics *in vivo*: Virion clearance rate, infected cell life span, and viral generation time, *Science* **271**:1582–1586.

73. Ogg, G. S., Jin, X., Bonhoeffer, S., Dunbar, P. R., Nowak, M. A., Monard, S., Segal, J. P., Cao, Y., Rowland-Jones, S. L., Cerundolo, V., Hurley, A., Markowitz, M., Ho, D. D., Nixon, D. F., and McMichael, A. J., 1998, Quantitation of HIV-1 specific cytotoxic T lymphocytes and plasma viral RNA, *Science* **279**:2103–2106.

74. McMichael, A. J., and Phillips, R. E., 1997, Escape of HIV from immune control, *Annu. Rev. Immunol.* **15**:271–296.

75. Nixon, D. F., Huet, S., Rothbard, J., Kieny, M. P., Delchambre, M., Thiriart, C., Rizza, C. R., Gotch, F. M., and McMichael, A. J., 1990, An HIV-1 and HIV-2 cross-reactive cytotoxic T-cell epitope, *AIDS* **4**:841–845.

76. Buseyne, F., Janvier, G., Fleury, B., Schmidt, D., and Riviere, Y., 1994, Multispecific and heterogenous recognition of the gag protein by cytotoxic T lymphocytes (CTL) from HIV infected patients: Factors other than the MHC control the epitope specificities, *Clin. Exp. Immunol.* **97**:353–360.

77. Rowland-Jones, S., Sutton, J., Ariyoshi, K., Dong, T., Gotch, F., McAdam, S., Whitby, D., Sabally, S., Gallimore, A., Corrah, T., Takiguchi, M., Schultz, T., McMichael, A., and Whittle, H., 1995, HIV-specific cytotoxic T-cells in HIV-exposed but uninfected Gambian women, *Nature Med.* **1**:59–64.

78. Voss, G., and Letvin, N., 1996, Definition of human immunodeficiency virus type 1 gp120 and gp41 cytotoxic T lymphocyte epitopes and their restricting major histocompatibility complex class I alleles in simian-human immunodeficiency virus infected rhesus monkeys, *J. Virol.* **70**:7335–7340.

79. Bertoletti, A., Cham, F., McAdam, S., Rostron, T., Rowland-Jones, S., Sabally, S., Corrah, T., Ariyoshi, K., and Whittle, H., 1998, Cytotoxic T cells from human immunodeficiency virus type 2 infected patients frequently cross-react with different human immunodeficiency virus type 1 clades, *J. Virol.* **72**:2439–2448.

80. Ariyoshi, K., Cham, F., Berry, N., Jaffar, S., Sabally, S., Corrah, T., and Whittle, H., 1995, HIV-2 specific cytotoxic T-lymphocyte activity is inversely related to proviral load, *AIDS* **9**:555–559.

81. Lucchiari-Hartz, M., Bauer, M., Niedermann, G., Maier, B., Meyerhans, A., and Elchmann, K., 1996, Human immune response to HIV-1 Nef. II. Induction of HIV-1/HIV-2 Nef cross-reactive cytotoxic T lymphocytes in peripheral blood lymphocytes on non-infected healthy individuals, *Int. Immunol.* **8**:588–584.

82. Gotch, F., McAdam, S. N., Allsopp, C. E., Gallimore, A., Elvin, J., Kieny, M. P., Hill, A. V., McMichael, A. J., and Whittle, H. C., 1993, Cytotoxic T cells in HIV-2 seropositive Gambians. Identification of a virus specific MHC restricted peptide epitope, *J. Immunol.* **151**:3361–3369.

83. Voss, G., Nick, S., Otteken, A., Luke, W., Stahl-Henning, C., and Hunsmann, G., 1992, Virus-specific cytotoxic T-lymphocytes in HIV-2 infected cynomolgus macaques, *AIDS* **6**:1077–1083.

84. Voss, G., and Hunsmann, G., 1993, Cellular immune response to SIVmac and HIV-2 in macaques: Model for the human HIV-1 infection, *J. Acquired Immune Defic. Syndr.* **6**:969–976.

85. Abimiku, A., Franchini, G., Aldrich, K., Myagkikh, M., Markham, P., Gard, E., Gallo, R., and Robert-Guroff, M., 1995, Humoral and cellular immune responses in rhesus macaques infected with human immunodeficiency virus type 2, *AIDS Res. Hum. Retrovir.* **11**:383–393.

86. Brinchmann, J. E., Gaudernack, G., and Vartdal, F., 1991, CD8+ T cells inhibit HIV replication in naturally infected CD4+ T cells: Evidence for a soluble inhibitor, *J. Immunol.* **144:**2961–2966.

87. Mackewicz, C., and Levy, J. A., 1992, CD8+ cell anti HIV activity: Non-lytic suppression of virus replication, *AIDS Res. Hum. Retrovir.* **8:**1039–1050.

88. Walker, C. M., Moody, D., Stites, D. P., and Levy, J. A., 1986, CD8+ lymphocytes can control HIV-1 infection by suppressing viral replication, *Science* **234:**1563–1566.

89. Walker, C. M., and Levy, J. A., 1989, A diffusible lymphokine produced by CD8+ T lymphocytes suppresses HIV replication, *Immunology* **66:**628–630.

90. Blackbourn, D. J., Locher, C. P., Ramachandran, B., Barnett, S. W., Murthy, K. K., Carey, K. D., Brasky, K. M., and Levy, J. A., 1997, CD8+ cells from HIV-2-infected baboons control HIV replication, *AIDS* **11:**737–746.

91. Locher, C. P., Blackbourn, D. J., Barnett, S. W., Murthy, K. K., Cobb, E. K., Rouse, S., Greco, G., Reyes-Teran, G., Brasky, K. M., Carey, K. D., and Levy, J. A., 1997, Superinfection with human immunodeficiency virus type 2 can reactivate virus production in baboons but is contained by a CD8 T cell antiviral response, *J. Infect. Dis.* **176:**948–959.

92. Rosok, B., Voltersvik, P., Larsson, B. M., Albert, J., Brinchmann, J. E., and Åsjö, B., 1997, CD8+ cells from HIV-1 type seronegative individuals suppress virus replication in acutcly infected cells, *AIDS Res. Hum. Retrovir.* **13:**79–85.

93. Maggi, E., Mazzetti, M., Ravina, A., Annunziato, F., deCarli, M., Piccini, M. P., Manetti, R., Carbonari, M., Pesce, A. M., del Prete, G., and Romagnani, S., 1994, Ability of HIV to promote a TH1 to Th0 shift and to replicate preferentially in TH2 and TH0 cells, *Science* **265:**244–248.

94. Travers, K., Mboup, S., Marlink, R., Guèye-Ndiaye, A., Siby, T., Thior, I., Trore, I., Dieng-Sarr, A., Sankalé, J.-L., Mullins, C., Ndoye, I., Hsieh, C.-C., Essex, M., and Kanki, P., 1995, Natural protection against HIV-1 infection provided by HIV-2, *Science* **268:**1612–1615.

8

HIV Mucosal Vaccines

HERMAN F. STAATS and JERRY R. McGHEE

1. INTRODUCTION

Human immunodeficiency virus (HIV) infection and acquired immune deficiency syndrome (AIDS) continue to be a major public health concern. Recent WHO estimates predict that 40 million men, women, and children will be infected with HIV by the year 2000, with more than 90% of the infections occurring in developing countries.[1] Other estimates predict that AIDS deaths will peak at 1.7–1.9 million deaths/year between the years 2006 and 2012 and continue at the rate of 0.8–1.8 million deaths/year through the year 2020.[2] By the year 2020, HIV infection is predicted to be the ninth leading cause of death in the world.[2] Although treatment of HIV-infected individuals with highly active antiretroviral therapy (HAART) may drastically reduce the amount of HIV detectable in peripheral blood, lymphoid tissues, and semen,[3–5] developing countries will not be able to afford the cost of these powerful drugs. Additionally, drug-resistant mutants can evolve that may be transmitted to persons who participate in high-risk behavior.[6] Therefore, prevention of HIV infection via vaccination remains the treatment of choice. Because HIV is most commonly transmitted after virus contact with a mucosal surface of the host,[7,8] vaccines that induce protective anti-HIV immune responses at mucosal surfaces are desirable, and possibly required, for vaccine-mediated protection against HIV infection. This chapter will discuss the importance of the mucosal immune system as it relates to

HERMAN F. STAATS • Departments of Medicine and Immunology, Center for AIDS Research, Duke University Medical Center, Durham, North Carolina 27710.   JERRY R. McGHEE • Department of Microbiology, Immunobiology Vaccine Center, University of Alabama at Birmingham, Birmingham, Alabama 35294.

Human Retroviral Infections, edited by Kenneth E. Ugen *et al.*
Kluwer Academic / Plenum Publishers, New York, 2000.

the development of an HIV vaccine that induces protective immune responses at mucosal tissues.

2. HIV INFECTION AT MUCOSAL SURFACES

Worldwide, HIV infection is a sexually transmitted disease spread by heterosexual contact.[8] In addition to heterosexual contact, HIV may be transmitted by the oral and anal routes. Simian immunodeficiency virus (SIV) and chimeric simian/human immunodeficiency virus (SHIV) infection of rhesus macaques provide a reliable model for human HIV infection and disease, and studies performed in this model indicate that infection may occur after exposure to SIV/SHIV by the vaginal, urethral, rectal, or oral routes.[9–18] The first cells to become infected with HIV at mucosal surfaces are likely to be mucosal epithelial cells or immune cells residing near the surface of mucosal tissues. $CD4^-$ mucosal epithelial cells may become infected with HIV via a mechanism involving the alternate gp120 receptor galactosylceramide[19–21] and CXCR4.[20] HIV may also be transported across the mucosal epithelial barrier via transcytosis through epithelial cells[22] or through M cells, specialized epithelial cells that cover mucosal lymphoid follicles.[23,24] Once past the epithelial barrier, lymphocytes, macrophages, or dendritic cells may become infected with HIV. Conversely, studies with intestinal explant cultures indicated that lymphocytes and macrophages became infected after exposure to HIV, while epithelial cells did not.[25] Indeed, dendritic cells in the lamina propria of the cervicovaginal mucosa were the first cells to be infected with SIV after intravaginal inoculation of rhesus macaques with SIV_{mac251}.[18] Therefore, HIV transmission at mucosal tissues may be initiated by a number of mechanisms, including (1) direct infection of mucosal epithelial cells,[19–21] (2) transcytosis of HIV through mucosal epithelial cells to initiate infection of dendritic cells in the lamina propria,[18,22] (3) transcytosis of HIV through mucosal M cells to initiate infection of underlying immune cells,[18,23,24] (4) transport through epithelium by, or direct infection of, Langerhans cells that reside in the mucosal epithelium,[18,26,27] and (5) trauma to the mucosal epithelium that allows HIV direct access to the blood (Fig. 1).[22]

Once infection occurs at the mucosal surface, HIV rapidly disseminates throughout the host. Vaginal infection of rhesus macaques with SIV_{mac251} resulted in SIV-infected dendritic cells in the lamina propria by 2 days and SIV-infected peripheral blood mononuclear cells (PBMC) by 5 days.[18] Likewise, by 2 days after intravaginal inoculation of pig-tailed macaques with $SHIV_{KU-1}$, cells in the submucosa of the vagina and uterus and cells in pelvic and mesenteric lymph nodes expressed viral RNA and protein, while the

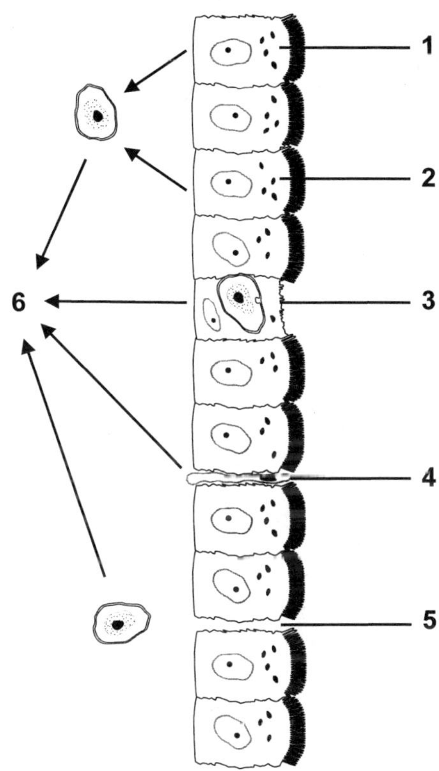

FIGURE 1. Potential mechanisms of HIV transmission at mucosal tissues. (1) Direct infection of mucosal epithelial cells. (2) Transcytosis of HIV through mucosal epithelial cells to initiate infection of dendritic cells in the lamina propria. (3) Trancytosis of HIV through mucosal M cells to initiate infection of underlying immune cells. (4) Transport of HIV through epithelium by, or direct infection of Langerhans cells that reside in the mucosal epithelium. (5) Trauma to the mucosal epithelium that allows HIV direct access to the blood. (6) HIV-infected macrophages, T cells, and/or dendritic cells migrate to regional lymph nodes, where HIV is disseminated throughout the host.

inguinal lymph node, spleen, and thymus remained negative.[28] By 4 days, cells expressing viral RNA and protein were also detected in the inguinal lymph node, spleen, and uterus, while all areas of the brain tested were polymerase chain reaction (PCR)-negative.[28] At 7 and 15 days after intravaginal infection, SHIV had spread to 3 of 13 and 10 of 15 tested regions in the brain, respectively.[28] These studies suggested that mucosal infection with HIV/SIV/SHIV proceeds in three phases. First, HIV/SIV/SHIV contacts the mucosal epithelium and infection is initiated.[18,28] Second, HIV/SIV/SHIV then spread to the regional lymph nodes via dendritic cells.[18,27–29] Lastly, in the regional lymph nodes, HIV/SIV/SHIV is transmitted to T cells and macrophages, expanded, and disseminated throughout the host.[27,29] Accordingly, HIV vaccines should induce immune responses that have an opportunity to (1) prevent HIV infection at the mucosal surface, (2) prevent the spread of HIV to regional lymph nodes, and/or (3) block the dissemination of HIV to circulating immune cells (Fig. 2).[15,30,31]

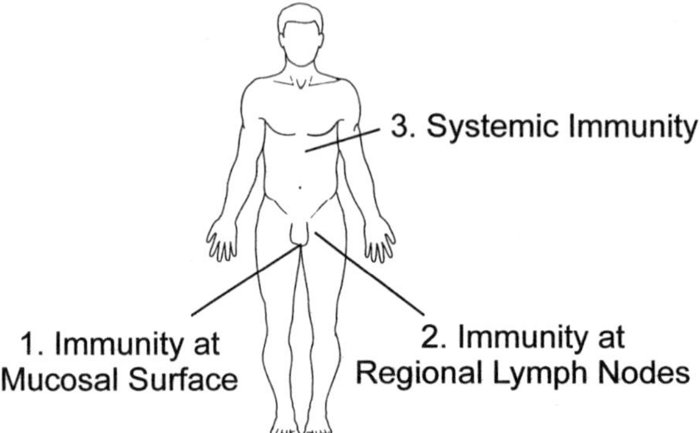

FIGURE 2. Effector sites of mucosal HIV-vaccine-induced immune responses for protection against mucosally transmitted HIV. HIV vaccines should induce immune responses that have an opportunity to (1) prevent HIV infection at the mucosal surface, (2) prevent the spread of HIV to regional lymph nodes, and/or (3) block the dissemination of HIV to circulating immune cells.

3. CORRELATES OF PROTECTION FOR MUCOSALLY TRANSMITTED HIV

3.1. HIV-Exposed but Uninfected

The exact type and magnitude of immune response required for protection against mucosally transmitted HIV are not known. It is important to distinguish between prevention of HIV infection and clearance or control of an established HIV infection since the immune response(s) required for prevention of HIV infection may not be the same as those needed to clear or control an established HIV infection. Studies of HIV-exposed uninfected individuals indicate that both humoral and cell-mediated immunity are involved with protection[32–34] (Table I). Indeed, HIV-specific immunoglobulin A (IgA), but not IgG, was detected in 82% of the vaginal wash samples of HIV-exposed, seronegative women.[32] HIV-specific PBMC were also identified in these HIV-exposed, seronegative women. However, PBMC production of the chemokines RANTES, MIP-1α, and MIP-1β was comparable between HIV-infected and uninfected individuals, suggesting that the production of chemokines did not play a role in resistance to HIV infection.[32] Cytotoxic T lymphocytes (CTL) appear to be involved in protection against HIV infection because uninfected children born to HIV-infected mothers had detectable CTL activity against a variety of HIV antigens including *env, gag/pol,* and *nef.*[33,34] Collectively, these results indicate that HIV-specific mucosal IgA responses and CTL correlate with protection against HIV infection, and vaccines designed to protect against mucosally transmitted HIV should be capable of inducing both mucosal anti-HIV IgA responses and HIV-specific CTL.

TABLE I
Antigens and Routes of Immunization Utilized for the Induction of Anti-HIV (SIV) Mucosal Immune Responses

Host	Antigen	Route of immunization	Immune responses induced[a]	Route of challenge	Protection against challenge	Ref.
HIV-exposed, uninfected humans	HIV	???	Vaginal anti-HIV IgA, HIV-specific proliferative PBMC and CTL	???	Yes	32–34
Rhesus macaques	Attenuated SIV	Intravenous	Rectal anti-SIV antibody, systemic CTL	Rectal SIV	4 or 4 protected from superinfection	35
Rhesus macaques	SIV gp120 and p27	TILN	Iliac anti-SIV IgA antibody-secreting cells, iliac lymph node cell production of CD8-suppressor factor, RANTES, and MIP-1α	Rectal SIV	4 of 7 protected	15
Cynomolgus macaques	HIV-2	Intravenous	SIV-specific CTL	Rectal SIV	2 of 4 protected	16
Rhesus macaques	SHIV	Vaginal	SIV-specific CTL in PBMC, SIV-specific vaginal IgG	Vaginal SIV	3 of 5 protected	36
Mice	HIV peptide	Intranasal, intragastric, intrarectal	HIV-specific systemic and mucosal IgG and IgA, systemic and mucosal CTL	N.D.	N.D.	37, 40, 41, 72, 87–89
Mice	HIV env	Intranasal	HIV-specific systemic and mucosal IgG and IgA	N.D.	N.D.	44, 85
Rhesus macaques	SIV VLP	Vaginal, rectal, oral, urethral	SIV-specific systemic and mucosal IgG and IgA, SIV-specific lymphocytes in regional lymph nodes	N.D.	N.D.	30, 31, 80, 81
Rhesus macaques, chimpanzees	r adenovirus HIV	Intranasal	HIV-specific systemic and mucosal IgG and IgA, proliferative responses	Vaginal SIV	2 of 6 uninfected, 2 of 6 naive uninfected	90, 91
Mice	r *Salmonella* HIV, r BCG SIV	Intragastric	Mucosal anti-HIV IgA, ASC, SIV-specific proliferative and CTL	N.D.	N.D.	98, 99
Mice	DNA HIV	Intranasal	anti-HIV systemic and mucosal IgG and IgA, systemic CTl	N.D.	N.D.	102

[a]For protection studies, only those immune response that correlate with protection are listed. N.D., Not done.

3.2. Mucosal SIV Challenge in Nonhuman Primates

Studies in nonhuman primates have also provided valuable information concerning protection against mucosally transmitted retroviral infections. HIV-2 exposed, seronegative macaques or macaques infected with live, attenuated SIV_{mac} or macaques previously infected with SHIV were partially protected against mucosal (rectal, vaginal) infection with pathogenic SIV.[15,35,36] However, it remains unclear what immune response (if any) was responsible for protection against mucosal SIV infection. Two studies determined that protection was associated with serum and mucosal anti-SIV antibodies and peripheral blood CTL,[35] while another study determined that only SIV-specific CTL were associated with protection.[16]

Vaccine-induced protection against rectal SIV infection has also been reported. Rhesus macaques were protected against rectal challenge with SIV by targeted iliac lymph node (TILN) immunization with SIV gp120 and p27.[15] Protection was correlated with an increased number of IgA antibody-secreting cells (ASC), increased production of CD8-suppressor factor, and increased production of the chemokines RANTES and MIP-1β by iliac lymph node cells.[15]

Unfortunately, not all studies mentioned above monitored for the presence of both systemic and mucosal antibody (IgG and IgA) and CTL responses. To better understand the immune responses required for protection against mucosal HIV/SIV/SHIV infection, all possible immune responses must be monitored in both the systemic and mucosal compartments. Also, mucosal vaccines designed to induce only antibody, only CTL, or both could be utilized to determine whether antibody, CTL, or both are required for protection. Indeed, HIV peptide immunogens have been found to induce only CTL or both antibody and CTL after mucosal and systemic administration.[37–41] The use of peptide immunogens may be useful as tools to induce restricted immune responses and identify the type of immune responses required for vaccine-mediated protection against mucosal HIV infection. Because protection against mucosal challenge may be easier to achieve than protection against a systemic challenge (due to the barrier functions provided by the mucosal epithelium) mucosally and systemically immunized macaques should be challenged with SIV/SHIV by both intravenous and mucosal routes.

4. MUCOSAL HIV VACCINE NEEDED

Although no consistent correlate for protection against mucosal or systemic HIV infection has been identified, the induction of mucosal and systemic humoral and cell-mediated immune responses is thought to be important for vaccine-induced protection against HIV infection. The hallmark of a mucosal immune response is the detection of antigen-specific secretory-IgA (S-IgA) in mucosal secretions and the presence of antigen-specific T and B cells in mucosal effector tissues.[42,43] Systemic immunization rarely if ever induces mucosal immune responses, while mucosal immunization has the advantage of inducing both systemic and mucosal immune responses.[30,31,37,40,44] Therefore, if mucosal immune responses are required for prevention of mucosal HIV infection, HIV vaccines will need to be administered via a mucosal route. The following sections will briefly review the organization of the mucosal immune system and discuss the protective mechanisms employed by the humoral and cellular arms of the mucosal immune system.

4.1. Organization of the Mucosal Immune System

The mucosal immune system may be divided into two functional components, inductive and effector tissues.[42,45,46] Inductive sites include tissues where antigen-specific B and T cells are first sensitized to specific antigen, while effector sites are those sites where antigen-specific immune responses are detected. Inductive tissues are identified as organized lymphoid tissues adjacent to mucosal surfaces and include (in the mouse) the Peyer's patch for the gastrointestinal tract and the nasal-associated lymphoreticular tissue (NALT) for the upper respiratory tract. In humans, the tonsils appear to serve as the inductive sites for mucosal immune responses active in the upper respiratory tract and oral cavity. Inductive tissues contain all the cellular components required for the induction of antigen-specific B and T (CD4$^+$ and CD8$^+$) cells.[47–51] At the inductive site, antigen is sampled from the environment via specialized antigen-sampling cells known as M cells, processed by antigen-presenting cells (APC), and presented to T and B lymphocytes. Antigen-specific T and B lymphocytes then leave the inductive site, enter the circulation, and eventually extravasate to mucosal effector tissues to perform their respective functions (i.e., T-helper, CTL, or antibody production). Mucosal effector tissues include the lamina propria of the gastrointestinal, genitourinary, and upper respiratory tracts.[42,45,46] Glandular tissues such as the mammary gland are also considered effector tissues of the mucosal immune system. Because immunization at one mucosal site commonly gives rise to antigen-specific mucosal immune responses at the site of immunization as well as other mucosal sites, the mucosal immune system is commonly referred to as the "common mucosal immune system."[46,52]

4.2. Mucosal Antibody Responses: Secretory-IgA

Antibodies in mucosal secretions may play a crucial role in protecting the mucosal surfaces of the host against HIV infection.[32] The antibodies in mucosal secretions are predominantly of the IgA class[53–55] and are primarily produced by local plasma cells which infiltrate the lamina propria regions of mucosal effector tissues.[56] Polymeric IgA, but not monomeric IgA or IgG, may be actively transported through mucosal epithelial cells into mucosal secretions after interaction with the polymeric immunoglobulin receptor (pIgR) produced by specialized epithelial cells.[57] IgA that is transported to the mucosal surfaces via this mechanism is referred to as secretory-IgA (S-IgA). S-IgA is structurally and functionally distinct from the immunoglobulins which predominate in the serum (IgG, monomeric and polymeric IgA).[58] S-IgA exists as an IgA polymer that is complexed with a portion of the pIgR known as secretory component (SC). SC protects S-IgA against proteolytic degradation.[59] This may provide S-IgA an advantage over other forms of immunoglobulin when functioning in mucosal secretions that contain proteolytic enzymes.

4.2.1. Protective Mechanisms Employed by S-IgA

Mechanisms that may be employed by S-IgA for protection against HIV infection at mucosal surfaces include immune exclusion, intracellular virus neutralization, and transepithelial transport of immune complexes.[60] Antibody-mediated protection against viral infection at a mucosal surface by immune exclusion is characterized by virus-specific antibody binding to the virus and preventing it from contacting the mucosal surface of the host, thereby preventing infection. Passive transfer studies with antigen-specific IgA have provided convincing data that pathogen-specific IgA alone prevented infection of animals challenged via gastric or nasal routes.[61–65] Although both IgG and IgA may protect against infection at mucosal surfaces by immune exclusion, S-IgA is better suited to protecting mucosal surfaces because (1) S-IgA, but not IgG, is actively transported into mucosal secretions by epithelial cells that express the pIgR, (2) S-IgA is more resistant to proteolytic enzymes present in the mucosal secretions, and (3) S-IgA is polymeric, and therefore has four (or more) antigen-binding sites, while IgG is monomeric with only two antigen-binding sites. Indeed, anti-gp160 S-IgA was superior to anti-gp160 IgG when tested for its ability to prevent HIV transcytosis through an epithelial cell layer.[66]

Intracellular virus neutralization is another protective mechanism utilized by S-IgA. With intracellular virus neutralization, virus-specific S-IgA is transported into the epithelial cell by the pIgR. If the cell is infected with the corresponding virus, virus-specific S-IgA can bind to the virus and neutralize infectivity intracellularly. Experiments *in vivo* with the rotavirus system have supported the concept of intracellular virus neutralization.[67] An interesting observation made during this study was that the protective S-IgA was specific for an inner capsid protein and did not exhibit *in vitro* neutralization activity.[67] Therefore, "nonneutralizing" anti-HIV S-IgA may neutralize HIV intracellularly if pIgR⁺-epithelial cells are infected with HIV. Recent data suggest that cervical epithelial cells may be infected with HIV after contact with HIV-infected cells.[68] Therefore, S-IgA antibodies specific for several HIV antigens, notably nonenvelope gene products, may be capable of mediating intracellular neutralization of HIV and may be an important component of a protective anti-HIV immune response. It is important to mention that IgG does perform intracellular virus neutralization because it does not interact with pIgR and therefore is not actively transported through epithelial cells.

The final mechanism that S-IgA may participate in that would be beneficial for protection against HIV infection is transepithelial transport of immune complexes.[60] In this mechanism, S-IgA binds to virus present in the lamina propria region of mucosal tissues and forms immune complexes. The immune complexes are then transported out of the host into the mucosal secretions by cells expressing the pIgR. Although transepithelial transport of immune complexes would not be a mechanism that would prevent infection, it may reduce the infectious inoculum to a point low enough to allow the host to mount an immune response able to clear the infection. Because IgG is not actively transported by pIgR, IgG does not appear capable of mediating transepithelial transport of immune complexes.

4.2.2. Anti-HIV IgA Effector Functions

Mucosal anti-HIV IgA may play a crucial role in protection against mucosal HIV infection. The most convincing evidence supporting this conclusion comes from the observation that HIV exposed but uninfected women had urinary and vaginal anti-HIV IgA, but not IgG, responses in the absence of serum anti-HIV antibodies.[32] Cervical inoculation with HIV that did not result in seroconversion has also been reported to induce cervical anti-HIV IgA responses in chimpanzees.[69] Although some of the exposed but seronegative women also had detectable HIV-specific PBMC proliferative responses, that anti-HIV IgA played a protective role in resistance to repeated HIV exposures cannot be ruled out.[32] Indeed, anti-HIV S-IgA was superior to anti-HIV IgG as determined by its ability to block HIV transcytosis through epithelial cells.[66] Additionally, anti-HIV IgA isolated from the sera of HIV-infected persons or from the fecal extracts of vaccinated mice had HIV-1-neutralizing capabilities *in vitro*.[70–72] Therefore, anti-HIV IgA induced by natural exposure or vaccination may have HIV-neutralizing activity and may play a role in resistance to mucosal HIV infection.

IgA has also been found to enhance HIV infection. IgA purified from the serum of HIV-infected individuals enhanced HIV infection of U937 cells, primary human blood monocytes, and intestinal lamina propria mononuclear cells.[73,74] Nevertheless, when enhancing IgA was mixed with physiologic concentrations of nonenhancing IgG, enhancement of HIV infection was not observed, suggesting that IgG may be able to block the enhancing effect of IgA *in vivo* in locations where IgG is abundant.[74] It remains to be determined what role, if any, IgA plays in the enhancement of HIV infection *in vivo*. Further studies are needed to identify the epitope recognized by enhancing IgA antibodies and to determine if vaccine-induced anti-HIV S-IgA antibodies are beneficial or detrimental to the host.

4.3. Mucosal Cell-Mediated Immunity: Cytotoxic T Lymphocytes

Cytotoxic T lymphocytes (CTL) represent another immune effector mechanism that may be capable of protecting the mucosal surfaces of the host against HIV infection.[75] Although mucosal CTL were not measured, systemic HIV-specific CTL were detected in uninfected children born to HIV-infected mothers.[33,34] However, virus-specific CTL have been detected in the cervix and vagina of HIV-infected women and SIV-infected rhesus macaques, respectively.[76,77] These results support the conclusion that the appropriate HIV vaccines may be able to induce HIV-specific CTL that home to and reside in the mucosal tissues of the host and protect against mucosally transmitted HIV. Results from other virus models suggest that mucosal CTL will be beneficial for protection against mucosally acquired HIV infection. For example, herpes simplex virus (HSV)-specific CTL induced by vaginal inoculation with an attenuated HSV-2 protected mice against a lethal HSV-2 vaginal infection.[78] Additional studies are required to identify immunization protocols that induce anti-HIV CTL that reside in mucosal tissues, as these may play a pivotal role in protection against mucosally transmitted HIV.

4.4. Routes of Immunization for the Induction of Anti-HIV Mucosal Immune Responses

Numerous routes of immunization have been evaluated for their ability to induce mucosal immune responses in humans and nonhuman primates. These include immunization via the nasal, oral (gastric), rectal, vaginal, urethral, and tracheal routes.[10,30,31,79–83] Although parenteral immunizations generally do not induce mucosal immune responses, a specialized intramuscular immunization known as targeted iliac lymph node immunization has been reported to induce mucosal immune responses in nonhuman primates.[15] This route of immunization is thought to induce mucosal immune responses because the antigen is injected near the iliac lymph node that drains the genitorectal mucosa and is therefore processed as a mucosally administered antigen.[15] Exactly what route of immunization will be optimal for the induction of mucosal anti-HIV S-IgA and CTL responses and protection against mucosal HIV infection remains to be determined. Specific results concerning the induction of anti-HIV mucosal immune responses will be discussed below according to the type of immunogen used.

5. MUCOSAL HIV VACCINE STRATEGIES

A wide variety of vaccine strategies have been evaluated for their ability to induce anti-HIV mucosal immune responses. These include HIV protein subunits, HIV peptides, virus-like particles (VLP), recombinant viral vectors, recombinant bacterial vectors, attenuated HIV (SIV), and DNA vaccines. The type of anti-HIV mucosal immune responses induced by the various vaccine strategies will be discussed below according to the form of antigen utilized as the vaccine.

5.1. Protein Subunits

Major viral proteins have been tested extensively for their ability to induce systemic and mucosal immune responses against HIV/SIV. Oral (intragastric) immunization of rhesus macaques with SSIV p55^{gag} and cholera toxin induced serum and mucosal (rectal and salivary) anti-p55 IgG and IgA responses.[84] Unfortunately, vaginal IgA, which may be required for protection against sexually transmitted HIV/SIV (see above), was not detected after oral immunization with p55^{gag} and cholera toxin. These results are consistent with results obtained in humans after oral immunization with CT-B as a protein antigen; oral immunization of humans did not induce consistent vaginal IgA responses, while intravaginal immunization did.[79] Intranasal immunization of mice with HIV gp160 has been found to induce systemic and mucosal anti-HIV IgG and IgA responses.[44,85] Intranasal immunization of mice with gp160 formulated with proteosomes and/or emulsomes and/or CT-B was superior to intranasal immunization with gp160 in saline for the induction of serum anti-gp160 IgG and serum and mucosal (intestinal, vaginal, and lung) anti-gp 160 IgA.[85] More recently, intranasal immunization of mice with oligomeric HIV-1 gp160 (o-gp160) formulated with a number of adjuvants including liposomes, liposomes and monophosphoryl lipid A (MPL), MPL-AF (AF: surfactant 1,2-dipalmitoyl-SN-glycero-3 phosphocholine), proteosomes, emulsomes, or proteosomes and emulsomes was evaluated.[44] When 50 μg of o-gpl60 was intranasally administered to mice at 0, 3, and 6 weeks in saline, a peak serum IgG titer of 1:820,000 was detected. When the same amount of antigen was administered with MPL-AF or proteosomes and emulsomes as an adjuvant, a peak serum IgG titer of 1:12,800,000 was detected.[44] Nasal immunization with o-gp160 with these adjuvants also induced a serum anti-gp160 IgA titer of 1:51,200. Anti-gp160 IgA responses were also detected in vaginal, lung, intestinal, and fecal mucosal samples. In Fact, vaginal and lung samples from mice intranasally immunized with o-gp160 and proteosomes and emulsomes was able to neutralize HIV-1$_{MN}$ *in vitro.*[44] However, it was not determined if the IgG, IgA, or both were responsible for neutralization. Combined, these studies indicate that not only does the form of antigen and route of immunization affect the immune response induced, but also the adjuvant used may play a crucial role in the induction of biologically active anti-HIV mucosal and systemic antibody responses. Similar conclusions were obtained after parenteral immunization of mice with o-gp160 and various adjuvants.[86] Future protection studies in nonhuman primates may wish to evaluate intranasal immunization with HIV antigens for its ability to induce protective immunity since intranasal immunizations of humans and rhesus macaques induced both antigen-specific IgG in the systemic compartment and antigen-specific IgA at numerous mucosal sites including salivary, nasal, and genital secretions.[82,83]

Targeted iliac lymph node (TILN) immunizations with SIV gp120 and p27 induced systemic and mucosal anti-SIV immune responses and protected macaques against intrarectal challenge with SIV$_{macJ5}$.[15] Immune responses induced by TILN immunization with SIV gp120 and p27, but not immunization by the intradermal route or by a combination of routes including nasal, rectal, and intramuscular, completely protected 4 of 7 macaques against intrarectal SIV challenge. Immune responses that correlated with protection included anti-p27 IgA antibody-secreting cells in the iliac lymph node and iliac lymph node cell production of CD8-suppressor factor and chemokines RANTES and MIP-1β.[15] Mucosal and parenteral immunization studies with HIV/SIV/SHIV proteins should be performed in parallel and concluded with all immunized animals being challenged by mucosal routes with HIV/SIV/SHIV. This would identify the specific immune response(s) that correlate with protection against a mucosal HIV/SIV/SHIV infection.

5.2. Peptide Vaccines

Synthetic peptides that correspond to T_{helper}, neutralizing B cell, or CTL epitopes from HIV glycoproteins represent another HIV vaccine strategy for the induction of anti-HIV mucosal immune responses. Oral immunization with the HIV-1 gp120 B1M-p3C synthetic lipopeptide was found to induce serum IgG and IgA and salivary IgA antipeptide antibody responses as well as splenic antipeptide and anti-HIV gp160 CTL responses.[87] Oral immunization with a macromolecular, multicomponent peptide vaccine with or without cholera toxin as a mucosal adjuvant induced serum and fecal antipeptide IgG and IgA responses.[72] Serum IgG titers peaked at 1:256, while fecal IgA titers reached a maximum of 1:2048. This peptide antigen, composed of HIV-1 sequences from the third variable region of gp120, a CD4-binding site, and a *gag* region, induced anti-HIV-1 S-IgA responses that were able to neutralize $HIV-1_{IIIB}$, $HIV-1_{SF2}$, and $HIV-1_{MN}$ *in vitro*.[72] The ability of this immunization protocol to induce anti-HIV IgA in the reproductive tract was not determined.

Intranasal immunization of mice with C4-V3 peptide which contains a T_{helper} epitope, a neutralizing B cell epitope, and a CTL epitope from HIV-1 gp120 was able to induce serum $HIV-1_{MN}$-neutralizing antibody responses when intranasally administered with cholera toxin.[40] In addition to high-titered serum IgG responses (1:131,072 in BALB/c mice and 1:524,288 in C57BL/6 mice), vaginal antipeptide IgG and IgA responses were also induced. Vaginal antipeptide antibody responses were associated with SC, suggesting that the anti-HIV IgA responses represented polymeric IgA that was transported across the mucosal epithelium to the vaginal surface. Intranasal priming followed by intranasal boosting with C4-V3 peptide was compared to intranasal priming followed by intravaginal, intragastric, or intrarectal boosting for its ability to induce serum antipeptide IgG and vaginal antipeptide IgA. Intranasal immunization was superior to all combinations tested as determined by (1) the amount of antigen used for immunization (200 µg peptide for intranasal immunization as compared to 2100 µg for intranasal + intragastric immunization), (2) the magnitude of serum IgG induced, and (3) the magnitude of vaginal IgA induced.[88]

Peptides that correspond to defined HIV CTL epitopes only or peptides containing both HIV CTL and T_{helper} epitopes induced peptide-specific CTL after intranasal, intragastric, or intrarectal immunization.[37,41] Intranasal immunization with HIV CTL epitope peptides induced peptide-specific CTL in the cervical lymph node (CLN), spleen, mesenteric lymph node (MLN), and lung.[37,89] Cholera toxin was required as a mucosal adjuvant for the induction of peptide-specific CTL after intranasal immunization with CTL epitope peptide.[37] In contrast, mucosal immunization with a peptide containing both a T_{helper} and CTL epitope was able to induce peptide-specific CTL in the absence of cholera toxin, although cholera toxin enhanced the induction of CTL.[41] Peptide-induced CTL recognized target cells expressing HIV-1 gp160 and protected mice against intrarectal challenge with a recombinant vaccinia virus expressing HIV-1 gp160.[41] Therefore, mucosally administered HIV peptide vaccines are capable of inducing both systemic and mucosal antibody (IgG and IgA) and CTL responses. To determine if peptide-induced mucosal immune responses are capable of protecting against mucosal challenge with SHIV, rhesus macaques are being mucosally immunized with HIV peptide immunogens and challenged with SHIV 89.6 (H. F. Staats, work in progress).

5.3. Virus-like Particles

Virus-like particles (VLP) containing SIV *gag* p27 have been extensively studied for their ability to induce p27-specific immune responses after mucosal immunization in female and male macaques.[30,31,80,81] Vaginal or rectal immunization with SIV p27-VLP followed by oral immunization with SIV p27-*gag* VLP was found to be an effective immunization protocol for the induction of anti-p27 IgA responses in vaginal and rectal secretions, respectively.[30,80] The combined vaginal–oral immunizations with SIV p27-*gag* VLP also induced p27-specific lymphocyte-proliferative responses in cells isolated from the blood, spleen, genital lymph nodes, and iliac lymph nodes, but not the mesenteric, bronchial, or axillary lymph nodes.[30] The combination of oral and rectal immunization followed by an intramuscular boost with SIV p27-*gag* VLP induced p27-specific lymphocyte-proliferative responses in the blood, spleen, internal iliac, inferior mesenteric, and iliac-paraaortic lymph nodes, but not in the superior mesenteric, bronchial, or axillary lymph nodes.[30] Topical urethral immunization of male macaques followed by oral immunization with SIV p27-*gag* VLP also induced p27-specific lymphocytes in the blood, spleen, internal iliac, inferior mesenteric, and iliac paraaortic lymph node.[31] In contrast, intramuscular immunization of female or male macaques induced p27-specific lymphocyte proliferative responses in the spleen and blood only.[30,31] Therefore, mucosal immunization was required for the induction of SIV-specific lymphocytes that reside in lymph nodes that drain mucosal tissues near the genitourinary tract. The observation that mucosal immunization induced compartmentalized mucosal and systemic immune responses led these investigators to hypothesize that the induction of mucosal and systemic immune responses may prevent mucosally transmitted HIV infection at three different states by (1) preventing HIV infection at the mucosal surface, (2) preventing the spread of HIV to regional lymph nodes, and/or (3) blocking the dissemination of HIV to circulating immune cells (Fig. 2).[15,30,31]

5.4. Recombinant Viral Vectors

Viral vectors that express portions of the HIV genome have been investigated as an antigen delivery system to induce systemic and mucosal anti-HIV immune responses. Intranasal immunization of chimpanzees with recombinant adenoviruses (Ad) expressing HIV-1$_{IIIB}$ gp160 or *gag* induced serum anti-HIV neutralizing antibody responses.[90] Intranasal immunization with this recombinant Ad induced serum, salivary, nasal, and vaginal IgG responses, with low IgA responses detected in salivary and nasal secretions, but not in vaginal or rectal secretions. Additional studies in the macaque model have determined that mucosal immunization with an adenovirus type 5 vector producing SIV$_{sm}$ *env* (Ad-SIV) and parenteral boosting with SIV gp120 was not able to prevent vaginal infection of macaques with SIV$_{mac251}$.[91] Immunization of macaques with the Ad-SIV by the nasal and oral routes at week 0 with an intratracheal boost at 12 weeks followed by paren teral immunization with SIV$_{mac251}$ gp120 at 24 and 36 weeks induced both systemic IgG and mucosal (nasal and rectal) IgA as well as proliferative responses specific for SIV.[91] SIV-specific CTL were not observed.[91] Upon vaginal challenge with SIV$_{mac251}$, SIV was isolated from four of six Ad-SIV immunized and four of six control macaques. Although the Ad-SIV immunizations did not prevent vaginal SIV infection, immunization was associated with a reduction in plasma SIV RNA during the primary infection.[91]

Subcutaneous immunization of mice with attenuated Venezuelan equine encephalitis virus (VEEV) expressing the HIV-1 matrix/capsid (MA/CA) coding region induced systemic and mucosal anti-HIV immune response including anti-HIV CTL, serum IgG and IgA, and vaginal IgA.[92] The unexpected result obtained with this construct was that parenteral immunization was able to induce both systemic and mucosal anti-HIV antibody responses. This observation may be associated with the ability of VEEV to replicate in inductive tissues of the mucosal immune system.[93] A potential drawback to the use of attenuated virus vectors is that repeated use of a vector may be prohibited if the vector is immunogenic enough so that the host develops an antivector immune response that prevents infection at subsequent immunizations (i.e., preexisting immunity).

5.5. Recombinant Bacterial Vectors

Recombinant bacterial vectors are also being investigated as HIV mucosal vaccine delivery vehicles. Prime candidates for use as recombinant bacterial vectors to deliver HIV antigens to the mucosal immune system are *Salmonella* spp. and Bacille Calmette-Guerin (BCG).[94–96] BCG, used to immunize against tuberculosis, is the most widely used vaccine in the world, is associated with a low occurrence of serious side effects, and can be engineered to express foreign antigens.[97] BCG genetically engineered to express the SIV *nef* protein induced *nef*-specific proliferative and CTL responses in systemic and mucosal tissues after oral inoculation of mice.[98] A single oral immunization of mice with attenuated, recombinant *Salmonella typhimurium* expressing a truncated form of HIV-1 gp120 induced gp120-specific IgA-antibody secreting cells in the lamina propria and mesenteric lymph nodes after a single immunization.[99] As with viral vectors, live bacterial vectors may be limited in their use due to preexisting immunity.

5.6. DNA Vaccines

DNA vaccines constitute another vaccine mechanism for the induction of HIV-specific mucosal immune responses.[100–102] Both intranasal and intravaginal routes of immunization have been utilized for the induction of anti-HIV mucosal immune responses.[101,102] Intranasal immunization with a mixture of DNA vectors encoding HIV *env* and *rev* genes formulated with liposomes induced HIV-specific serum IgG, fecal and vaginal IgA, and splenic CTL.[102] Inclusion of DNA vectors expressing cytokines (interleukin-4 [IL-4], IL-12, or granulocyte-macrophage colony-stimulating factor [GM-CSF]) into the HIV DNA vaccine-liposome formulation enhanced the induction of serum IgG and fecal IgA antibody (IL-4, GM-CSF) and splenic CTL responses (IL-12, GM-CSF).[102]

5.7. Live Attenuated SIV

A final mucosal HIV vaccine strategy that will be mentioned is attenuated HIV. To evaluate the safety and efficacy of attenuated HIV as a vaccine, attenuated SIV has been utilized as a representative model. The fact that attenuated SIV was pathogenic after oral inoculation to neonatal macaques indicates that attenuated HIV will not be safe for use in the human population.[13] However, mucosal inoculation with attenuated SIV may provide a means to identify the type of immune responses needed for protection against mucosally transmitted SIV if mucosal challenge studies are performed. Vaginal inoculation of rhesus macaques with attenuated SIV induced the production of low-level serum and vaginal IgG and IgA as well as anti-SIV CTL activity in peripheral blood cells.[103] Although the attenuated SIV induced humoral and CTL responses, the ability of these immune responses to protect against virulent mucosal challenge with pathogenic SIV was not determined. In a similar study, vaginal inoculation of rhesus macaques with simian/human immunodeficiency virus (SHIV) led to transient infection that induced both humoral and cell-mediated immunity to SHIV.[36] Vaginal challenge of these macaques with pathogenic SIV_{mac239} resulted in three of five macaques remaining virus isolation negative after two exposures to SIV.[36] Protection was associated with the detection of SIV-specific CTL and vaginal IgG.[36] Although total IgA was detected in the vaginal secretions, SIV-specific IgA was not detected at the time of pathogenic challenge.

6. MUCOSAL ADJUVANTS

Mucosal immunization in the absence of adjuvants may induce local humoral immunity in the presence of systemic immunological tolerance.[104] This phenomenon is referred to as oral tolerance. For optimal induction of systemic and mucosal immune responses after mucosal immunization, a mucosal adjuvant is required.[37,40,105,106] The most commonly utilized mucosal adjuvant is cholera toxin, although heat-labile enterotoxin (LT) and pertussis toxin (PT) also possess mucosal adjuvant activity.[106] Even though CT is a potent mucosal adjuvant, as little as 5 µg administered to humans via the gastric route may cause severe diarrheal disease.[107] Because the use of these molecules as mucosal adjuvants in humans may be precluded by their toxic nature, a number of groups have succeeded in mutating CT, LT, or PT so that the toxicity is reduced or absent while the adjuvanticity is maintained.[108–113] Additionally, since CT, LT, and/or PT are very immunogenic, preexisting immunity to these proteins may limit their use as mucosal adjuvants. Other reagents such as liposomes, monophosphoryl lipid A, proteosomes, and emulsomes have exhibited mucosal adjuvant activity when intranasally administered with HIV oligomeric gp160.[44] Additional studies must be performed to identify ad adjuvant that exhibits adjuvant activity in the absence of adverse reactions. The identification of a safe mucosal adjuvant is critical to the development of effective HIV vaccines.

7. CONCLUSIONS

HIV infection and AIDS represent a major public health problem that is primarily transmitted via heterosexual contact. HIV infection may be initiated at the mucosal surfaces of the host by a number of potential mechanisms including (1) direct infection of mucosal epithelial cells, (2) transcytosis of HIV through mucosal epithelial cells to initiate infection of dendritic cells in the lamina propria, (3) transcytosis of HIV through mucosal M cells to initiate infection of underlying immune cells, (4) transport through epithelium by direct infection of Langerhans cells that reside in the mucosal epithelium, and (5) trauma to the mucosal epithelium that allows HIV direct access to the blood. After HIV infection is initiated at the mucosal surface, HIV spreads to the regional lymph nodes where circulating T cells, macrophages, and/or dendritic cells become infected and disseminate the infection throughout the host. Accordingly, HIV vaccines should induce mucosal and systemic immune responses capable of (1) preventing HIV infection at the mucosal surface, (2) preventing the spread of HIV to regional lymph nodes, and/or (3) blocking the dissemination of HIV to circulating immune cells. Although the immune responses required for protection against mucosally transmitted HIV are not known, a variety of mucosal vaccination strategies, including protein subunits, peptides, attenuated viral and bacterial vectors, and DNA vaccines are able to induce both mucosal and systemic humoral and cell-mediated immune responses to HIV. Whether vaccines are administered systemically or mucosally, it seems as if challenge via the mucosal route is easier to protect against than systemic challenge. Although this may represent a viral dose issue, we know that natural challenge via heterosexual contact must be low. By searching for a vaccine that protects against systemic challenge, we may be setting ourselves up for failure. To better mimic the natural conditions, we need to test vaccines for their ability to protect against mucosal challenge. Although protection against a mucosal challenge may not protect against systemic challenge, a vaccine that protects against mucosal challenge would be beneficial to the vast majority of people who will become infected with HIV.

ACKNOWLEDGMENTS. The authors would like to thank Dr. Barton F. Haynes, Dr. Dani P. Bolognesi, and Dr. Kent J. Weinhold for helpful discussions and comments. We thank Susan C. Hellenbrand for the figures. H.F.S. is supported by grants DAMD17-94-J-4467, 2 P30 AI28662, and 5 UO1 AI35351-04, and by a grant from the Veterans Affairs Research Center on AIDS and HIV Infection, Durham, North Carolina.

REFERENCES

1. Mertens, T. E., and Low-Beer, D., 1996, HIV and AIDS: Where is the epidemic going? *Bull. WHO* **74:**121–129.
2. Murray, C. J. L., and Lopez, A. D., 1997, Alternative projections of mortality and disability by cause 1990–2020—Global burden of disease study, *Lancet* **349:**1498–1504.
3. Anonymous, 1997, Viramune triple therapy shows suppression of HIV after 1 year, 7 months, *AIDS Patient Care STDS* **11:**456–457.

4. Notermans, D. W., Jurriaans, S., Dewolf, F., Foudraine, N. A., Dejong, J. J., Cavert, W., Schuwirth, M., Kauffmann, R. H., Meenhorst, P. L., McDade, H., Goodwin, C., Leonard, J. M., Goudsmit, J., and Danner, S. A., 1998, Decrease of HIV-1 RNA levels in lymphoid tissue and peripheral blood during treatment with ritonavir, lamivudine and zidovudine, *AIDS* **12:**167–173.

5. Gupta, P., Mellors, J., Kingsley, L., Riddler, S., Singh, M. K., Schreiber, S., Cronin, M., and Rinaldo, C. R., 1997, High viral load in semen of human immunodeficiency virus type 1-infected men at all stages of disease and its reduction by therapy with protease and nonnucleoside reverse transcriptase inhibitors, *J. Virol.* **71:**6271–6275.

6. Imrie, A., Beveridge, A., Genn, W., Vizzard, J., and Cooper, D. A., 1997, Transmission of human immunodeficiency virus type 1 resistant to nevirapine and zidovudine, *J. Infect. Dis.* **175:**1502–1506.

7. CDC, 1997, Update: Trends in AIDS incidence, deaths, and prevalence—United States, 1996, *Morbid. Mortal. Weekly Rep.* **46:**165–173.

8. Mestecky, J., and Jackson, S., 1994, Reassessment of the impact of mucosal immunity in infection with the human immunodeficiency virus (HIV) and design of relevant vaccines, *J. Clin. Immunol.* **14:**259–272.

9. Miller, C. J., Marthas, M., Torten, J., Alexander, N. J., Moore, J. P., Doncel, G. F., and Hendrickx, A. G., 1994, Intravaginal inoculation of rhesus macaques with cell-free simian immunodeficiency virus results in persistent or transient viremia, *J. Virol.* **68:**6391–6400.

10. Marx, P. A., Compans, R. W., Gettie, A., Staas, J. K., Gilley, R. M., Mulligan, M. J., Yamschikov, G. V., Chen, D., and Eldridge, J. H., 1993, Protection against vaginal SIV transmission with microencapsulated vaccine, *Science* **260:**1323–1327.

11. Miller, C. J., Alexander, N. J., Sutjipto, S., Lackner, A. A., Hendrickx, A. G., Gettie, A., Lowenstine, L. J., Jennings, M., and Marx, P. A., 1989, Genital mucosal transmission of simian immunodeficiency virus: Animal model for heterosexual transmission of human immunodeficiency virus, *J. Virol.* **63:**4277–4284.

12. Baba, T. W., Koch, J., Mittler, E. S., Greene, M., Wyand, M., Penninck, D., and Ruprecht, R. M., 1994, Mucosal infection of neonatal rhesus monkeys with cell-free SIV, *AIDS Res. Hum. Retrovir.* **10:**351–357.

13. Baba, T. W., Jeong, Y. S., Penninck, D., Bronson, R., Greene, M. F., and Ruprecht, R. M., 1995, Pathogenicity of live, attenuated SIV after mucosal infection of neonatal macaques, *Science* **267:**1820–1825.

14. Baba, T. W., Trichel, A. M., An, L., Liska, V., Martin, L. N., Murphey-Corb, M., and Ruprecht, R. M., 1996, Infection and AIDS in adult macaques after nontraumatic oral exposure to cell-free SIV, *Science* **272:**1486–1489.

15. Lehner, T., Wany, Y., Cranage, M., Bergmeier, L. A., Mitchell, E., Tao, L., Hall, G., Dennis, M., Cook, N., Brookes, R., Klavinskis, L., Doyle, C., and Ward, R., 1996, Protective mucosal immunity elicited by targeted iliac lymph node immunization with a subunit SIV envelope and core vaccine in macaques, *Nature Med.* **2:**767–775.

16. Putkonen, P., Makitalo, B., Bottiger, D., Biberfeld, G., and Thorstensson, R., 1997, Protection of human immunodeficiency virus type 2-exposed seronegative macaques from mucosal simian immunodeficiency virus transmission, *J. Virol.* **71:**4981–4984.

17. Lu, Y., Brosio, P., Lafaile, M., Li, J., Collman, R. G., Sodroski, J., and Miller, C. J., 1996, Vaginal transmission of chimeric simian/human immunodeficiency viruses in rhesus macaques, *J. Virol.* **70:**3045–3050.

18. Spira, A. I., Marx, P. A., Patterson, B. K., Mahoncy, J., Koup, R. A., Wolinsky, S. M., and Ho, D. D., 1996, Cellular targets of infection and route of viral dissemination after an intravaginal inoculation of simian immunodeficiency virus into rhesus macaques, *J. Exp. Med.* **183:**215–225.

19. Furuta, Y., Eriksson, K., Svennerholm, B., Fredman, P., Horal, P., Jeansson, S., Vahlne, A., Holmgren, J., and Czerkinsky, C., 1994, Infection of vaginal and colonic epithelial cells by the human immunodeficiency virus type 1 is neutralized by antibodies raised against conserved epitopes in the envelope glycoprotein gp120, *Proc. Nat. Acad. Sci. USA* **91:** 12559–12563.

20. Delezay, O., Koch, N., Yahi, N., Hammache, D., Tourres, C., Tamalet, C., and Fantini, J., 1997, Co-expression of CXCR4/fusin and galactosylceramide in the human intestinal epithelial cell line Ht-29, *AIDS* **11:**1311–1318.

21. Fantini, J., Yahi, N., Baghdiguian, S., Spitalnik, S. L., and Gonzalez-Scarano, F., 1994 Human immunodeficiency virus infection of human intestinal epithelial cells, *Mucosal Immunol. Update* **2:**1–13.

22. Bomsel, M., 1997, Transcytosis of infectious human immunodeficiency virus across a tight human epithelial cell line barrier, *Nature Med.* **3:**42–47.

23. Amerongen, H. M., Weltzin, R., Farnet, C. M., Michetti, P., Haseltine, W. A., and Neutra, M. R., 1991, Transepithelial transport of HIV-1 by intestinal M cells: A mechanism for transmission of AIDS, *J. Acquired Immune Defic. Syndr.* **4:**760–765.

24. Neutra, M. R., Frey, A., and Kraehenbuhl, J.-P., 1996, Epithelial M cells: Gateway for mucosal infection and immunization, *Cell* **86:**345–348.

25. Batman, P. A., Fleming, S. C., Sedgwick, P. M., MacDonald, T. T., and Griffin, G. E., 1994, HIV infection of human fetal intestinal explant cultures induces epithelial cell proliferation, *AIDS* **8:**161–167.

26. Blauvelt, A., and Katz, S. I., 1995, The skin as target, vector, and effector organ in human immunodeficiency virus disease, *J. Inv. Dermatol.* **105:**S122–S126.

27. Pope, M., Elmore, D., Ho, D., Marx, P., 1997, Dendritic cell–T cell mixtures, isolated from the skin and mucosae of macaques, support the replication of SIV, *AIDS Res. Hum. Retrovir.* **13:**819–827.

28. Joag, S. V., Adany, I., Li, Z. A., Foresman, L., Pinson, D. M., Wang, C. Y., Stephens, E. B., Raghavan, R., and Narayan, O., 1997, Animal model of mucosally transmitted human immunodeficiency virus type 1 disease—Intravaginal and oral deposition of simian/human immunodeficiency virus in macaques results in systemic infection, elimination of CD4(+) T Cells, and AIDS, *J. Virol.* **71:**4016–4023.

29. Ayehunie, S., Groves, R. W., Bruzzese, A. M., Ruprecht, R. M., Kupper, T. S., and Langhoff, E., 1995, Acutely infected Langerhans cells are more efficient than T cells in disseminating HIV type 1 to activated T cells following a short cell–cell contact, *AIDS Res. Hum. Retrovir.* **11:**877–884.

30. Lehner, T., Bergmeier, L. A., Panagiotidi, C., Tao, L., Brookes, R., Klavinskis, L. S., Walker, P., Walker, J., Ward, R. G., Hussain, L., Gearing, A. J. H., and Adams, S. E., 1992, Induction of mucosal and systemic immunity to a recombinant simian immunodeficiency viral protein, *Science* **258:**1365–1369.

31. Lehner, T., Tao, L., Panagiotidi, C., Klavinskis, L. S., Brookes, R., Hussain, L., Meyers, N., Adams, S. E., Gearing, A. J., and Bergmeier, L. A., 1994, Mucosal model of genital immunization in male rhesus macaques with a recombinant simian immunodeficiency virus p27 antigen, *J. Virol.* **68:**1624–1632.

32. Mazzoli, S., Trabattoni, D., Caputo, S. L., Piconi, S., Ble, C., Meacci, F., Ruzzante, S., Salvi, A., Semplici, F., Longhi, R., Fusi, M. L., Tofani, N., Biasin, M., Villa, M. L., Mazzotta, F., and Clerici, M., 1997, HIV-specific mucosal and cellular immunity in HIV-seronegative partners of HIV-seropositive individuals, *Nature Med.* **3:**1250–1257.

33. De Maria, A., Cirillo, C., and Moretta, L., 1994, Occurrence of human immunodeficiency virus type 1 (HIV-1)-specific cytolytic T cell activity in apparently uninfected children born to HIV-1-infected mothers, *J. Infect. Dis.* **170:**1296–1299.

34. Rowland-Jones, S. L., Nixon, D. F., Aldhous, M. C., Gotch, F., Ariyoshi, K., Hallam, N., Kroll, J. S., Froebel, K., and McMichael, A., 1993, HIV-specific cytotoxic T-cell activity in an HIV-exposed but uninfected infant, *Lancet* **341**:860–861.

35. Cranage, M. P., Whatmore, A. M., Sharpe, S. A., Cook, N., Polyanskaya, N., Leech, S., Smith, J. D., Rud, E. W., Dennis, M. J., and Hall, G. A., 1997, Macaques infected with live attenuated SIVmac are protected against superinfection via the rectal mucosa, *Virology* **229**:143–154.

36. Miller, C. J., McChesney, M. B., Lu, X. S., Dailey, P. J., Chutkowski, C., Lu, D., Brosio, P., Roberts, B., and Lu, L. C., 1997, Rhesus macaques previously infected with simian/human immunodeficiency virus are protected from vaginal challenge with pathogenic Sivmac239, *J. Virol.* **71**:1911–1921.

37. Porgador, A., Staats, H. F., Faiola, B., Gilboa, E., and Palker, T. J., 1997, Intranasal immunization with CTL epitope peptides from HIV-1 or ovalbumin and the mucosal adjuvant cholera toxin induces peptide-specific CTLs and protection against tumor development *in vivo, J. Immunol.* **158**:834–841.

38. Hart, M. K., Weinhold, K. J., Scearce, R. M., Washburn, E. M., Clark, C. A., Palker, T. J., and Haynes, B. F., 1991, Priming of anti-human immunodeficiency virus (HIV) CD8+ cytotoxic T cells *in vivo* by carrier-free HIV synthetic peptides, *Proc. Nat. Acad. Sci. USA* **88**: 9448–9452.

39. Haynes, B. F., Torres, J. V., Langlois, A. J., Bolognesi, D. P., Gardner, M. B., Palker, T. J., Scearce, R. M., Jones, D. M., Moody, M. A., McDanal, C., *et al.*, 1993, Induction of HIVMN neutralizing antibodies in primates using a prime-boost regimen of hybrid synthetic gp120 envelope peptides, *J. Immunol.* **151**:1646–1653.

40. Staats, H. F., Nichols, W. G., and Palker, T. J., 1996, Mucosal immunity to HIV-1: Systemic and vaginal antibody responses after intranasal immunization with the HIV-1 C4/V3 peptide T1SP10 MN(A), *J. Immunol.* **157**:462–472.

41. Belyakov, I. M., Derby, M. A., Ahlers, J. D., Kelsall, B. L., Earl, P., Moss, B., Strober, W., and Berzofsky, J. A., 1998, Mucosal immunization with HIV-1 peptide vaccine induces mucosal and systemic cytotoxic T lymphocytes and protective immunity in mice against intrarectal recombinant HIV-vaccinia challenge, *Proc. Nat. Acad. Sci. USA* **95**:1709–1714.

42. Staats, H. F., Jackson, R. J., Marinaro, M., Takahashi, I., Kiyono, H., and McGhee, J. R., 1994, Mucosal immunity to infection with implications for vaccine development, *Curr. Opin. Immunol.* **6**:572–583.

43. Staats, H. F., and McGhee, J. R., 1996, Application of basic principles of mucosal immunity to vaccine development, in: *Mucosal Vaccines* (H. Kiyono, P. L. Ogra, and J. R. McGhee, eds.), Academic Press, New York, pp. 17–39.

44. VanCott, T. C., Kaminski, R. W., Mascola, J. R., Kalyanaraman, V. S., Wassef, N. M., Alving, C. R., Ulrich, J. T., Lowell, G. H., and Birx, D. L., 1998, HIV-1 neutralizing antibodies in the genital and respiratory tracts of mice intranasally immunized with oligomeric gp160, *J. Immunol.* **160**:2000–2012.

45. McGhee, J. R., Mestecky, J., Dertzbaugh, M. T., Eldridge, J. H., Hirasawa, M., and Kiyono, H., 1992, The mucosal immune system: From fundamental concepts to vaccine development, *Vaccine* **10**:75–88.

46. McGhee, J. R., Xu-Amano, J., Miller, C. J., Jackson, R. J., Fujihashi, K., Staats, H. F., and Kiyono, H., 1994, The common mucosal immune system: From basic principles to enteric vaccines with relevance for the female reproductive tract, *Reprod. Fertil. Dev.* **6**:369–377.

47. Wu, H.-Y., Nikolova, E. B., Beagley, K. W., and Russell, M. W., 1996, Induction of antibody secreting cells and T-helper and memory cells in murine nasal lymphoid tissue, *Immunology* **88**:493–500.

48. Wu, H.-Y., Nikolova, E. B., Beagley, K. W., Eldridge, J. H., and Russell, M. W., 1997,

Development of antibody-secreting cells and antigen-specific T cells in cervical lymph nodes after intranasal immunization, *Infect. Immunity* **65**:227–235.

49. Sarsfield, P., Rinne, A., Jones, D. B., Johnson, P., and Wright, D. H., 1996, Accessory cells in physiological lymphoid tissue from the intestine: An immunohistochemical study, *Histopathology* **28**:205–211.

50. Kelsall, B. L., and Strober, W., 1996, Distinct population of dendritic cells are present in the subepithelial dome and T cell regions of the murine Peyer's patch, *J. Exp. Med.* **183**:237–247.

51. Everson, M. P., McDuffie, D. S., Lemak, D. G., Koopman, W. J., McGhee, J. R., and Beagley, K. W., 1996, Dendritic cells from different tissues induce production of different T cell cytokine profiles, *J. Leuk. Biol.* **59**:494–498.

52. McDermott, M. R., and Bienenstock, J., 1979, Evidence for a common mucosal immunologic system. I. Migration of B immunoblasts into intestinal, respiratory, and genital tissues, *J. Immunol.* **122**:1892–1898.

53. Tomasi, T. B., and Zigelbaum, S., 1963, The selective occurrence of a globulin in certain body fluids, *J. Clin. Inv.* **42**:1552–1560.

54. Tomasi, T. B., Tam, E. M., Solomon, A., and Pendergrast, R. A., 1965, Characteristics of an immune system common to certain external secretions, *J. Exp. Med.* **121**:101–124.

55. Hanson, L.-A., 1961, Comparative immunologic studies of the immune globulins of human milk and the blood stream, *Int. Arch. Allergy Immunol.* **18**:241–267.

56. Brandtzaeg, P., 1974, Mucosal and glandular distribution of immunoglobulin components: Immunohistochemistry with a cold ethanol-fixation technique, *Immunology* **26**: 1101–1104.

57. Brandtzaeg, P., Krajci, P., Lamm, M. E., and Kaetzel, C. S., 1994, Epithelial and hepatobiliary transport of polymeric immunoglobulins, in: *Handbook of Mucosal Immunology* (P. L. Ogra, J. Mestecky, M. E. Lamm, S. Warren, J. R. McGhee, and J. Bienenstock, eds.), Academic Press, San Diego, California, pp. 113–126.

58. Mestecky, J., and McGhee, J. R., 1987, Immunoglobulin A (IgA): Molecular and cellular interactions involved in IgA biosynthesis and immune response, *Adv. Immunol.* **40**: 253–245.

59. Renegar, K. B., Jackson, G. D. F., and Mestecky, J., 1998, *In vitro* comparison of the biologic activities of monoclonal monomeric IgA, polymeric IgA, and secretory IgA, *J. Immunol.* **160**:1219–1223.

60. Mazanec, M. B., Nedrud, J. G., Kaetzel, C. S., and Lamm, M. E., 1993, A three-tiered view of the role of IgA in mucosal defense, *Immunol. Today* **14**:430–435.

61. Michetti, P., Mahan, M. J., Slauch, J. M., Mekalanos, J. J., and Neutra, M. R., 1992, Monoclonal secretory immunoglobulin A protects mice against oral challenge with the invasive pathogen *Salmonella typhimurium*, *Infect. Immunity* **60**:1786–1792.

62. Renegar, K. B., and Small, P. A. J., 1991, Passive transfer of local immunity to influenza virus infection by IgA antibody, *J. Immunol.* **146**:1972–1978.

63. Lee, C. K., Weltzin, R., Soman, G., Georgakopoulos, K. M., Houle, D. M., and Monath, T. P., 1994, Oral administration of polymeric immunoglobulin A prevents colonization with *Vibrio cholerae* in neonatal mice, *Infect. Immunity* **62**:887–891.

64. Winner, L., Mack, J., Weltzin, R., Mekalanos, J. J., Kraehenbuhl, J.-P., and Neutra, M. R., 1991, New model for analysis of mucosal immunity: Intestinal secretion of specific monoclonal immunoglobulin A from hybridoma tumors protects against *Vibrio cholerae* infection, *Infect. Immunity* **59**:977–982.

65. Czinn, S. J., Cai, A., and Nedrud, J. G., 1993, Protection of germ-free mice from infection by *Helicobacter felis* after active oral or passive IgA immunization, *Vaccine* **11**:637–642.

66. Hocini, H., Belec, L., Iscaki, S., Garin, B., Pillot, J., Becquart, P., and Bomsel, M., 1997,

High-level ability of secretory IgA to block HIV type 1 transcytosis—Contrasting secretory IgA and IgG responses to glycoprotein 160, *AIDS Res. Hum. Retrovir.* **13**:1179–1185.

67. Burns, J. W., Siadat-Pajouh, M., Krishnaney, A. A., and Greenberg, H. B., 1996, Protective effect of rotavirus VP6-specific IgA monoclonal antibodies that lack neutralizing activity, *Science* **272**:104–107.

68. Tan, X., and Phillips, D. M., 1996, Cell-mediated infection of cervix derived epithelial cells with primary isolates of human immunodeficiency virus, *Arch. Viol.* **141**:1177–1189.

69. Black, K. P., Fultz, P. N., Girard, M., and Jackson, S., 1997, IgA immunity in HIV type 1-infected chimpanzees 2. Mucosal immunity, *AIDS Res. Hum. Retrovir.* **13**:1273–1282.

70. Burnett, P. R., VanCott, T. C., Polonis, V. R., Redfield, R. R., and Birx, D. L., 1994, Serum IgA-mediated neutralization of HIV type 1, *J. Immunol.* **152**:4642–4648.

71. Kozlowski, P. A., Chen, D., Eldridge, J. H., and Jackson, S., 1994, Contrasting IgA and IgG neutralization capacities and responses to HIV type 1 gp120 V3 loop in HIV-infected individuals, *AIDS Res. Hum. Retrovir.* **10**:813–822.

72. Bukawa, H., Sekigawa, K. I., Hamajima, K. I., Fukushima, J., Yamada, Y., Kiyono, H., and Okuda, K., 1995, Neutralization of HIV-1 by secretory IgA induced by oral immunization with a new macromolecular multicomponent peptide vaccine candidate, *Nature Med.* **1**: 681–685.

73. Janoff, E. N., Wahl, S. M., Thomas, K., and Smith, P. D., 1995, Modulation of human immunodeficiency virus type 1 infection of human monocytes by IgA, *J. Infect. Dis.* **172**: 855–858.

74. Kozlowski, P. A., Black, K. P., Shen, L., and Jackson, S., 1995, High prevalence of serum IgA HIV-1 infection-enhancing antibodies in HIV-infected persons. Masking by IgG, *J. Immunol.* **154**:6163–6173.

75. London, S. D., 1994, Cytotoxic lymphocytes in mucosal effector sites, in: *Handbook of Mucosal Immunology* (P. L. Ogra, J. Mestecky, M. E. Lamm, W. Strober, J. R. McGhee, and J. Bienenstock, eds.), Academic Press, San Diego, California, pp. 325–332.

76. Lohman, B. L., Miller, C. J., and McChesney, M. B., 1995, Antiviral cytotoxic T lymphocytes in vaginal mucosa of simian immunodeficiency virus-infected rhesus macaques, *J. Immunol.* **155**:5855–5860.

77. Musey, L., Hu, Y., Eckert, L., Christensen, M., Karchmer, T., and McElrath, M. J., 1997, HIV-1 induces cytolytic T lymphocytes in the cervix of infected women, *J. Exp. Med.* **185**: 293–303.

78. McDermott, M. R., Goldsmith, C. H., Rosenthal, K. L., and Brais, L. J., 1989, T lymphocytes in genital lymph nodes protect mice from intravaginal infection with herpes simplex virus type 2, *J. Infect. Dis.* **159**:460–466.

79. Kozlowski, P. A., Cu-Uvin, S., Neutra, M. R., and Flanigan, T. P., 1997, Comparison of the oral, rectal, and vaginal immunization routes for induction of antibodies in rectal and genital tract secretions of women, *Infect. Immunity* **65**:1387–1394.

80. Lehner, T., Panagiotidi, C., Bergmeier, L. A., Ping, T., Brookes, R., and Adams, S. E., 1992, A comparison of the immune responses following oral, vaginal, or rectal route of immunization with SIV antigens in nonhuman primates, *Vaccine Res.* **1**:319–330.

81. Lehner, T., Brookes, R., Panagiotidi, C., Tao, L., Klavinskis, L. S., Walker, J., Walker, P., Ward, R., Hussain, L., Gearing, J. H., Adams, S. E., and Bergmeier, L. A., 1993, T- and B-cell functions and epitope expression in nonhuman primates immunized with simian immunodeficiency virus antigen by the rectal route, *Proc. Nat. Acad. Sci. USA* **90**:8638–8642.

82. Russell, M. W., Moldoveanu, Z., White, P. L., Sibert, G. J., Mestecky, J., and Michalek, S. M., 1996, Salivary, nasal, genital, and systemic antibody responses in monkeys immunized intranasally with a bacterial protein antigen and the cholera toxin B subunit, *Infect. Immunity* **64**:1272–1283.

83. Bergquist, C., Johansson, E.-L., Lagergard, T., Holmgren, J., and Rudin, A., 1997, Intranasal vaccination of humans with recombinant cholera toxin B subunit induces systemic and local antibody responses in the upper respiratory tract and the vagina, *Infect. Immunity* **65**:2676–2684.

84. Kubota, M., Miller, C. J., Imaoka, K., Kawabata, S., Fujihashi, K., McGhee, J. R., and Kiyono, H., 1997, Oral immunization with simian immunodeficiency virus P55(Gag) and cholera toxin elicits both mucosal IgA and systemic IgG immune responses in nonhuman primates, *J. Immunol.* **158**:5321–5329.

85. Lowell, G. H., Kaminski, R. W., VanCott, T. C., Slike, B., Kersey, K., Zawoznik, E., Loomisprice, L., Smith, G., Redfield, R. R., Amselem, S., and Birx, D. L., 1997, Proteosomes, emulsomes, and cholera toxin B improve nasal immunogenicity of human immunodeficiency virus Gp160 in mice—Induction of serum, intestinal, vaginal, and lung IgA and IgG, *J. Infect. Dis.* **175**:292–301.

86. VanCott, T. C., Mascola, J. R., Kaminski, W., Kalyanaraman, V., Hallberg, P. L., Burnett, P. R., Ulrich, J. T., Rechtman, D. J., and Birx, D. L., 1997, Antibodies with specificity to native gp120 and neutralization activity against primary human immunodeficiency virus type 1 isolates elicited by immunization with oligomeric gp160, *J. Virol.* **71**:4319–4330.

87. Nardelli, B., Haser, P. B., and Tam, J. P., 1994, Oral administration of an antigenic synthetic lipopeptide (MAP-P3C) evokes salivary antibodies and systemic humoral and cellular responses, *Vaccine* **12**:1335–1339.

88. Staats, H. F., Montgomery, S. P., and Palker, T. J., 1997, Intranasal immunization is superior to vaginal, gastric, or rectal immunization for the induction of systemic and mucosal anti-HIV antibody responses, *AIDS Res. Hum. Retrovir.* **13**:945–952.

89. Staats, H. F., 1997, Unpublished observations.

90. Lubeck, M. D., Natuk, R. J., Chengalvala, M., Chanda, P. K., Murthy, K. K., Murthy, S., Mizutani, S., Lee, S. G., Wade, M. S., Bhat, B. M., Dheer, S. K., Eichberg, J. W., Davis, A. R., and Hung, P. P., 1994, Immunogenicity of recombinant adenovirus–human immunodeficiency virus vaccines in chimpanzees following intranasal administration, *AIDS Res. Hum. Retrovir.* **10**:1443–1449.

91. Buge, S. L., Richardson, E., Alipanah, S., Markham, P., Cheng, S., Kalyan, N., Miller, C. J., Lubeck, M., Udem, S., Eldridge, J., and Robert-Guroff, M., 1997, An adenovirus-simian immunodeficiency virus env vaccine elicits humoral, cellular, and mucosal immune responses in rhesus macaques and decreases viral burden following vaginal challenge, *J. Virol.* **71**:8531–8541.

92. Caley, I. J., Betts, M. R., Irlbeck, D. M., Davis, N. L., Swanstrom, R., Frelinger, J. A., and Johnston, R. E., 1997, Humoral, mucosal, and cellular immunity in response to a human immunodeficiency virus type 1 immunogen expressed by a Venezuelan equine encephalitis virus vaccine vector, *J. Virol.* **71**:3031–3038.

93. Charles, P. C., Brown, K. W., Davis, N. L., Hart, M. K., and Johnston, R. E., 1997, Mucosal immunity induced by parenteral immunization with a live attenuated Venezuelan equine encephalitis virus vaccine candidate, *Virology* **228**:153–160.

94. Doggett, T. A., and Brown, P. K., 1996, Attenuated *Salmonella* as vectors for oral immunization, in: *Mucosal Vaccines* (H. Kiyono, P. L. Ogra, and J. R. McGhee, eds.), Academic Press, San Diego, California, pp. 105–118.

95. Langermann, S., 1996, Recombinant BCG as vector for mucosal immunity, in: *Mucosal Vaccines* (H. Kiyono, P. L. Ogra, and J. R. McGhee, eds.), Academic Press, San Diego, California, pp. 129–136.

96. Hoft, D. F., and Gheorghiu, M., 1996, Mucosal immunity induced by oral administration of Bacille Calmette-Guerin, in: *Mucosal Vaccines* (H. Kiyono, P. L. Ogra, and J. R. McGhee, eds.), Academic Press, San Diego, California, pp. 269–279.

97. Stover, C. K., de la Cruz, V. F., Fuerst, T. R., Burlein, J. E., Benson, L. A., Bennett, L. T., Bansal, G. P., Young, J. F., Lee, M. H., Hatfull, G. F., Snapper, S. B., Barletta, R. G., Jacobs, Jr., W. R., and Bloom, B. R., 1991, New use of BCG for recombinant vaccines, *Nature* **351**:456–460.

98. Lagranderie, M., Balazuc, A. M., Gicquel, B., and Gheorghiu, M., 1997, Oral immunization with recombinant *Mycobacterium bovis* BCG simian immunodeficiency virus Nef induces local and systemic cytotoxic T-lymphocyte responses in mice, *J. Virol.* **71**:2303–2309.

99. Wu, S. G., Pascual, D. W., Lewis, G. K., and Hone, D. M., 1997, Induction of mucosal and systemic responses against human immunodeficiency virus type 1 glycoprotein 120 in mice after oral immunization with a single dose of a *Salmonella*-HIV vector, *AIDS Res. Hum. Retrovir.* **13**:1187–1194.

100. Ulmer, J. B., Donnelly, J. J., Shiver, J. W., and Liu, M. A., 1996, Prospects for induction of mucosal immunity by DNA vaccines, in: *Mucosal Vaccines* (H. Kiyono, P. L. Ogra, and J. R. McGhee, eds.), Academic Press, San Diego, California, pp. 119–127.

101. Wang, B., Dang, K., Agadjanyan, M. G., Srikantan, V., Li, F., Ugen, K. E., Boyer, J., Merva, M., Williams, M. V., and Weiner, D. B., 1997, Mucosal immunization with a DNA vaccine induces immune responses against HIV-1 at a mucosal site, *Vaccine* **15**:821–825.

102. Okada, E., Sasaki, S., Ishii, N., Aoki, I., Yasuda, T., Nishioka, K., Fukushima, J., Miyazaki, J., Wahren, B., and Okuda, K., 1997, Intranasal immunization of a DNA vaccine with IL-12- and granulocyte-macrophage colony-stimulating factor (GM-CSF)-expressing plasmids in liposomes induces strong mucosal and cell-mediated immune responses against HIV-1 antigens, *J. Immunol.* **159**:3638–3647.

103. Lohman, B. L., McChesney, M. B., Miller, C. J., Otsyula, M., Berardi, C. J., and Marthas, M. L., 1994, Mucosal immunization with a live, virulence-attenuated simian immunodeficiency virus (SIV) vaccine elicits antiviral cytotoxic T lymphocytes and antibodies in rhesus macaques, *J. Med. Primatol.* **23**:95–101.

104. Mowat, A. M., 1994, Oral tolerance and regulation of immunity to dietary antigens, in: *Handbook of Mucosal Immunology* (P. L. Ogra, M. E. Lamm, J. R. McGhee, J. Mestecky, W. Strober, and J. Bienenstock, eds.), Academic Press, San Diego, California, pp. 185–201.

105. Elson, C. O., 1996, Cholera toxin as a mucosal adjuvant, in: *Mucosal Vaccines* (H. Kiyono, P. L. Ogra, and J. R. McGhee, eds.), Academic Press, New York, pp. 59–72.

106. Elson, C. O., and Dertzbaugh, M. T., 1994, Mucosal adjuvants, in: *Handbook of Mucosal Immunology* (P. L. Ogra, M. E. Lamm, J. R. McGhee, J. Mestecky, W. Strober, and J. Bienenstock, eds.), Academic Press, San Diego, California, pp. 391–402.

107. Levine, M. M., Kaper, J. B., Black, R. E., and Clements, M. L., 1983, New knowledge on pathogenesis of bacterial enteric infections as applied to vaccine development, *Microbiol. Rev.* **47**:510–550.

108. Dickinson, B. L., and Clements, J. D., 1995, Dissociation of *Escherichia coli* heat-labile enterotoxin adjuvanticity from ADP-ribosyltransferase activity, *Infect. Immunity* **63**:1617–1623.

109. Douce, G., Fontana, M., Pizza, M., Rappuoli, R., and Dougan, G., 1997, Intranasal immunogenicity and adjuvanticity of site-directed mutant derivatives of cholera toxin, *Infect. Immunity* **65**:2821–2828.

110. Yamamoto, S., Kiyono, H., Yamamoto, M., Imaoka, K., Yamamoto, M., Fujihashi, K., Van Ginkel, F. W., Noda, M., Takeda, Y., and McGhee, J. R., 1997, A nontoxic mutant of cholera toxin elicits Th2-type responses for enhanced mucosal immunity, *Proc. Nat. Acad. Sci. USA* **94**:5267–5272.

111. Yamamoto, S., Takeda, Y.., Yamamoto, M., Kurazono, H., Imaoka, K., Yamamoto, M., Fujihashi, K., Noda, M., Kiyono, H., and McGhee, J. R., 1997, Mutants in the ADP-

ribosyltransferase cleft of cholera toxin lack diarrheagenicity but retain adjuvanticity, *J. Exp. Med.* **185**:1203–1210.

112. Roberts, M., Bacon, A., Rappuoli, R., Pizza, M., Cropley, I., Douce, G., Dougan, G., Marinaro, M., McGhee, J., and Chatfield, S., 1995, A mutant pertussis toxin molecule that lacks ADP-ribosyltransferase activity, PT-9K/129G, is an effective mucosal adjuvant for intranasally delivered proteins, *Infect. Immunity* **63**:2100–2108.

113. O'Hagan, D. T., 1997, Recent advances in vaccine adjuvants for systemic and mucosal adjuvants, *J. Pharmacy Pharmacol.* **49**:1–10.

9

Nucleic Acid Vaccination against HIV-1

AMI R. SHAH, DAVID B. WEINER, and JEAN D. BOYER

1. INTRODUCTION

The human immunodeficiency virus-1 (HIV-1) is an RNA retrovirus belonging to the lentivirus family that selectively infects and kills CD4$^+$ T cells and macrophages resulting in immune system failure. HIV-1 is transmitted from individual by blood and sexual contact. The spread of HIV-1 infection continues to be a major public health concern worldwide. The United Nations Joint Programme on AIDS reports that an estimated 8500 people become infected with HIV daily and that 90% of these newly infected individuals reside in developing countries.[1] Recent breakthroughs in chemotherapeutic regimens using three or more different antiretroviral agents have generated optimism regarding the control of HIV-1 infection *in vivo*.[2] However, global access to these pharmaceutical agents is extremely limited due to cost and the strict nature of the administration regimens. These restrictions indicate the need for alternatives. Therefore, the most effective solution to the worldwide spread of HIV-1 infection will most likely rely upon a protective vaccination strategy.

Indeed, the World Health Organization estimates that vaccination against tetanus, diphtheria, whooping cough, measles, polio, and tuberculosis prevents about 3 million deaths per year, making it the most effective public health measure in reducing human morbidity and mortality. Tradi-

AMI R. SHAH • Rollins School of Public Health, Emory University, Atlanta, Georgia, 30322. DAVID B. WEINER and JEAN D. BOYER • Stellar-Chance Laboratories, Department of Pathology and Laboratory Medicine, University of Pennsylvania, Philadelphia, Pennsylvania 19104-6100.

Human Retroviral Infections, edited by Kenneth E. Ugen *et al.*
Kluwer Academic / Plenum Publishers, New York, 2000.

tional vaccines fall into two categories: attenuated, nonpathogenic, live infectious material; and killed, inactivated, or subunit preparations. Live, attenuated vaccines lead to the production of the most potent and enduring cellular and humoral immune responses. However, a live, attenuated HIV-1 vaccine presents several safety concerns including the risk of reversion to a more pathogenic form of virus during replication in the host. Also, in immunodeficient individuals, even weak attenuated viruses may cause disease. Whole, inactivated vaccines can induce protection, but only generate T helper and humoral responses via major histocompatibility class (MHC) class II-restricted presentation of the antigens. Since there is no intracellular production of viral antigens within the host, no significant cytotoxic T cell responses are induced, and, as shown later in this chapter, the cellular response appears to be important with respect to HIV-1 vaccine development. Further, this strategy may require repeated boosting to achieve life-long immunity. The effectiveness of these two HIV-1 vaccine approaches for achieving protection in animal models is unclear and alternative approaches are clearly needed for vaccine development. Nucleic acid vaccination is one of the new vaccination technologies that induce both cellular and humoral responses without the safety concerns of live, attenuated vaccines and therefore it warrants significant attention in the quest for a prophylactic vaccine strategy against HIV-1 (Table I and Fig. 1).

2. OBSTACLES TO HIV-1 VACCINE DEVELOPMENT

2.1. Introduction

The development of a vaccine for HIV-1 has been hindered by several factors. First, HIV-1 is subject to a significant amount of sequence divergence

TABLE I
Vaccination Strategies

Vaccine type	Immune response		
	Helper T cells	Cytotoxic T cells	Antibody
Attenuated (live)	+	+	+
Inactivated (whole)	+	−	+
Subunit/peptide	+	−/+	+
Recombinant	+	−/+	+
Plasmid DNA	+	+	+
Canarypox	+	+	+

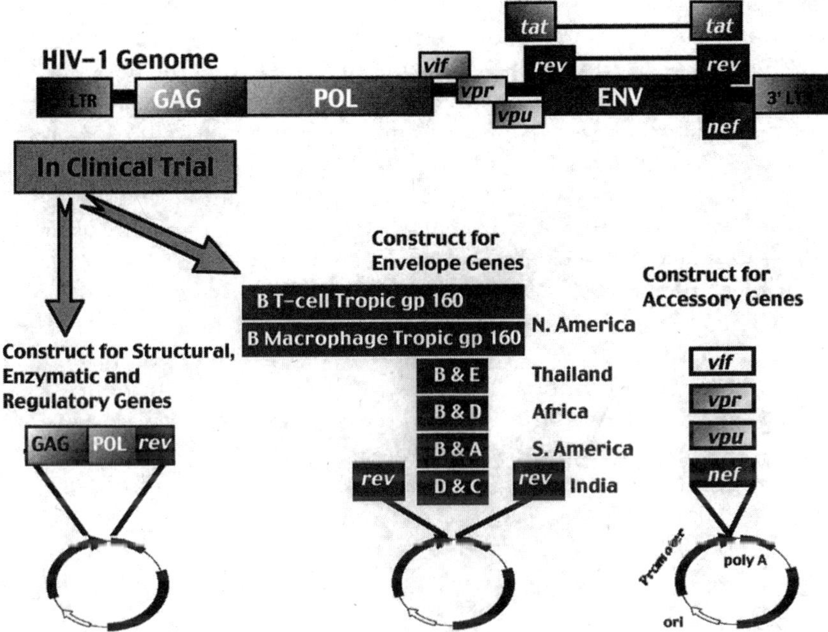

FIGURE 2. Potential immunologic targets for DNA vaccination against HIV-1. The HIV-1 genome is organized into three major structural and enzymatic genes, two regulatory genes, and four accessory genes. The immunologic targets include *env, gag, pol,* and the four accessory genes. Furthermore, the high variability between the envelopes of various HIV-1 clades suggests that different envelope subtypes might be needed for an effective vaccine against HIV-1. The probability of the host to respond to any vaccine increases as the number of immunologic targets is increased.

attenuated within a potential vaccine. The *rev* protein, through its interaction with the Rev-response element (RRE), increases the number of unspliced RNAs from the nucleus by its displacement of host splicing factors which would prevent RNA transport from the nucleus to the cytoplasm. Rev also plays a major role in the production of structural proteins of HIV-1; therefore immune responses directed against the *rev* gene itself could serve to effectively interfere with the viral life cycle.[26] Additionally, the fact that Rev is produced early suggests that immune responses targeting here could be especially important as well as to those targeted against the *nef* protein (see below).

The accessory genes (*vif, vpr, vpu,* and *nef*) serve as potential targets for DNA vaccination as well. These four genes are termed "accessory" because upon their deletion from the HIV-1 genome, *in vitro* viral replication remains intact. The *vif* (viral infectivity factor) protein, which is located in the plasma membrane, is important for the production of infectious virions in certain cells.[28] Experiments using *vif* mutants elucidated the necessity of *vif* expression for infectivity of peripheral blood mononuclear cells and macrophages.[27–29] The *vpu* (viral protein u) protein plays a role in the downregulation of CD4[+] cells and the increase of viral transcription.[30] The *vpr* (viral protein r) protein plays several roles including increasing viral replication, arresting host cell division, and reactivating virus from a latent phase.[31–33] While the *nef* (negative factor) protein has not been shown to play a critical role in viral infection of cell lines *in vitro*,[34] experiments conducted with the related simian immunodeficiency virus have shown a decrease in viral replication in rhesus macaques infected with virus with a deletion in the *nef* gene.[35] More importantly, follow-up examination found evidence of decreased pathogenesis of SIV infection. This observation was in sharp contrast to animals infected with either wild-type *nef* or virus containing a stop codon with the *nef* gene. These latter animals produced significantly higher levels of virus, and 5 of 12 subsequently developed immunodeficiency and died. Finally, *nef* protein appears to provide an important target for cytotoxic T-lymphocyte activity. In summary, it must be reiterated that the accessory genes include properties that are not necessarily desirable in a putative vaccine. For example, Nef and Vpr both have been reported to have negative effects on the immune response *in vitro*.[36] Therefore engineering these vaccine targets to decrease function while maintaining immunogenicity is an important issue.

Most vaccines against HIV-1 have encoded the *env* gene in particular and to a lesser extent the *gag/pol* genes. The *env* gene codes for a large glycoprotein, gp160, which is cleaved into two smaller molecules, gp120 and gp41, by a host protease, and was the first major vaccine target. This protein is the target of neutralizing antibody responses as well as a target for cellular immunity. The two genes *gag* and *pol* also serve as targets for vaccination. The *gag* gene codes for the viral core proteins, while *pol* codes for the enzymatic proteins RT, Int, and protease (Pro). These antigens similarly are targets for cellular immunity and have now entered into the arena of clinical trials.

An additional important fact to consider in the design of a successful vaccine against HIV-1 is the heterogeneity of the envelopes of the different subtypes. Therefore, multicomponent vaccines consisting of DNA encoding for several envelopes may be better suited to mount more effective immune responses against several subtypes of HIV-1.

An important vaccine technology which combines the ability to molecularly engineer viral genes to eliminate pathogenic features while maximizing the number of targets and the features associated with a live vaccine (induction of cellular and humoral immune responses) is nucleic acid immunization.

4. DNA-BASED VACCINES

Nucleic acid immunization, also known as genetic or polynucleotide immunization, is based on the injection of nucleic acid sequences specific for a pathogen directly into host target tissue.[37-40] The injection of cDNA expression cassettes results in *in vivo* expression of the encoded proteins with the development of specific humoral and cellular immune response against the encoded antigens (Fig. 1).

The mechanism of immune induction by DNA vaccination has been examined in a number of studies.[41] Among the issues studied is the induction of B cell- and CD4-mediated responses. Synthesized proteins are released extracellularly and these proteins prime the induction of the humoral response in addition to a CD4 cellular response. The CD4 or the T helper response involves the MIIC class II restricted pathway and professional antigen-presenting cells (APCs) which take up the foreign antigen. With regard to the induction of a CTL MHC class I restricted response, antigen is processed intracellularly via the proteosome pathway and presented to the CD8 lymphocytes. To prime efficiently for a CD8 cytotoxic response costimulation molecules (B7-1 and B7-2, found predominantly on professional APCs) are required. While DNA has been delivered intramuscularly, intradermally, as well as mucosally, it is the dermis and mucosa that have a preponderance of professional APCs. Intramuscular injection has, however, provided a challenge to immunologists studying the mechanism of lymphocyte priming. The myocytes predominantly express the foreign antigen following DNA injection. Although myocytes do express MHC class I molecules, they do not express the costimulatory molecules. Recent observations by Chattergoon *et al.*[42] demonstrated that following intramuscular injection, macrophages and dendritic cells are in fact transfected *in vivo* with the plasmids. The APCs appeared to home to the lateral lymph nodes, where efficient T cell priming may occur. Earlier work suggested that antigen itself is transferred to bone marrow-derived APCs and the protein crosses over into the MHC class I pathway.[43] Nevertheless, a cellular CD8 MHC class I restricted response is induced following DNA vaccination. Thus, DNA inoculation mimics the effects of live, attenuated vaccines in that the host cell is directed to produce the antigenic protein without the risk for reversion to the virulent form of the virus from the injected DNA.

Nucleic acid immunization offers pharmaceutical as well as immuno-

TABLE II
Advantages of Nucleic Acid Immunization

1. Plasmids can be rapidly and inexpensively constructed
2. Large-scale manufacturing procedures are available
3. Better stability and storage capability compared to other vaccine preparations
4. Small quantities of vector are sufficient in producing effective immune responses
5. Protection can be achieved in large-primate models
6. Several antigens can be administered in a single vaccine construct
7. DNA vaccines have been shown to produce cellular and humoral immune responses

logical advantages (Table II). DNA can be produced on a large scale with great purity without the risk of contamination with potentially dangerous agents. In addition, DNA is extremely stable relative to proteins and other biological polymers. This approach allows specific genes to be expressed in nonreplicating vectors with the ability to manipulate the sequences to accommodate different subtypes and/or parts of the genome of the organism of interest. Genes which lead to undesired immunologic inhibition or cross-reactivity (autoimmunity) may either be altered or deleted. In this way, genes which encode important immunologic epitopes can be included, while those that confer pathogenicity or virulence can be excluded. From the immunologic perspective, the ability to maintain DNA in an episomal form provides the potential for significant antigen expression. This has implications for the duration of immunologic memory responses, an important element of protective vaccine strategies. A number of laboratories have been vigorously pursuing the development of a prophylactic vaccine strategy for HIV-1 by the use of expression vectors that code for proteins of HIV-1.[44-65]

5. PROGRESS OF NUCLEIC ACID IMMUNIZATION AGAINST HIV-1

5.1. Murine Studies

5.1.1. Immune Responses

Wang and colleagues demonstrated the first immune responses *in vivo* to a DNA-based vaccine against HIV-1. Here we briefly describe these ground-breaking studies. A more thorough description of Wang's work is provided by Bagarazzi.[44] In an initial study, a group of 10 mice received a series of four intramuscular injections of a construct encoding HIV-1 *env* gp160-*rev* at 2-week intervals. Sera obtained from the 10 mice reacted with recombinant gp160 protein as determined by enzyme-linked immunosorbent assay (ELISA). In addition, neutralizing antibodies were induced against both homologous and heterologous isolates. Seroconversion was also noted in 100% of mice after a single immunization with the HIV-1 *gag/pol* construct. In a related study, mice were injected intramuscularly with a plasmid encoding gp160 of HIV-1$_{Z6}$ or vector alone followed by excision of the injected muscle 72 h later. Muscle sections were treated with a pool of murine anti-HIV gp160 and gp41 monoclonal antibodies and specific reactivity was determined by indirect immunohistochemistry with a horseradish peroxidase (HRP)-conjugated detection system. Reactivity to both gp160 and gp41 was observed along needle tracks through the muscle, demonstrating that gp160-specific antigen can be expressed in myocytes transfected with HIV-1 gene constructs. Finally, immunoglobulin isotyping studies of the gp160-specific antibodies elicited by envelope DNA vaccination revealed a predominance of IgG isotypes, with 19% IgG1, 51% IgG2, 16% IgG3, 10% IgM, and 5% IgA responses measured. These results are consistent with a secondary immune response suggesting helper T cell stimulation (Th1 particularly) and antiviral activity through nucleic acid-based immunization. It should also be noted the envelope construct induced seroconversion against gp160 in increasing numbers of mice with each boost, with 95% of the animals responding by the fourth injection. The *env* construct also contained sequences coding for Tat and Rev proteins and specific antibody formation to these proteins was also demonstrated by ELISA.

Studies in other laboratories verified Wang's initial findings. DNA-based constructs encoding *env* and core proteins were coadministered and shown to produce antibody responses to more than one antigen (transient anti-*env* and persistent p24).[45] Okuda and colleagues reported a similar boosting effect in the rabbit model. Rabbits were injected with a plasmid combination[46] and responses were seen to multiple antigens. They also found that the second *rev* construct could augment anti-*env* responses above what was found when the *env* construct was injected alone. This may represent intracellular cooperation between exogenously delivered gene products. These results cumulatively and independently illustrate the ability of DNA-based vaccines to elicit specific immune responses to multiple viral antigens using single or multiple constructs. Indeed, these early studies demonstrating that DNA immunization could induce antibody responses against HIV-1 antigens built a solid foundation for the future of genetic immunization against HIV-1.

It was further observed in a number of studies that DNA vaccines can induce cellular responses.[44] Splenocytes from HIV-envelope-inoculated mice were compared to vector-inoculated controls for their ability to proliferate in response to specific stimulation with recombinant gp120. Subsequent studies have broadened these results with other constructs in response to gp120, gp160, gp41 external domain, and several peptides of gp120 and gp41 and found that all antigens were specifically responded to in this assay. The induction of CTLs in the murine model after vaccination with envelope DNA constructs has been reported. Constructs designed from the HIV-1$_{NL4-3}$ isolate were also shown to elicit CTL activity. Specific CTLs were also reported after a single epidermal immunization with a HIV-1 gp120 construct delivered via the Accell® gene delivery system.[47] Lastly, cytokine responses have also been measured. The higher levels of interferon-γ (IFN-γ) compared to interleukin-4 (IL-4) released from stimulated splenocytes revealed IFN-γ levels that resembled the CTL activity.

Murine studies of potential HIV-1 vaccine strategies are limited. Mice are not susceptible to infection with HIV-1 and the subsequent immunodeficiency, and therefore this model is unable to investigate the immunogenicity. However, Wang *et al.* developed a lymphoma HIV-antigen-expressing cell-challenge model to test protection. Indeed, the model assesses the ability of nucleic acid-based immunization to protect against relevant HIV-1 cellular challenge.[48] A clonal population of lymphoma cells (SP2/0) was transfected with HIV-1 gp160 envelope and was shown to express the transfected antigen in association with the H-2^d class I MHC antigens. All naive and vector-immunized control mice subsequently challenged with gp160 expressing SP2/0 cells developed tumors within 13 days and died within 9 weeks. In contrast, most animals previously immunized with the HIV-1$_{Z6}$ envelope construct survived the same lethal challenge. Since cellular immunity is a critical component of tumor rejection, one can project that successful protection is likely a direct results of *in vivo*-induced cell-mediated responses directed against the gp160 antigen on the SP2/0 cells.

5.1.2. Studies of Mucosal Immunity

DNA immunization at mucosal sites is another area of active research. Mucosal immunity is of utmost importance because a predominant mode of HIV-1 transmission is through the mucosa. Wang *et al.* demonstrated that intravaginal delivery of an HIV-1-based DNA vaccine can lead to the elicitation of local vaginal immunoglobulins.[49] Klavinskis *et al.*, in an effort to improve upon mucosal delivery of their HIV-1-based vaccines, further demonstrated that incorporating DNA encoding the luciferase gene into liposomes with cationic lipids enhanced luciferase expression in nasal tissue.[50] In addition, this was associated with the induction of a humoral immune response in serum and vaginal fluids and also a proliferative and cytotoxic T lymphocyte response in the spleen and iliac lymph nodes draining the genital–rectal mucosa. In a similar study Hamajima *et al.* determined that a polymer, carboxymethylcellulose sodium salt, had adjuvant effects when coadministered with an HIV-1 *env* DNA-based vaccine, either intranasally or intramuscularly.[51] Finally, Winchell *et al.* delivered HIV-1 DNA plasmids directly to the Peyer's patches of rabbits.[52] This delivery method, although seemingly cumbersome, resulted in long-lived mucosal and serum antibody levels. Protection from mucosal viral challenge remains to be seen, but it seems likely that these findings will serve to strengthen the immune response generated by systemic immunization.

5.1.3. Modulation of the Immune Response

Cytokines have been postulated to play a direct role as immune modulators. They are involved in the inflammatory and immune responses as the initiators and regulators of the immune system, and enhancement of the overall immune response continues to be the subject of experiments. Since the report from Raz and colleagues in 1993 that the injection of cytokine genes into muscle resulted in the biological action of these cytokines *in vivo*,[53] a number of studies have used coinjection with cytokine genes to enhance the immune response. Xiang and Ertl[54] examined the effects of granulocyte-macrophage colony-stimulating factor (GM-CSF), IL-4, and IFN-γ on the immunization of mice with vectors under the control of the SV40 promoter. They found that GM-CSF enhanced the initial T helper and B cell responses, but over a period of time antibody titers fell below levels detected in the mice not receiving any lymphokines. On the other hand, IL-4 induced an initial drop in both T helper and B cell responses, followed by antibody titers steadily rising to levels that surpassed the control mice after 5 months.[54] Kim *et al.*[55] also studied coimmunization with cytokine genes and found that coinjection of plasmids encoding cytokine genes with HIV-1 DNA vaccines can drive the immune response either toward a TH1- or TH2-type response. In this study, coinjection with IL-12 enhanced cytotoxic T lymphocyte activity, while coinjection with GM-CSF enhanced proliferation.[55] Similarly, Prayaga *et al.* demonstrated that coinjection with TH1-type cytokine, IL-2, IL-7, and IL-12 blocked TH2-like immune responses.[56] In a related study, coinjection with antiinterferon-γ antibody showed that the cell-mediated immune response was severely inhibited.

CD4 cells are a major source of cytokines, although CD8 cells can also produce various cytokines. However, it now appears that the production of chemokines by CD8 cells (as well as other cell types) also plays a significant role in the recruitment of APCs and lymphocytes to the site of an immune response. Chemokines are divided into three subgroups, alpha, beta, and gamma chemokines. The chemokines, like the cytokines, have overlapping functions. Kim *et al.* found, by coinjecting chemokine-encoding plasmids with HIV-1 expression vectors, that they could enhance the immune response to the HIV-1 antigens.[57] They in fact demonstrated MIP1-α and IL-8 enhanced antibodies, while RANTES and MCP-1 enhanced CTL responses.

In addition to these cytokines and chemokines, a number of cell surface markers affect the induction of immune responses. For example, CD28 is a marker of naive T cells which acts as a receptor for the costimulatory markers, the B7-1 (CD80) and B7-2 (CD86) molecules, which are expressed on the surface of antigen-presenting cells. Indeed, the B7-2 costimulatory molecules have been studied the most thoroughly with regard to DNA-based vaccines. An enhancement of the cellular immune responses, particularly CTLs, was attained from the coinjection of plasmids encoding the B7-2 costimulatory molecules. This work was later supported by the findings of Horspool *et al.*,[60] who also demonstrated the distinct roles of B7-1 and B7-2 and T cell stimulation through the CD28 molecule resulting again in an enhancement of the cellular immune response. Many costimulatory molecules are currently under investigation by incorporating into expression plasmids.

With regard to the enhancement and improvement of the immune response following DNA vaccination, there are improvements with regard to the expression vector itself that can be made, i.e., increased expression. Tsuji *et al.* demonstrated that coinjection of the TCA3 expression plasmid with the pCMV160/REV plasmid significantly enhanced cell-mediated immunity as assessed by CTL and delayed type hypersensitivity (DTH) assays.[61] Similar to the studies on mucosal immunization, use of reagents such as bupivicaine can enhance *in vivo* transfection of cells. This was demonstrated in work by Toda *et al.*,[62] where DNA inoculated with mannan-coated liposomes also demonstrated enhanced cell-mediated immunity.

The strategy of coadministration of cytokine/chemokine genes and costimulatory molecules could further refine current DNA vaccination technology by allowing for control of the magnitude and type of immune response elicited. For example, in the case when a T cell-mediated response is important, IL-12 or CD86 genes could be chosen to modulate the immune response by coimmunization with a DNA antigen. Alternatively, coadministration with TH2-type cytokines could be used in the design of a vaccine strategy in which humoral responses are needed to resist infection. For cases in which the correlates of protection are unknown, as in the case of HIV-1, a combination of strategies that induce both TH1- and TH2-type responses can be utilized. Being able to control the immune response in this directed manner has very positive implications for the next generation of genetic vaccine design.

5.1.4. CpG Motifs and Immunostimulatory Sequences

One aspect of genetic immunization that has recently received attention is the immunostimulatory activity of DNA itself. It has been observed that DNA from bacteria, but not from vertebrates, can induce a nonspecific immune response, possibly due to differences in the frequency of unmethylated CpG dinucleotides found in the two genomes. Krieg showed that oligonucleotides containing one or more CpG dinucleotides ("CpG motif" or "immunostimulatory sequence") could trigger B cell proliferation and immunoglobulin secretion.[66] The relevance of these findings for genetic immunization became clear when Sato *et al.*[67] reported that a DNA vaccine whose plasmid backbone contained CpG immunostimulatory sequences (ISSs) induced a more vigorous antibody and CTL response than an otherwise identical vaccine which did not contain the ISS, despite a higher level of gene expression produced by the latter plasmid. The response to the vaccine lacking the ISS could be restored to normal levels by co-injecting it with noncoding plasmid containing ISS. This study demonstrated that the plasmid vector itself could have a significant adjuvant effect on DNA vaccine-induced immunity, and suggests that this effect may be mediated by CpG-containing ISS.

Subsequent studies have confirmed that CpG motifs can enhance immunity to genetic immunization,[67–70] and have shown that they can qualitatively modify the immune response by preferentially inducing a T_h1 response. This phenomenon may be one reason why most DNA vaccines studied to date induce a predominantly T_h1 response when injected intramuscularly or intradermally.

It should be noted, however, that all of the data about CpG motifs and DNA vaccines reported thus far have been generated in rodents; it will be most interesting to see if a similar immunostimulatory effect is seen in primates.

5.1.5. Codon Optimization

The 5′ toward the 3′ sequence of a coding strand of DNA encodes amino acids as "codons," or groups of 3 nucleotides read together. Each of the 4 possible nucleotides can occupy each of the three positions of the codon, so there are 4^3 (64) possible trinucleotide sequences. Sixty-one of these codons specify amino acids while 3 encode stop codons specifying termination of protein synthesis.

The DNA strand coding sequence is converted to mRNA for translation into proteins by DNA polymerase II. On the ribosome, the link that translates the codon sequence encoded by mRNA into a full polypeptide—the tRNA—recognizes one codon sequence and transfers the specified single amino acid individually to the growing polypeptide chain. There are varying degrees of redundancy built into this system. For example, the mRNA sequences UCU, UCC, UCA, and UCG represent four different codons each with their own specific tRNA—each of these specify the same amino acid. However, there can be significant wobble or nonselectivity at the third position. Due to specific modifications of the tRNA anticodon or differences in tRNA pools, a particular codon family can be read by tRNAs with different anticodons in different organisms. This provides an opportunity to modify codon usage in order to utilize a more common tRNA, thus enhancing translation efficiency. In the case of HIV antigens, such optimization must be considered in light of the rev regulatory gene. Rev binds to unspliced or singly spliced env or gag/pol mRNAs and escorts them from the nucleus to the endoplasmic reticulum (ER), facilitating appropriate translation. Rev binds to env through the RRE and to gag/pol through undefined cryptic RREs in an mRNA sequence-dependent fashion. Changing codon usage may destroy the mRNA-dependent RREs and facilitate transport of inappropriately spliced HIV mRNA, resulting in a less than perfect translated product. Thus, the approaches of using a rev-enhanced expression system and optimizing codon sequences might actually work against each other. Both systems have apparent benefits and should be considered.

5.2. Primate Studies

There are three different primate models for HIV vaccine studies. They include the SIV and chimeric SIV/HIV-1 (SHIV-1) challenge models in macaques and the HIV challenge model in chimpanzees. Although the pathogenesis of SIV is similar to HIV, vaccines developed against HIV-1 cannot be tested in the SIV challenge model. However, macaques are also susceptible to infection with the chimeric SHIV-1 virus in which the SIV envelope genes are replaced by the HIV-1 envelope genes, while keeping the SIV *gag/pol* genes intact. The SHIV model is sufficient for testing HIV-1-envelope-based vaccines. Attention should also be drawn to the fact that the SIV and SHIV-1 viruses do not clearly simulate HIV-1 infection in humans in terms of disease pathogenesis and immunology. The infection of chimpanzees with virulent isolates is another representative animal model of HIV-1 infection and pathogenesis. It is known that chimpanzees are almost genetically identical to humans and a host of human biochemical, molecular, and immunologic reagents as well as T cell repertoire reagents are cross-reactive in these species. In addition, the progression of disease in these animals is very similar to that in humans. For example, viral replication persists even in the face of neutralizing antibody and cellular immune responses in both chimpanzees and humans infected with HIV-1. Despite the associated issues, each of these models has been used to test candidate HIV-1 vaccines including HIV-1 DNA-based vaccines.

Immune responses to HIV-1 antigens have been generated in macaques using DNA immunization, and HIV-1 DNA vaccines have successfully protected macaques from a viral challenge using a SIV/HIV-1 chimera. In an early study, one of four cynomolgus macaques were protected from intravenous challenge with 50 $TCID_{50}$ of SHIV-1$_{HlxB2}$.[65] This study showed that DNA vaccination could lead to the control of viral replication *in vivo.* Protection was also achieved by a so-called prime–boost immunization regimen, where animals were preliminarily immunized with an HIV-1 *env* DNA vaccine and then boosted with recombinant gp160 protein.[72] Finally, a study by Lu *et al.* in which DNA vaccines encoding SIV *env* (four different plasmids) and SIV *gag* (one plasmid) were injected into rhesus macaques provided evidence to the contrary.[73] After receiving six injections, those monkeys were subsequently challenged with SIV$_{mac}$251. None of the vaccinated animals were protected from viral infection despite the fact that the animals had demonstrated cellular and humoral immune responses.

A study in the chimpanzee model also reported on the ability of DNA to protect against challenge with virus. Using the HIV model, three chimpanzees were inoculated with an HIV-1$_{MN}$ envelope construct. These animals developed both cellular and humoral immune responses to specific envelope antigens.[74] Antibody responses to the cyclic V3-loop peptide of HIV-1$_{MN}$ were measured for each animal by the third boost in comparison to a vector-immunized animal that never developed measurable antibody titers. Vaccination also induced each animal to develop antibodies to gp120. These animals were also immunized with a *gag/pol* construct and one of the animals developed measurable antibodies to Pr55 core protein. Following boosting, serum from all three immunized animals was ultimately able to neutralize HIV-1$_{MN}$ *in vitro* to varying levels; however, only one animal demonstrated a significant titer of neutralizing antibodies. CTL activity was measured in two of the three immunized animals. One chimpanzee developed activity to *gag/pol* and *env* targets and the other showed activity against *gag/pol* predominantly. Two of the experimental animals and one control animal were subsequently challenged intravenously with a high dose (250 chimpanzee ID$_{50}$) of a heterologous stock of HIV-1$_{SF-2}$.

Both vaccinated chimpanzees challenged in this manner were protected from infection with HIV-1$_{SF-2}$.[74] The animals chosen were the animal with the highest antibody response and the animal with the highest CTL response as well as the control animal. The control vector-immunized animal became infected by the identical challenge. This result represents the first example of protection from viral challenge by a DNA vaccine in the chimpanzee model. Interestingly, of the two animals that were protected, one displayed substantial neutralization activity, while the other had significant CTL responses. This demonstration of prophylactic vaccine activity in chimpanzees is consistent with other reports of protection in that the correlates of protection differed from animal to animal.

DNA immunization has been evaluated further for its safety and its ability to control viral replication in infected chimpanzees. A previously HIV-1-infected adult chimpanzee was vaccinated with the same envelope construct given to the naive chimpanzees mentioned above. Antibodies to peptides within both gp120 and gp41 increased after vaccination. Measurements of viral load by both reverse transcriptase-polymerase chain reaction and peripheral blood mononuclear cell coculture methods exhibited a decrease to undetectable levels 20 weeks after immunization.[75] In a follow-up study, a second chimpanzee was vaccinated as above and similarly exhibited a sustained decrease in viral burden in its peripheral compartment after a delay following immunization. The results of the two studies coupled with the lack of any significant demonstrable toxicity in the vaccinated chimpanzees would suggest that DNA vaccines should be explored in humans.

5.3. Human Studies

Studies directed by Sandra Calarota at the Swedish Institute for Infectious Disease Control tested the effects of DNA constructs encoding the *nef*, *rev*, and *tat* regulatory genes of HIV-1 in nine HIV-1-seropositive patients.[76] Three patients received HIV-1 *nef* cDNA, three received HIV-1 *rev* cDNA, and three received HIV-1 *tat* cDNA, at days 0, 6, and 180. Preliminary results revealed no significant clinical or laboratory effects measured in any of the patients. In addition, specific cellular responses were observed. All three patients immunized with *nef* cDNA mounted strong cytotoxic T lymphocyte responses, whereas two of the patients immunized with *rev* cDNA showed enhanced cellular responses, and all three of the *tat* cDNA immunized patients showed a low-level CTL response through the course of the study. In addition, immunization was accompanied by an increased frequency of precursor CTLs, suggesting that even immunodeficient individuals have the capacity of forming new HIV-1-specific memory cells. The induction of cellular responses in humans after genetic immunization is a promising step toward DNA-based vaccines against HIV-1.

MacGregor *et al.* also reported on results in a therapy trial.[77] Sequential groups of five subjects received three intramuscular injections of an HIV-1 *env/rev* DNA construct at the same concentration (30, 100, or 300 µg) at 10-week intervals. Both antibody and cell-mediated immune responses were studied. Geometric mean antibody titers against gp120 as well as specific CTL and lymphoproliferative responses from the subjects were measured at various points: immediately before initiating the three-dose series of injections, immediately after the third dose (week 21), and 4 months later (week 36). Vaccine administration was well tolerated and CD4/CD8 lymphocyte levels and plasma HIV-1 levels remained relatively constant throughout the course of the study. Although the sample sizes were small, interesting differences were noted among the three dose groups. For example, there was a lack of apparent boosting in the mean antibody titers in the 30-μg dose group, while the 100-μg and 300-μg dose groups displayed statistically significant rises in geometric mean titers by the 36-week follow-up point. MacGregor is now evaluating the effects of the HIV-1 *env/rev* and *gag/pol* DNA vaccines in HIV-1-infected subjects who are concurrently on highly active antiretroviral therapy. This study will help to examine whether more potent immune responses can be achieved when viral replication is controlled and there is the absence of the immunopathogenicity associated with HIV-1.

A Phase I trial using the same HIV-1 envelope construct has now been initiated in high-risk seronegative individuals (Table III). This study is designed to determine the safety and immunogenicity of the HIV-1 *env/rev* construct mentioned above when administered to healthy, HIV-seronegative subjects in a dose-escalation scheme. Sequential groups of six subjects received three doses of vaccine at the same concentration (100 μg, 300 μg, or 1 mg) at entry, week 4, and week 8, with a booster injection scheduled for 6 months post initial immunization. Preliminary results show that the *env/rev* DNA-based vaccine is safely tolerated at all doses. Furthermore, this

TABLE III

Lymphocyte Proliferation to HIV-1 Antigen following Immunization of HIV-1 Seronegative Volunteers with an HIV-1 DNA Based Vaccine[72] [a]

Cohort	No. tested	No. responders/No. tested	No. multiple responders[b]
gp160			
100 μg	4	3/4	0
300 μg	6	5/6	2
rev protein			
100 μg	4	1/4	1
300 μg	6	4/6	4

[a]HIV seronegative subjects received 3 doses of vaccine at the same concentration (100 mg or 300 mg) at entry, week 4, week 8, and week 26. Subjects were tested for their ability to proliferate to the HIV-1 antigens—gp160 and rev—in a standard proliferation assay. Four of six subjects were tested in the 100 μg dose group while all six subjects were tested in the 300 μg dose group.

[b]Positive proliferative responses at more than one time point.

vaccine has induced HIV-1-specific cellular immune responses in several subjects, demonstrating that DNA can induce antigen-specific immune responses in HIV-seronegative humans. Additionally, a second Phase I trial to evaluate a *gag/pol* construct has also gotten underway and demonstrated the induction of cellular responses. These studies are pioneering the use of DNA vaccines for clinical evaluation in humans and it will be seen in the next year or so if the potent immune responses generated by DNA vaccination against HIV-1 in mice and nonhuman primates can be seen also in humans.

6. CONCLUSION

Research investigating DNA-based vaccines as an effective strategy for worldwide prevention of HIV-1 has only just begun. From the rational design of the vaccine constructs to clinical trials in humans, this vaccine strategy provides a great deal of promise for providing a safe and inexpensive method of worldwide HIV-1 prevention. However, the search for the optimal approach to nucleic acid vaccination against HIV-1 is still ongoing. The injection of foreign genes by genetic immunization has resulted in specific immune responses with characteristics of protective immunity against a number of infectious diseases in small animals, primates, and even humans. Until recently, the majority of the work on DNA vaccines has focused on the testing of single-gene constructs. Recently we have initiated clinical studies of a multiplasmid vaccine cocktail. These expression cassettes were designed with the hope of stimulating immunity against multiple cellular and humoral epitopes. There is also the prospect for developing multiple envelope-subtype constructs in an attempt to address the antigenic diversity of the virus throughout the world. The demonstrated power of coadministration with cytokine and costimulatory molecule genes to direct and enhance cellular and humoral immune responses generates optimism regarding a combination approach to nucleic acid vaccination against HIV-1. These recent developments suggest many new possibilities for the future and perhaps will lead the way to overcoming HIV-1, a daunting challenge for vaccine development.

REFERENCES

1. HIV/AIDS, 1996, UNAIDS Fact Sheet, *UNAIDS/WHO.*
2. Ho, D. D., 1996, Therapy of HIV infections: Problems and prospects, *Bull. N.Y. Acad. Med.* **73:**37–45.

3. Walker, B. D., 1994, The rationale for immunotherapy in HIV-1 infection, *J. Acquired Immune Defic. Syndr.* **7**:S6–S13.

4. Cao, Y., Qin, L., Zhang, L., Safrit, J., and Ho, D. D., 1995, Virologic and immunologic characterization of long-term survivors of human immunodeficiency virus type-1 infection, *N. Engl. J. Med.* **332**:201–208.

5. Ugen, K. E., Srikantan, V., Goedert, J. J., Nelson, R. P., Williams, W. V., and Weiner, D. B., 1997, Vertical transmission of immunodeficiency virus type 1: Seroreactivity by maternal antibodies to the carboxy region of the gp41 envelope, *J. Infect. Dis.* **175**:63–69.

6. Ugen, K. E., Goedert, J. J., Boyer, J., Refaeli, Y., Frank, I., Williams, W. V., Willoughby, A., Landesman, S., Mendez, H., Rubinstein, A., Kieber-Emmons, T., and Weiner, D. B., 1992, Vertical transmission of HIV infection. Reactivity of maternal sera with glycoprotein 120 and 41 peptides from HIV type 1, *J. Clin. Invest.* **89**:1923–1930.

7. Rossi, P., Moschese, V., Broliden, P., Fundaro, C., Quintl, I., Plebani, A., Giaquinto, C., Tovo, P. A., Ljunggren, K., Rosen, J., Wigzell, H., Jondal, M., and Wahren, B, 1989, Presence of maternal antibodies to human immunodeficiency virus 1 envelope glycoprotein gp120 epitopes correlates with the uninfected status of children born to seropositive mothers, *Proc. Natl. Acad. Sci. USA* **86**:8055–8058.

8. Cheingsong-Papov, R., Callow, D., Beddows, S., Shaunak, S , Wasi, C , Kaleebu, P., Gilks, C., Petrascu, I. V., Garaev, M. M., Watts, D. M., *et al.*, 1992, Geographic diversity of human immunodeficiency virus type-1: Serologic reactivity to *env* envelopes and relationship to neutralization, *J. Infect. Dis.* **165**:256–261.

9. Emini, E., Schleif, W. A., Nunberg, J. H., Conley, A. J., Eda, Y., Tokiyoshi, S., Putney, S. D., Matsushita, S., Cobb, K. E., Jett, C. M., *et al.*, 1992, Prevention of HIV-1 infection in chimpanzees by gp120 V3 domain-specific monoclonal antibody, *Nature* **355**:728–730.

10. Putkonen, P., Thorstensson, R., Ghavamzadeh, L., Albert, J., Hild, K., Bibefeld, G., and Norrby, E., 1991, Prevention of HIV-2 and SIV_{SM} infection by passive immunization in cynomolgus monkeys, *Nature* **352**:436–438.

11. Graham, B., and Wright, P. F., 1995, Candidate AIDS vaccines, *N. Engl. J. Med.* **33**:1331–1339.

12. Schwartz, D., Gorse, G., Clements, M. L., Belshe, R., Izu, A., Duliege, A. M., Berman, P., Twaddell, T., Stablein, D., Sposto, R., Siliciano, R., and Matthews, T., 1993, Induction of HIV-1 neutralising and syncytium-inhibiting antibodies of uninfected recipients of HIV-1 (IIIB) rgp 120 subunit vaccine, *Lancet* **342**:69–73.

13. Berman, P., Gregory, T. J., Riddle, L., Nakamura, G. R., Champe, M. A., Porter, J. P., Wurm, F. M., Hershberg, R. D., Cobb, E. K., and Eichberg, J. W., 1990, Protection of chimpanzees from infection by HIV-1 after vaccination with recombinant glycoprotein gp120 but not gp160, *Nature* **345**:622–625.

14. Bruck, C., Thiriat, C., Delers, A., *et al.*, 1993, Comparison of vaccine protection in chimpanzees immunized with two different forms of recombinant HIV-1 envelope glycoprotein, *AIDS Res. Hum. Retrovir.* **9**:S110.

15. Girard, M., Meigner, B., Barre-Sinoussi, F., Kieny, M. P., Matthews, T., Muchmore, E., Nara, P. L., Wei, Q., Rimsky, L., Weinhold, K., and Fultz, P. N., 1995, Vaccine-induced protection against infection by heterologous regions in human immunodeficiency virus type-I, *J. Virol.* **69**:6239–6248.

16. Pinto, L., Sullivan, J., Berzofsky, J. A., Clerici, M., Kessler, H. A., Landay, A. L., and Shearer, G. M., 1995, Env-specific cytotoxic T lymphocyte responses in HIV seronegative health care workers occupationally exposed to HIV-contaminated body fluids, *J. Clin. Invest.* **96**: 867–876.

17. Rowland-Jones, S., Sutton, J., Ariyoshi, K., Dong, T., Gotch, F., McAdam, S., Whitby, D., Sabally, S., Allimore, A., Corrah, T., Takiguchi, M., McMichael, A., and Whittle, H., 1995, HIV-specific T-cells in HIV-exposed but uninfected Gambian women, *Nature Med.* **1**:59–64.

18. Borrow, P., Lewicki, H., Hahn, B. H., Shaw, G. M., and Oldstone, M. B., 1994, Virus-specific CD8[+] cytotoxic T-lymphocyte activity associated with control of viremia in primary human immunodeficiency virus type-I infection, *J. Virol.* **68:**6103–6110.

19. Koup, R., Safrit, J. T., Cao, Y., *et al.*, 1994, Temporal association of cellular immune responses with the initial control of viremia in primary human immunodeficiency type-I syndrome, *J. Virol.* **64:**621–629.

20. Hulskotte, E. G., Geretti, A. M., Siebelink, K. H., van Amerongen, G., Cranage, M. P., Rud, E. W., Norley, S. G., de Vries, P., and Osterhaus, A. D., 1995, Vaccine-induced virus neutralizing antibodies and cytotoxic T cells do not protect macaques from experimental infection with simian immunodeficiency virus $SIV_{mac32H(J5)}$, *J. Virol.* **69:**6289–6296.

21. Rosenberg, E. S., Billings, J. M., Caliendo, A. M., *et al.*, 1997, Vigorous HIV-1 specific CD4 T cell responses associated with control of viremia, *Science* **278:**1447–1450.

22. Cocchi, R., DeVico, A. L., Garzino-Demo, A., Arya, S. K., Gallo, R. C., and Lusso, P., 1995, Identification of RANTES, MIP-1 alpha and MIP-1 beta as the major HIV-1 suppressive factors produced by CD8[+] T cells, *Science* **270:**1811–1815.

23. Walker, C. M., Moody, D. J., Stites, D. P., and Levy, J. A., 1989, CD8[+] lymphocytes can control HIV infection *in vitro* by suppressing virus replication, *Science* **66:**1663–1666.

24. Mackewicz, C. E., Ortega, H. W., and Levy, J. A., 1991, CD8 activity correlates with the clinical state of the infected individual, *J. Clin. Invest.* **87:**1462–1466.

25. Heeney, J. L., Teeuwsen, V. J., van Gils, M., Bogers, W. M., DeGiuli Morghen, C., Radaelli, A., Barnett, S., Morein, B., Akerblom, L., Wang, Y., Lehner, T., and David, D., 1998, Beta chemokines and neutralizing antigenicity titers correlate with sterilizing immunity generated in HIV-1 vaccinated macaques, *Proc. Natl. Acad. Sci. USA* **18:**10803–10808.

26. Malim, M., Hauber, J., Le, S. Y., Maizil, J. V., and Cullen, B. R., 1989, The HIV-1 *rev* trans-activator acts through a structured target sequence to activate a nuclear export of un-spliced viral mRNA, *Nature* **338:**254–257.

27. Fan, L., and Peden, K., 1992, Cell-free transmission of *vif*-mutants of HIV-1, *Virology* **190:** 19–29.

28. Gabduza, D., Lawrence, K., Langhoff, E., Terwilliger, E., Dorfman, T., Haseltine, W., and Sodrowski, J., 1992, Role of *vif* in replication of human immunodeficiency virus type-1 transmission in CD4[+] lymphocytes, *J. Virol.* **66:**6489–6495.

29. Von Schwedler, U., Song, J., Aiken, C., and Trono, D., 1993, Vif is crucial for human immunodeficiency virus type-1 proviral DNA synthesis in infected cells, *J. Virol.* **67:**4945–4955.

30. Klimkait, T., Strebel, K., Hoggan, M. D., Martin, M. A., and Orenstein, J. M., 1990, The human immunodeficiency virus type-1 specific protein Vpu is required for efficient virus mutation and release, *J. Virol.* **64:**621–629.

31. Levy, D. N., Fernandes, L. S., Williams, W. V., and Weiner, D. B., 1993, Induction of cell differentiation by human immunodeficiency virus-1 Vpr, *Cell* **72:**541–550.

32. Levy, D. N., Refaeli, Y., MacGregor, R. R., and Weiner, D. B., 1994, Serum Vpr regulates productive infection and latency of human immunodeficiency virus type-1, *Proc. Natl. Acad. Sci. USA* **91:**10873.

33. Levy, D. N., Refaeli, Y., and Weiner, D. B., 1994, Extracellular Vpr protein increases cellular permissiveness to HIV replication and reactivates virus from latency, *J. Virol.* **69:**1243–1252.

34. Kim, S., Ikeuchi, K., Byrn, R., Groopman, J., and Baltimore, D., 1989, Lack of negative influence on viral growth by the *nef* gene of human immunodeficiency virus type-1, *Proc. Natl. Acad. Sci. USA* **86:**9544–9548.

35. Kestler, H., Ringler, D. J., Mori, K., Panicalli, D. L., Sehgai, P. K., Daniel, M. D., and Desrosiers, R. C., 1991, Importance of the *nef* gene for maintenance of high virus loads and the development of AIDS, *Cell* **65:**651–662.

36. Ayyavoo, V., Nagashunmugan, N., Boyer, J., Mahalingam, S., Fernandes, L. S., LePhong,

L. J., Hgun, C., Chattergoon, M., Goedert, J. J., Friedman, H., and Weiner, D. B., 1997, Development of genetic vaccines for pathogenic gene construction of attenuated *vif* immunization cassettes, *AIDS* **11**:1433–1444.

37. Tang, D., Devit, M., and Johnston, S. A., 1992, Genetic immunization is a simple method for eliciting an immune response, *Nature* **356**:152–154.

38. Ulmer, J., Donnelly, J., Parker, S. E., Rhodes, G. H., Felgner, P. L., Dwarki, V. L., *et al.*, 1993, Heterologous protection against influenza by injection of a DNA encoding a viral protein *Science* **259**:1745–1749.

39. Wang, B., Ugen, K. E., Srikantan, V., Agadjanyan, M. G., Dang, K., Refaeli, Y., Sato, A. I., Boyer, J. D., Williams, W. V., and Weiner, D. B., 1993, Gene inoculation generates immune responses against human immunodeficiency virus type 1, *Proc. Natl. Acad. Sci. USA* **90**:4156–4160.

40. Davis, H., Michel, M. L., and Whalen, R. G., 1993, DNA-based immunization induces continuous secretion of hepatitis B surface antigen and high levels of circulating antibody, *Hum. Mol. Genet.* **2**:1847–1851.

41. Cohen, A., Boyer, J. D., and Weiner, D. B., 1998, DNA vaccination, *FASEB* **12**:1611–1626.

42. Chattergoon, M., Robinson, T., Boyer, J. D., and Weiner, D. B., 1998, Specific immune induction following DNA-based immunization through *in vivo* transfection and activation of macrophages, *J. Immunol.* **169**:5707–5718.

43. Corr, M., Lee, D. J., Carson, D. A., and Tighe, H., 1996, Gene vaccination with naked plasmid DNA: Mechanisms of CTL priming, *J. Exp. Med.* **184**:1555–1560.

44. Bagarazzi, M. L., Boyer, J. D., Ayyavoo, V., and Weiner, D. B., 1998, Nucleic acid based vaccines as an approach to immunization against human immunodeficiency virus type 1, *Curr. Top. Microbiol. Immunol.* **226**:107–144.

45. Lu, S., Santoro, J. C., Fuller, D. H., Haynes, J. R., and Robinson, H. L., 1995, Use of DNA expressing HIV-1 *env* and non-infectious HIV-1 particles to raise antibody responses in mice, *Virology* **209**:147–154.

46. Okuda, K., Bukawa, H., Hamajima, K., Kawamoto, S., Sekigawa, K. I., Yamada, Y., Tanaka, S. I., Ishii, N., Aoki, I., Nakamura, M., Yamamoto, H., Cullen, B. R., and Fukushima, J., 1995, Induction of potent humoral and cell-mediated immune responses following direct injection of DNA encoding the HIV type I *env* and *rev* products, *AIDS Res. Hum. Retrovir.* **11**:933–943.

47. Haynes, J., Fuller, D. H., Eisenbraun, M. D., Ford, M. J., and Pertmner, T. M., 1994, Accell particle medicated immunization elicits humoral, cytotoxic and protective responses, *AIDS Res. Hum. Retrovir.* **10**:S43–S45.

48. Wang, B., Merva, M., Dang, K., Ugen, K. E., Boyer, J. D., Williams, W. V., and Weiner, D. B., 1994, DNA inoculation induces protective *in vivo* immune responses against cellular challenge with HIV-1 antigen expressing cells, *AIDS Res. Hum. Retrovir.* **10**:S35–S41.

49. Wang, B., Dang, K., Agadjanyan, M. G., Srikantan, V., Li, F., Ugen, K. E., Boyer, J. D., Merva, M., Williams, W. V., and Weiner, W. V., 1997, Mucosal immunization with a DNA vaccine induces immune responses against HIV-1 at a mucosal site, *Vaccine* **15**:818–820.

50. Klavinskis, L. S., Gao, L., Barnfeld, C., Lehner, T., and Parker, S, 1997, Mucosal immunity with DNA–liposome complexes, *Vaccine* **15**:818–820.

51. Hamijima, K., Sasaki, S., Fukushima, J., Kaneko, T., Xin, K. Q., Kusoh, I., and Okuda, K., 1998, Intranasal administration of HIV-1 DNA vaccine formulated with a polymer, carboxymethylcellulose, augments mucosal antibody production and cell-mediated immune response, *Clin. Immunol. Immunopathol.* **88**:205–210.

52. Winchell, J. M., Routray, S., Betts, P. W., Kruingen, H. J., and Silbart, L. K., 1998, Mucosal and systemic antibody response to a C4/V3 construct following DNA vaccination of rabbits via Peyers patch, *J. Infect. Dis.* **178**:850–853.

53. Raz, E., Watanabe, A., Baird, S. M., Eisenberg, R. A., Parr, T. B., Lotz, M., Kipps, T. J., and Carson, D. A., 1993, Systemic immunological effects of cytokine genes injected into skeletal muscle, *Proc. Natl. Acad. Sci. USA* **90:**4523–4527.

54. Ziang, Z. Q., and Ertl, H. C. J., 1995, Manipulation of the immune response to plasmid-encoded viral antigen by coinoculation with plasmids expressing cytokines, *Immunity* **2:** 129–137.

55. Kim, J., Ayyavoo, V., Bagarazzi, M. L., Chattergoon, M. A., Dang, K., Wang, B., Boyer, J. D., and Weiner, D. B., 1997, *In vivo* engineering of a cellular immune response by coadministration of IL-12 expression vector with a DNA immunogen, *J. Immunol.* **158:**816–826.

56. Prayaga, S. K., Ford, M. J., and Haynes, J. R., 1997, Manipulation of HIV-1 gp120 specific immune responses elicited via gene gun based DNA immunization, *Vaccine* **15:**1349–1352.

57. Kim, J. J., Nottingham, L. K., Sin, J. I., Tsai, A., Morrison, L., Dang, K., Hu, Y., Kazahaya, K., Bennett, M., Dentchev, T., Wilson, D. M., Boyer, J. D., Agadjanyan, M. G., Chalian, A. A., and Weiner, D. B., 1998, CD8$^+$ T cells-expressed chemokines expand and modulate antigen-specific immune responses in the periphery, *J. Clin. Invest.* **102:**1112–1124.

58. Kim, J., Bagarazzi, M. L., Trivedi, N., Hu, Y., Kazahaya, K., Wilson, D., Ciccarelli, R., Chattergoon, M., Dang, K., Mahalingam, S., Chalian, A., Agadjanyan, M. G., Boyer, J. D., Wang, B., and Weiner, D., 1997, Engineering of *in vivo* immune responses to DNA immunization via codelivery of costimulatory molecule genes, *Nature Biotechnol.* **15:**641–646.

59. Iwaski, A., Nicla-Stiernholm, B. J., Cahn, A. K., Birinstein, N. L., and Barber, B. H., 1997, Enhanced CTL responses mediated by plasmid DNA immunogens encoding costimulatory molecules and cytokines, *J. Immunol.* **158:**4591–4601.

60. Horspool, J. H., Perrin, P. J., Woodcock, J. B., Cox, J. H., June, C. H., Harlan, D. M., St. Louie, D. C., and Lee, K. P., 1998, Nucleic acid vaccine-induced immune response require CD28 costimulation and are regulated by CTLA4, *J. Immunol.* **160:**2706–2714.

61. Tsuji, T., Fukushima, J., Hamajima, I., Ishii, N., Aoki, I., Bukawa, H., Ishigatsubo, Y., Tani, K., Okubo, T., Dorf, M. E., and Okuda, K., 1997, HIV-1 specific cell-mediated immunity is enhanced by co-inoculation of TCA3 expression plasmid with DNA vaccine, *Immunology* **90:**1–6.

62. Toda, S., Ishii, N., Okada, E., Kusakabe, K. I., Arai, H., Hamajima, K., Gorai, I., Nishioka, K., and Okuda, K., 1997, HIV-1 specific cell mediated immune responses induced by DNA vaccination were enhanced by mannan-coated liposomes and inhibited by anti-interferon gamma antibody, *Immunology* **90:**1–6.

63. Boyer, J. D., Wang, B., Ugen, K. E., Agadjanyan, M. G., Javadian, M. A., Frost, P., Dang, K., Carrano, R. A., Ciccarelli, R., Coney, L., Williams, W. V., and Weiner, D. B., 1996, *In vivo* protective anti-HIV immune responses in non-human primates through DNA immunization, *J. Med. Primatol.* **25:**242–250.

64. Fynan, E., Webster, R., Fuller, D., Haynes, J., Santoro, J., and Robinson, H., 1993, DNA vaccines: Protective immunizations by parenteral, mucosal and gene-gun inoculations, *Proc. Natl. Acad. Sci. USA* **90:**11478–11482.

65. Wang, B., Boyer, J., Srikantan, V., Ugen, K., Gilbert, L., Phan, C., Dang, K., Merva, M., Agadjanyan, M. G., Newman, M., Carrano, R., McCallus, D., Coney, L., Williams, W. V., and Weiner, D. B., 1995, Induction of humoral and cellular immune responses to the human immunodeficiency type 1 virus in non human primates by *in vivo* inoculation, *Virology* **221:** 102–112.

66. Krieg, A. M., 1996, An innate immune defense mechanism based on the recognition of CpG motifs in microbial DNA, *J. Lab. Clin. Med.* **128:**128–133.

67. Sato, Y., Roman, M., Tighe, H., Corr, M., Nguyen, M. D., Silverman, G. J., Lotz, M., Carson, D. A., and Raz, E., 1996, Immunostimulatory DNA sequences necessary for effective intradermal gene immunization, *Science* **273:**352–354.

68. Klinman, D. M., Yamshchikov, G., and Ishigatsubo, Y., 1997, Contribution of CpG motifs to the immunogenicity of DNA vaccines, *J. Immunol.* **158:**3635–3639.

69. Leclerc, C., Deriaud, E., Rojas, M., and Whalen, R. G., 1997, The preferential induction of a T_h1 immune response by DNA-based immunization is mediated by the immunostimulatory effect of plasmid DNA, *Cell. Immunol.* **179:**97–106.

70. Roman, M., Martin-Orozco, E., Goodman, J. S., Nguyen, M. D., Sato, Y., Renaghy, A., Kornbluth, R. S., Richman, D. D., Carson, D. A., and Raz, E., 1997, Immunostimulatory DNA sequences function as T helper-1 promoting adjuvants, *Nature Med.* **3:**849–854.

71. Boyer, J. D., Cohen, A., Schumann, K., Nath, B., Lacy, K., bagarazzi, M., Higgins, T., Baine, Y., Ciccarelli, R., Ginsberg, R., MacGregor, R. R., and Weiner, D. B., in press, Vaccination with seronegative volunteers with an HIV-1 env/rev DNA vaccine induces proliferation.

72. Letvin, N. L., Montefiori, D. C., Yasutomi, T., *et al.*, 1997, Potent, protective anti-HIV immune responses generated by bimodal HIV envelope DNA plus protein vaccination, *Proc. Natl. Acad. Sci. USA* **94:**9373–9383.

73. Lu, S., Arthos, J., Montefiori, D. C., Yasutomi, Y., Manson, K., Mustafa, F., Johnason, E., Santoro, J. C., Wissink, J., Mullins, J. I., Haynes, J. R., Letvin, N. L., Wynand, M., and Robinson, H. L., 1996, Simian immunodeficiency virus DNA vaccine trial in macaques, *J. Virol.* **70:**3978–3991.

74. Boyer, J. D., Ugen, K. E., Wang, B., Agadjanyan, M. G., Gilbert, L., Bagarazzi, M. L., Chattergoon, M. A., Frost, P., Javadian, M. A., Williams, W. V., Refaeli, Y., Ciccarelli, R. B., McCallus, D., Coney, L., and Weiner, D. B., 1997, Protection of chimpanzees from high dose heterologous HIV-1 challenge by DNA vaccination, *Nature Med.* **3:**526–532.

75. Boyer, J. D., Ugen, K. E., Chattergoon, M., Wang, B., Shah, A., Agadjanyan, M. G., Bagarazzi, M. L., Javadian, M. A., Williams, W. V., and Weiner, D. B., 1997, DNA vaccination as anti-human immunodeficiency virus immunotherapy in infected chimpanzees, *J. Infect. Dis.* **176:**1501–1509.

76. Calarota, S., Bratt, G., Nordlund, S., Hinkula, J., Leandersson, A. C., Sandstrom, F., and Wahren, B., 1998, Cellular cytotoxic response induced by DNA vaccination in HIV-1 infected patients, *Lancet* **351:**1320–1325.

77. MacGregor, R. R., Boyer, J. D., Ugen, K. E., Lacy, K. E., Bagarazzi, M. L., Chattergoon, M. A., Baine, Y., Higgins, T. J., Ciccarelli, R. B., Coney, L. R., Ginsberg, R. S., and Weiner, D. B., 1998, First human trial of a DNA-based vaccine against HIV-1 infection: Safety and host responses, *J. Infect. Dis.* **178:**92–100.

10

Passive Immunotherapy against HIV-1

Current Status and Potential

JOSEPH P. COTROPIA and KENNETH E. UGEN

1. INTRODUCTION

Passive antibody therapy has been demonstrated to provide protection against a number of bacterial and viral pathogens.[1-8] Specifically, antibodies have an established role in preventing infection by a number of viruses even though it is acknowledged that cytotoxic T lymphocytes (CTL) are critical for the elimination of infection by many viruses. A good example of the role of humoral immunity in mediating protection against a viral infection is hepatitis B. In this case the recombinant hepatitis B surface antigen vaccine induces neutralizing antibodies which are protective against infection.[8]

From animal studies in the mid-1970s, antibodies directed against retroviral surface glycoproteins were demonstrated to suppress viremia and provide protection against viral-induced disease.[9] Therefore, initial experimental models provided evidence that passive immunization could be effective in the prophylaxis and treatment of human immunodeficiency virus type 1 (HIV-1). In fact, there is also considerable evidence that passively administered antibodies can play an important role in preventing or modulating infection by retroviruses including HIV-1, simian immunodeficiency virus (SIV), feline immunodeficiency virus (FIV), and human T cell leukemia

JOSEPH P. COTROPIA • BioClonetics, Inc., Philadelphia, Pennsylvania 19147, and Department of Medicine, University of Pennsylvania School of Medicine, Philadelphia, Pennsylvania 19104.   KENNETH E. UGEN • Department of Medical Microbiology and Immunology, University of South Florida College of Medicine, Tampa, Florida 33612.

Human Retroviral Infections, edited by Kenneth E. Ugen *et al.*
Kluwer Academic / Plenum Publishers, New York, 2000.

virus (HTLV). However, as the acquired immune deficiency syndrome (AIDS) epidemic has continued from the 1980s into the 1990s, a protective role for the humoral immune response against HIV-1 has yet to be clearly defined.

The purpose of this review is to summarize the evidence for and potential of passive immunotherapy against HIV-1. First, a brief description of the immune responses which normally occur during infection against gp120 and gp41 will be given. The use of both polyclonal antibodies as well as human monoclonal antibodies against HIV-1 for prophylaxis and/or therapy will be discussed.

2. HUMORAL IMMUNE RESPONSES AGAINST HIV-1

Shortly after the primary infection in humans, HIV-1 stimulates a humoral immune response which results in the production of antibodies directed against most of the viral structural components. A particular subset of antibodies is directed against HIV-envelope gp120 and gp41, which are the surface (SU) and transmembrane (TM) glycoproteins, respectively. These important proteins may be involved in the induction of active immunity, as revealed by the production of neutralizing, cell-fusion-inhibiting and antibody-dependent cellular-cytotoxicity (ADCC)-mediating antibodies.

Both polyclonal and monoclonal antibodies have been demonstrated to react with surface components of cells infected by HIV primary isolates, indicating that targeted cells express surface antigens which can induce and, moreover, bind antibodies.[10]

In vitro assays of antibody-dependent cell-mediated (ADCM) cytolysis demonstrate that HIV-1-infected cells express surface viral antigens.[11–14]

Additionally, in early studies oligomeric envelope components have been demonstrated to be immunogenic. In fact, human monoclonal antibody binding to oligomeric forms of gp41 on Western blot have been described.[15,16]

Acute infections with HIV-1 lead to a rapid and extensive phase of virus proliferation, which generates massive numbers of cell-free virus. In addition, with the concomitant appearance of a serologic response, the plasma viremia generally decreases.

Also, importantly, HIV-1 has been demonstrated to mutate with high frequency. As a consequence of rapid and characteristic unfaithful replication, since the duplicating reverse transcriptase (RT) enzyme lacks a proof-reading function, resultant conformational variations and antigenic changes expressed in the progeny decrease the efficacy of antibodies generated earlier in infection to continue to protect the host.

3. HUMORAL IMMUNE RESPONSES AGAINST gp120

In the acute infection phase, the antibodies produced are HIV-1-type-specific, and in the years following initial infection, generally broader, but lower titered virus-neutralizing antibody responses develop which include HIV-1 strains with a divergent gp120 third hypervariable domain (V3). The V3 loop within the carboxy-terminal half of gp120 contains a major neutralization domain, the principal neutralizing determinant (PND), so named according to the results obtained from the first *in vitro* neutralization studies. The early neutralization assays were conducted utilizing HIV-1 isolates passaged multiple times through T cell lines. Such HIV-1 strains are referenced as lab strains or T cell (T lymphocyte)-adapted strains, i.e., T cell-line-adapted (TCLA).

Antigenic epitopes on the viral envelope surface (SU) glycoprotein (gp120) were identified using HIV-1-seropositive sera and a large panel of human monoclonal antibodies generated *in vitro*.

A region corresponding to the CD4-binding domain (CD4bd) and epitopes within the first and fifth conserved regions of gp120 (C1/C5) are recognized by HIV-1 isolate cross-reactive antibodies prevalent in the serum of HIV-1-seropositive individuals. Additional studies indicated that monoclonal antibody IgG1b12, specific for CD4bd, reacts with structures present on oligomeric envelope.[17]

It is important to note that anti-V3-loop monoclonal antibodies significantly neutralized lab strains of HIV-1 passaged in T cells.[18–20] The V3 loop in primary strains does not appear accessible on the native surface envelope gp120, and therefore antibodies directed against the V3 region may not significantly neutralize HIV-1 field isolates.

4. HUMORAL IMMUNE RESPONSES AGAINST gp41

From the beginning of the AIDS epidemic the preponderance of studies concerning the HIV-1 envelope have focused on antibodies against the surface gp120 component. Over the past several years, however, investigations have now shifted to include studies that demonstrate the importance of the transmembrane gp41 structure.

With regard to gp41, antigenic epitopes are clustered in two regions. Cluster I antibodies are directed to an immunodominant region that can potentially form a disulfide loop; cluster II antibodies map to a region that includes a distal helix on the gp41 ectodomain.

Neutralizing as well as nonneutralizing and *in vitro* enhancing antibodies directed against a series of overlapping epitopes within the immuno-dominant domain of gp41 (cluster I) are produced in human sera in response to HIV-1 infection.

Immunoreactive regions of HIV type 1 gp41 have been finely mapped by reacting (polyclonal) HIV-1 antibody-positive human sera with sequential overlapping synthetic peptides which covered the transmembrane protein.[21] Immunoreactive regions of gp41 have been reported by Horal *et al.*[22] Within cluster I, three immunoreactive domains were identified, and five different and partially overlapping epitopes recognized by HIV-1-positive human sera were found within one immunodominant region, DQQLLGIWGCSGKLIC-TTAVPWN. The five different and partially overlapping epitopes were delineated as LLGIWG, WGCSGK, GKLICT, ICTTAV, and TAVPWN. It is important to note that no single serum sample demonstrated reactivity to all individual epitopes.

Human sera fractionated to obtain anti-HIV-1 antibodies and human monoclonal antibodies derived from immortalized B cells obtained from HIV-1-infected individuals have demonstrated binding to cluster I and cluster II within the transmembrane gp41.[23,24]

Antigenic epitopes within gp41 have been characterized using human monoclonal antibodies generated from individuals with high-titer HIV-1-seropositive serum.

Immunological binding to the immunodominant region of HIV-1 transmembrane gp41 ([579]RILAVERYLKDQQLLGIWGCSGKLICTT[606Myers]) and biological reactivity against HIV-1 are presented in Table I.

The human monoclonal antibody, clone 3 antibody, which specifically binds to the gp41-conserved linear epitope LIC within the immunodominant region (cluster I), has been demonstrated to bind to both oligomer as well as monomeric transmembrane glycoprotein. In addition, the clone 3 antibody has been demonstrated to have neutralizing capacity against lab-adapted strains.[25,26] Also, clone 3 antibody has been demonstrated to have broad neutralizing capacity against primary HIV isolates in clades B, E, and F.[27]

Clone 3 antibody is the only human monoclonal antibody in Table I that specifically binds within the represented gp41 immunodominant region (cluster I) and also possesses neutralization activity against HIV-1. All other human monoclonal antibodies which have been produced against this region either possesses HIV-1-enhancing activity or no biological activity at all.

With regard to the cluster II region of gp41, the human monoclonal antibody 2F5 has been demonstrated to bind to the gp41 oligomeric structure and has been reported to have broad neutralizing capacity for lab as well as clinical HIV-1 isolates.[28] 2F5 binds to the epitope ELDKWA, a con-

TABLE I

**Human Monoclonal Antibody Immunological Binding
to the Immunodominant Region of HIV-1 Transmembrane gp41
([579]RILAVERYLKDQQLLGIWGCSGKLICTT[608Myers])
and Biological Reactivity Against HIV-1**

Human mAb	Class	Amino acid sequence (epitope)	Biological activity	References
V10-9	IgG1k	RILAVERYLKDQQLLGIW	Enhancing	71, 72
86	IgG1k	RILAVERYLKDQQLLGIW	Enhancing	71, 72
50-69	IgG2k	QQLLGIWG	Enhancing	24
246-D	IgG1k	QQLLGIWG	Enhancing	24
181-D	IgG2k	QLLGIWG	None	24
240-D	IgG1k	LLGIWGCSG	Enhancing	24
2F11	IgG1	DQQLLGIWGCSG	Enhancing	73
41-7	IgG1k	GCSGKLIC	None	74
F240	IgG1k	LLCIWCCSCKLICT	None/ enhancing	75
Clone 3	IgG1	GCSGKLICTT	Neutralizing	25

served peptide sequence near the transmembrane segment. Moreover and importantly, when combined with 2F5, clone 3 antibody has been shown to produce a synergistic neutralizing effect against clades B, E, and F.[27]

Investigations to characterize specifically which of the many antibodies induced after HIV-1 infection can actually provide a neutralizing function continue to be the intense focus for many research groups. Results from the Antibody Serological Project (ASP) suggest that the functional activity of HIV-specific antibodies most likely to predict therapeutic efficacy is the capacity to neutralize primary isolates.[29] Additionally, since a large proportion of primary isolates replicate only in peripheral blood mononuclear cells (PBMC), a PBMC neutralization assay appears essential to characterize the neutralizing ability of anti-HIV-1 monoclonal antibodies.

Early *in vitro* assays demonstrated that HIV-1 lab strains passaged in T cell lines were extremely sensitive to neutralization by the following: sera from HIV-1-infected individuals or sera from individuals who were vaccinees immunized with gp120, and reagents including soluble CD4, and monoclonal antibodies.[30–32]

Currently, PBMC-based neutralization assays utilizing low passaged primary HIV-1 isolates have identified four human monoclonal antibodies with the capability of neutralizing a broad range of field isolates.[25,27,29]

The four human monoclonal anti-HIV-1 antibodies with broad neutralizing effect against clinical isolates are listed along with immunological

TABLE II
**The Four Human Monoclonal Anti-HIV-1 Antibodies
with Broad Neutralizing Activity Against Clinical HIV-1 Isolates**[a]

Antibodies	Core epitope recognized	Neutralization		References
		TCLA strains	Primary strains	
b12 (IgG1k) Conformational	gp120 (CD4bd/V2 loop)	Yes	Yes	29, 76, 77
2G12 (IgG1k) Conformational	gp120 (C2/C3/V4)	Yes	Yes	78, 79
2F5 (IgG3k) Linear	gp41 cluster II elDKWA aa = 662–667	Yes	Yes	28, 79, 80
Clone 3 (IgG1) Linear	gp41 Cluster I gcsgkLICtt aa = 597–606	Yes	Yes	25–27

[a]TCLA, T cell-line-adapted; CD4bd, CD4-binding domain; aa, amino acids.

characteristics in Table II. The production and characterization of these human monoclonal antibodies directed against surface envelope components and other data provide evidence that virions as well as HIV-1-infected cells can induce a humoral immune response with a neutralizing capacity.

5. CONVERGENT/SYNERGISTIC PASSIVE IMMUNOTHERAPY

Enhanced or synergistic neutralizing effects *in vitro* have been reported utilizing human monoclonal antibodies in combinations, e.g., anti-V3 and CD4bd antibodies.[33–35]

Using a chimeric simian immunodeficiency virus (SHIV−*vpu*+) which expressed HIV-1IIIB envelope antigens, the lowest effective antibody concentration for 90% viral neutralization was achieved with combinations of monoclonal antibodies b12, 2F5, 2G12, and anti-V3 694/98D, and, depending on the combination regimen, the concentration of monoclonal antibodies required to reach 90% virus neutralization was reduced approximately 2- to 25-fold as compared to the dose requirement of individual monoclonal antibodies to produce an equivalent effect.[36] Therefore, combinations of monoclonal antibodies may have a role in passive immuno-prophylaxis against HIV-1, as indicated by the synergy noted in the combination regimens.

However, it must be noted that in another case of combined regimens a cocktail of 10 monoclonal antibodies could not neutralize one primary isolate, designated US4.[37] Additionally, this non-syncytium-inducing strain was neutralization-resistant when tested individually using a panel of 17 monoclonal antibodies which also included b12 and 2F5.

6. EVIDENCE FOR PROTECTIVE EFFICACY OF HUMORAL IMMUNITY AGAINST HIV-1

As evidenced in the acute phase of HIV-1 infection, the initial burst of viremia is indeed brought under control during the asymptomatic phase of the disease. The decrease in level of the virus during the seroconversion phase therefore indicates that effective immune responses can initially occur that are capable of containing the early virus.

As a relevant clinical correlation, in a group of HIV-1-infected infants born to seropositive mothers, those infants that lacked the presence of antibodies to a 9-mer peptide JB7 within gp41 (^{603}CSGKLICTT$^{611\text{Wain-Hobson}}$) had a rapid progression to symptomatic AIDS.[38] In addition, it has also been suggested by some studies that pregnant HIV-1-infected women who demonstrate a more vigorous humoral immune response against envelope glyco-proteins are less likely to transmit infection to their offspring than those with less seroreactivity to these proteins. Specifically, reactivity differences to regions of gp41 have been noted.[39–44]

It is also of interest to note from the data of Vanini *et al.* that a decreasing titer of antibodies against CSGKLIC directly correlated with disease progression.[23] In addition, Loomis-Price *et al.* also recently demonstrated that high antibody reactivity to the peptide SGKLICTTAVFW is associated with slow progression to AIDS.[45] To that end it is important to note that clone 3, which can both prevent fusion of virus-infected cells with uninfected cells and neutralize infectivity of free viral particles as demonstrated in biological assays *in vitro*, binds to the decapeptide (10-mer) GCSGKLICTT.[25,26] These observations therefore indicate that antibodies produced against the clone 3 epitope may have some protective role.

Also, Boyer *et al.* demonstrated that vaccination of HIV-1-infected chimpanzees with DNA constructs which express the gp120 and gp41 envelope glycoproteins can result in boosted antibody responses to a peptide which contains the clone 3 epitope (GKLICTTVPWNASWSNKSL).[46] Such an observation indicates the potential for vaccination to induce or boost humoral immune responses to epitopes within the envelope glycoprotein which mediate neutralization.

In addition to studies of binding reactivity to the immunodominant loop of gp41, an important investigation has also suggested an association between high antibody reactivity to the epitope for the human monoclonal antibody 2F5 (ELDKWA) and lack of disease progression in children perinatally infected with HIV-1.[47] Taken together these studies indicate that humoral immune responses to neutralizing epitopes in gp41 may mediate protection against infection by HIV-1 or disease progression.

7. PASSIVE IMMUNOTHERAPY TRIALS OF CLINICAL RELEVANCE UTILIZING HIV IMMUNE SERUM GLOBULIN

Data from early clinical trials demonstrated that passive immunization improved the status of patients with advanced AIDS.[48,49] In those trials, passive immunization was accomplished by transfusing HIV-antibody-containing plasma obtained from asymptomatic HIV-1-infected individuals into symptomatic AIDS patients. In general, benefits derived from the studies were manifested by decreases in opportunistic infections. In these initial early studies the symptomatic AIDS recipients showed a decrease in p24 antigenemia and viral RNA by polymerase chain reaction (PCR) and also a delay in time to positivity for plasma and PBMC viral culture.

In a clinical trial by Jacobson *et al.* the results did not rule out the potential passive immunotherapeutic value of HIV immune serum globulin (HIVIG) which had been inactivated by 0.5% β-propriolactone and freezing.[50] However, overall, the study failed to demonstrate clinical benefit since there was a nonsignificant trend in delay to opportunistic infection and no change in plasma or PBMC viral culture. According to Karpas, β-propriolactone at 0.5% may partially inactivate the antibody-neutralizing activity contained in the transfused plasma.[51]

In a study by Levy *et al.* which utilized a higher monthly dose of HIVIG (500 ml), there was a differential benefit based on CD4 counts when comparing patients with $<50/mm^3$ versus $>50/mm^3$ as well as reduced mortality, but no change in incidence of opportunistic infections.[52] Patients also demonstrated a decrease in plasma p24 antigenemia.[52]

In a lengthy study by Vittecoq *et al.* in which infusions of HIVIG were provided twice a month for 1 year, no clinical benefit was demonstrated in the recipients.[53] Of promise, however, was the passive immunotherapy study involving newborn rhesus macaques by Van Rompay *et al.*[54] SIV hyperimmune serum given subcutaneously prior to oral SIV inoculation protected six newborns against infection.[54] This beneficial prophylactic result is in contrast to the case when SIV hyperimmune serum was given to three newborns 3 weeks after oral SIV inoculation. In this situation viremia was not reduced and all three infants developed AIDS and immune complex disease and died within 3 months of age. These data indicate that passively acquired anti-HIV IgG can decrease perinatal HIV transmission. However, anti-HIV IgG may not provide therapeutic benefit to infants after HIV infection has already been established. Trials with various HIVIG preparations are summarized in Table III.

8. MONOCLONAL ANTIBODIES UTILIZED IN PASSIVE IMMUNOTHERAPY

8.1. Chimpanzee Model

Data to support the efficacy of passive immunotherapy in chimpanzees have been published. In these studies it has been determined that neutralization of *in vivo* HIV-1 infectivity can be mediated by *in vitro* neutralizing antibody directed against the gp120 major, yet hypervariable, neutralizing epitope.[55] This has most clearly been demonstrated in additional specific chimpanzee studies.[56,57]

Specifically, the preexposure, Cβ1-treated (chimeric monoclonal anti-V3 antibody) chimpanzee remained negative for virus isolation and PCR-negative when challenged 24 h after administration of the antibody.[58]

In expanded studies in the chimpanzee, the protective efficacy of the anti-V3-domain antibody was again demonstrated as a prophylactic agent, thereby indicating that an antibody by itself, in the absence of other virus-specific immune responses, could be a postexposure prophylactic agent when administered within 10 min after HIV challenge.[58]

With regard to gp41, the neutralizing monoclonal antibody 2F5 was infused into two chimpanzees which were then challenged intravenously with a primary HIV-1 isolate (5016) passaged a limited number of times *in vitro* only in human PBMC cultures.[59] Serum RNA PCR tests were negative in one of the two antibody-infused animals until week 8 and in the other antibody-infused animal until week 12. Both animals seroconverted at week 14. These early effects differ from those in the two control animals, where infection was established immediately, demonstrated by positive cell-associated DNA PCR and serum RNA PCR tests within 1 week, seroconversion by 4 weeks after HIV inoculation, and subsequent development of lymphadenopathy in the acute phase.

In these primate HIV-1 studies, passive immunization resulted in antibody levels sufficient to successfully protect chimpanzees from virus chal-

TABLE III

Utilization of Hyperimmune Serum for Passive Immunotherapy in Retroviral Infections[a]

Host	Clinical state	Virus	Reagent	Dose	Challenge	Clinical response	References
Human (n = 42)	AIDS	HIV	HIVIG	300 ml q 2 wk × 1 yr	—	No clinical benefit	53
Human (n = 68)	Symptomatic HIV-1	HIV	HIVIG	500 ml q month × 1 yr	—	Reduced mortality and improved CD4 count	52
Human (n = 75)	Symptomatic HIV-1	HIV	HIVIG	250 ml q month × 1 yr	—	No clinical benefit	52
Human (n = 63)	AIDS	HIV	HIVIG	250 ml q 4 wk × 65 wk	—	No clinical benefit	50
Human (n = 10)	AIDS/ARC	HIV	HIVIG	500 ml q 4 wk × 12 wk	—	Improved lab and clinical parameters	49
Human (n = 6)	AIDS/ARC	HIV	HIVIG	50–250 ml	—	Improved lab and clinical parameters	48
Newborn rhesus macaques (n = 2)	—	SIV	SIV-HIS, 1 dose at birth	20 ml/kg SQ	Oral uncloned SIVmac251 (challenge was 2 days after SIV-specific serum)	2/2 protected	54
Newborn rhesus macaques (n = 4)	—	SIV	SIV-HIS doses at birth, 1 and 2 wk	20 ml/kg SQ	As above	4/4 protected	54
Newborn rhesus macaques (n = 3)	—	SIV	SIV-HIS, 1 dose at 3 wk	20 ml/kg SQ	Oral uncloned SIVmac251 (challenge was 3 wk before SIV-HIS)	0/3 protected	54

[a]ARC, AIDS-related complex; HIVIG, HIV immune serum globulin; SIV-HIS, simian immunodeficiency virus hyperimmune serum; q, every; SQ, subcutaneous.

lenge. The passive immunization studies cited above suggest that the characteristics of anti-HIV antibody—immunochemical specificity, affinity (avidity), and quantity—all contribute to preventing transmission of HIV-1.

The results of the studies reported thus far emphasize that antibody specificity, affinity, and quantity are all critical factors involved in the effective neutralization of the virus. In the potential use of these or other antibodies for passive immunization, all three factors must be taken into consideration.

In principle, then, human monoclonal antibodies with appropriate affinity and concentration can be administered to patients who lack neutralizing antibodies against neutralizable epitopes, thereby providing efficacious passive immunotherapy. A summary of the passive immunotherapy studies with human monoclonal antibodies utilizing the chimpanzee model system is presented in Table IV.

8.2. Human Trials

A mouse monoclonal antibody directed against the V3 loop was utilized in passive immunotherapy in one of the first human studies involving 11 late-stage HIV-infected patients.[60]

The viral RNA in plasma quantitated by PCR decreased in four cases, was stable in four others, and increased in three cases. A human anti-mouse antibody response developed in eight patients and antiidiotypic antibodies appeared in six. Antibodies inhibiting gp120 binding to CD4 became detectable or increased in six patients during immunotherapy. There was also a decreased total gp120 content in serum, permitting a better T cell activation.

In advancing protocol, a chimeric mouse–human monoclonal antibody (CGP 47 439) to V3 loop was utilized in a phase I/II study wherein escalation of antibody doses was examined.[61] The antibody was well tolerated and no toxicity was noted. In some patients a virus burden reduction was observed.

A small phase I clinical trial ($n = 4$) with the anti-HIV-1 gp41 human monoclonal antibody 2F5 has also been performed.[62] This safety trial utilized ascending doses of the antibody (100, 300, and 900 mg) administered at 4-week intervals. No adverse effects were noted. In addition, no anti-HIV-1 activity was detected. However, no HIV-1 quasispecies nor neutralization escape mutants appeared to be generated after the medicinal administration of 2F5.

Recently, another phase I dose escalation study was reported which examined the functional activity of the human monoclonal antibody F105 (IgG1kappa) in eight HIV-1-infected individuals without any AIDS-defining criteria. F105 is reactive with a discontinuous epitope that overlaps the CD4-binding domain.[63]

TABLE IV
Utilization of Monoclonal Antibodies for Passive Immunotherapy against HIV-1 Infection

Host	Disease state	Virus	Study	Reagent	Viral target	Dose	Viral challenge	Clinical response	References
Human ($n = 8$)	Non-AIDS	HIV-1	Phase I	F105, human	CD4bd	Single 100 or 500 mg/m^2	×	No anti-HIV-1 activity	63
Chimpanzee ($n = 2$)	—	HIV-1	—	2F5, human	ELDKWA (gp41)	15 mg/kg	HIV-1 (primary isolate)	Delayed seroconversion; reduced viral load	59
Human ($n = 4$)	Stage IV	HIV-1	Phase I	2F5, human	ELDKWA (gp41)	100, 300, or 900 mg	×	No adverse effects; no anti-HIV-1 activity	62
Human ($n = 12$)	Stage IV	HIV-1	Phase I/IIA	CGP47439, chimeric mouse/human	V3 loop (gp120)	1, 10, 25, 50, 100, 200 mg escalation	×	Virus burden reduction	61
Human ($n = 11$)	Stage IV	HIV-1	Phase I	F58/H3, P4/D10 mouse	V3 loop (gp120)	125 mg iv q 2 wk for 3 mo	×	Decrease in serum gp120	60
Chimpanzee ($n = 1$)	—	HIV-1	—	c_1, chimeric mouse/human	V3 loop (gp120	36 mg/kg iv	HIV-1 IIIB, + 24 h	1/1 protected	58
Chimpanzee ($n = 1$)	—	HIV-1	—		V3 loop (gp120)	36 mg/kg iv	HIV-1 IIIB, − 10 min	1/1 protected	58

A single dose of F105 at 100 or 500 mg/m^2 was administered intravenously. There was no evidence of anti-HIV-1 activity following a single antibody infusion, and no toxicity was recorded. A summary of the human clinical trials is presented in Table IV.

8.3. hu-PBL-SCID Mouse Model

To characterize the reactivity of antibodies against HIV-1 and to develop monoclonal antibodies as therapeutic reagents that might be used in passive immunotherapeutic modalities, a comparatively inexpensive and generally available model for initial laboratory evaluation is the mouse with severe combined immunodeficiency (SCID) which is transplanted with human peripheral blood lymphocytes (hu-PBL) by peritoneal injection.

Neutralizing monoclonal antibodies, either the murine (BAT123) or its mouse–human chimeric form (CGP 47 439), directed against the V3 region of HIV-1IIIB were given intraperitoneally to hu-PBL-SCID mice at a dose of 40 mg/kg.[64] The mice were then challenged intraperitoneally with 10 mouse infectious doses of HIV-1IIIB. HIV-1 could not be recovered from spleen cells or peritoneal lavage collected at 3 weeks as measured by PCR methods for the detection of HIV-1 DNA.

In additional studies using BAT123, animals were protected against subsequent infection with LAI strain.[65] However, as expected, the animals were not protected against the hetcrologous virus strains (SF2, JR-CSF, AD6) when the mouse monoclonal antibody (1 mg/kg; 25 μg/mouse) was given 1 h before challenge inoculation. This dose resulted in a peak serum antibody concentration of 16 μg/ml 24 h after the injection, 50-fold greater than the *in vitro* ID$_{90}$. A dose threefold greater than the *in vitro* ID$_{90}$ only protected approximately 57% (four of seven) hu-PBL-SCID mice. These results indicate antibody concentrations adequate to neutralize virtually 100% of virus infectivity *in vitro* are required to obtain sterilizing immunity *in vivo*.

Postexposure protection (96%; 23 of 24) was observed when the antibody was given within 4 h of virus inoculation. No protection was achieved, however, when BAT123 was administered 2 weeks after infection had been established.

A human monoclonal Fab fragment (Fab b12) derived from a combinatorial antibody library prepared from bone marrow of a long-term, asymptomatic HIV-1-seropositive donor has been shown to be potent in the neutralization of HIV-1.[66] The hu-PBL-SCID mouse model was utilized to test the capacity for both the b12 Fab fragment and the whole immunoglobulin IgG1 b12 produced in Chinese hamster ovary cells in tissue culture to protect in passive immunotherapy by injecting the reconstituted mice with graded amounts of antibody prior to challenge with HIV-1$_{SF2}$. Fab b12 tested at a dose of 1.9 mg/kg was able to protect 25% of hu-PBL-SCID mice from HIV-1 infection. Complete protection was obtained (no mice infected of five challenged) with a regimen in which 100 μg of IgG1 b12 was injected per mouse (4.5–7 mg/kg) at 7 days and again at 1 day before the inoculum challenge with the virus.

Pre- and postexposure prophylaxis in the hu-PBL-SCID mouse model was studied using a human monoclonal antibody (694/98-D) directed against the V3 loop of HIV-1.[67] Preexposure administration of 1.32 mg/kg antibody produced a 50% protection against HIV-1LAI strain. Of interest, in a 10-fold higher dosing of antibody at 13.2 mg/kg, virus isolated from one mouse 3 weeks after passive immunization demonstrated resistance to subsequent *in vitro* neutralization by 694/98-D. Amino acid changes were revealed for the linear core epitope recognized by 694/98-D and one flanking amino acid, as demonstrated in the V3-loop sequence analysis of the cloned virus.

Postexposure prophylaxis in mice revealed that 694/98-D was effective when administered 15 min after virus, producing 100% protection, but the efficacy declined to 50% for treatment delayed to 1 h post virus inoculation.

In order to study the impact of neutralizing antibodies on the course of established infection with HIV-1 primary isolates in the hu-PBL-SCID mouse model, 1 week after infection with JR-CSF, mice were treated with 50 mg/kg of IgG1 b12, a combination of IgG1 b12, 2F5, and 2G12, or control antibody.[68] Using a dose two orders of magnitude higher than *in vitro* neutralization titers generating 90% inhibition as an indication of a minimal significant effect, a modification in the course of HIV infection was not achieved. Additionally, 70% of viruses isolated from IgG1 b12-treated mice were demonstrated to be neutralization escape mutants. In these analyses mutations in the CD4-binding domain (CD4bd), which is involved in the IgG1 b12 epitope, were detected by sequencing the envelope gene. Consequently, these preliminary experiments indicate that even a cocktail of monoclonal antibodies including b12 and the addition of two more potent antibodies, 2F5 and 2G12, was not able to control HIV infection in the hu-PBL-SCID mouse model.

One caveat regarding the hu-PBL-SCID mouse model is that the system utilizes an artificially engrafted chimeric animal in which the human cells are moderately activated.[69,70] Table V summarizes the passive immunotherapeutic studies with human monoclonal antibodies utilizing the hu-PBL-SCID model system.

9. SUMMARY

The role that neutralizing antibodies play in preventing or controlling HIV-1 infection in humans has not been established and is a source of contention among investigators. Nonetheless, it is clear that in animal

TABLE V

Passive Immunotherapy Utilizing Monoclonal Antibodies in the hu-PBL-SCID Mouse Model

Host	Virus	Reagent	Viral target[a]	Dose mAb	mAb schedule relative to viral challenge	Clinical response	References
hu-PBL-SCID mouse	JR-CSF or SF 162	Combination: b12, 2G12, 2F5	CD4bd, C2/C3, ELDKWA	50 mg/kg	[Established JR-CSF or SF-162]	Does not modify course; 70% of viruses rescued from b12-treated mice were neutralization-escape mutants	68
hu-PBL-SCID mouse	HIV-1 (LAI)	694/98-D	V3	1.32 mg/kg 40 mg/kg	−2 hr −2 hr	50% protection 100% protection	67
hu-PBL-SCID mouse	HIV-1 (LAI)	694/98-D	V3	40 mg/kg	+15 min +60 min	100% protection 50% protection	67
hu-PBL-SCID mouse	HIV-1 (SF2)	Fab b12 IgG b12	CD4bd CD4bd	1.9 mg/kp ip 4.5–7 mg/kg ip	−1 h and +1 h −7 and −1 day	25% protection 100% protection	66
hu-PBL-SCID mouse	HIV-1 (LAI)	BAT 123	V3	1 mg/kg ip	−1 h	100% protection	65
hu-PBL-SCID mouse	HIV-1 (LAI)	BAT 123	V3	1 mg/kg ip	+4 h +2 wk	96% protection 0% protection	65
hu-PBL-SCID mouse	HIV-1 (IIIB)	BAT 123 CGP47 439[b]	V3 V3	40 mg/kg ip 40 mg/kg ip	+1 h +1 h	100% protection 100% protection	64

[a]CD4bd, CD4-binding domain.
[b]Chimeric mouse/human.

model systems the administration of neutralizing antibody preparations (either monoclonal or polyclonal) can prevent infections by retroviruses including HIV-1. Therefore, the generation and study of novel anti-HIV-1 polyclonal and monoclonal antibody preparations is warranted. It is possible that if a clinical niche for such antibody preparations is established, it will likely consist of a combination of anti-HIV-1 human monoclonal and/or polyclonal antibodies. A combination approach with antibodies directed against conserved epitopes of HIV-1 will likely decrease the emergence of neutralization-resistant mutants. This is important since the expansion of immunoescape mutants would severely limit the clinical utility of such a regimen. Ongoing and expanded synergistic and convergent immuno-therapeutic studies must then more accurately establish the anti-HIV-1 role for antibodies which are provided either through passive administration or induced by active vaccination.

ACKNOWLEDGMENTS. Some of the research findings reported in this review was supported in part by an NIH grant from the National Heart Lung and Blood Institute to K.U. (RO1-HL59818).

REFERENCES

1. Robbins, J. B., Schneerson, R., and Szu, S. C., 1995, Perspective: Hypothesis: Serum IgG antibody is sufficient to confer protection against infectious diseases by inactivating the inoculum, *J. Infect. Dis.* **171:**1387–1398.
2. Groothuis, J. R., Simoes, E. A. F., Levin, M. J., Hall, C. B., Long, C. E., Rodriguez, W. J., Arrobio, J., Meissner, H. C., Fulton, D. R., and Welliver, R. C., 1993, Prophylactic administration of respiratory syncytial virus immune globulin to high-risk infants and young children, *N. Engl. J. Med.* **329:**1524–1530.
3. Santosham, M., Reid, R., Ambrosino, D. M., Wolff, M. C., Almeido-Hill, C., Moxon, R., Chir, B., and Siber, G. R., 1987, Prevention of *Haemophilus influenzae* type b infections in high-risk infants treated with bacterial polysaccharide immune globulin, *N. Engl. J. Med.* **317:**923–929.
4. Shurin, P. A., Rehmus, J. M., Johnson, C. E., Marchant, C. D., Carlin, S. A., Super, D. M., van Hare, G. F., Jones, P. K., Ambrosino, D. M., and Siber, G. R., 1993, Bacterial polysaccharide immune globulin for prophylaxis of acute otitis media in high-risk children, *J. Pediatr.* **123:**801–810.
5. Hemming, V. G., Rodriguez, W., Kim, W. H., Brandt, C. D., Parrott, R. H., Burch, B., Prince, G. A., Baron, P. A., Fink, R. J., and Reaman, G., 1987, Intravenous immunoglobulin treatment of respiratory syncytial virus infections in infants and young children, *Antimicrob. Agents Chemother.* **31:**1882–1886.
6. Beasley, R. P., Lin, C. C., Wang, K. Y., Hsieh, F. J., Hwang, L. Y., Stevens, C. E., Sun, T. S., and Szumuness, W., 1981, Hepatitis B immune globulin (HBIG) efficacy in the interruption of perinatal transmission of hepatitis B virus carrier state, *Lancet* **2:**388–393.
7. Beasley, R. P., Hwang, L. Y., Stevens, C. E., Lin, C. C., Hsieh, F. J., Wang, K. Y., Sun, T. S., and Szmuness, W., 1983, Efficacy of hepatitis B immune globulin for prevention of perinatal

transmission of the hepatitis B virus carrier state: Final report of a randomized double-blind, placebo-controlled trial, *Hepatology* **3**:135–141.

8. Krugman, S., 1994, Hepatitis B vaccine, in: *Vaccines* (S. A. Plotkin and E. A. Mortimer, eds.), Saunders, Philadelphia, pp. 419–438.

9. Gorny, M. K., Pinter, A., and Zolla-Pazner, S., 1989, Specific immunity to HIV and other retroviral infections, in: *Progress in AIDS Pathology* (H. Rotterdam and S. C. Sommers, eds.), Field and Wood, New York, pp. 181–199.

10. Zolla-Pazner, S., O'Leary, J., and Burda, S., 1995, Serotyping of primary human immunodeficiency virus type 1 isolated from diverse geographic locations by flow cytometry, *J. Virol.* **69**:3807–3815.

11. Tyler, D. S., Stanley, S. D., Zolla-Pazner, S., Gorny, M. K., Shadduck, P. P., Langlois, A. J., Matthews, T. J., Bolognesi, D. P., Palker, T. J., and Weinhold, K. J., 1990, Identification of sites within gp41 that serve as targets for antibody-dependent cellular cytotoxicity by using human monoclonal antibodies, *J. Immunol.* **145**:3276–3282.

12. Alsmadi, O., Herz, R., Pinter, A., and Tilley, S. A., 1997, A novel antibody-dependent cellular cytotoxicity epitope in gp120 is identified by two monoclonal antibodies isolated from a long-term survivor of human immunodeficiency virus type 1 infection, *J. Virol.* **71**:925–933.

13. Baum, L. L., Cassutt, K. J., and Knigge, K., 1996, HIV-1 gp120-specific antibody-dependent cell-mediated cytotoxicity correlates with rate of disease progression, *J. Immunol.* **157**:2168–2173.

14. Ljunggren, K., Moschese, V., Broliden, P.-A., Giaquinto, C., Quinti, I., Fenyö, E. M., Wahren, B., Rossi, P., and Jondal, M., 1990, Antibodies mediating cellular cytotoxicity and neutralization correlate with a better clinical stage in children born to human immunodeficiency virus-infected mothers, *J. Infect. Dis.* **161**:198–202.

15. Zolla-Pazner, S., Gorny, M. K., Pinter, A., and Honnen, W., 1989, Reinterpretation of human immunodeficiency virus Western blot patterns, *N. Engl. J. Med.* **320**:1280–1281.

16. Pinter, A., Honnen, W., and Tilley, S. A., 1989, Oligomeric structure of gp41, the transmembrane protein of HIV, *J. Virol.* **63**:2674–2679.

17. Moore, J. P., Cao, Y., Qin, L., Sattentau, Q. J., Pyati, J., Koduri, R., Robinson, J., Barbas, C. R., Burton, D. R., and Ho, D. D., 1995, Primary isolates of human immunodeficiency virus type 1 are relatively resistant to neutralization by monoclonal antibodies, *J. Virol.* **69**:101–109.

18. Javaherian, K., Langlois, A. J., McDanal, C., Ross, K. L., Eckler, L. I., Jellis, C. L., Profy, A. T., Rusche, J. R., Bolognesi, D. P., and Putney, S. D., 1989, Principal neutralizing domain of the human immunodeficiency virus type 1 envelope protein, *Proc. Natl. Acad. Sci. USA* **86**:6768–6772.

19. Javaherian, K., Langlois, A. J., LaRosa, G. J., Profy, A. T., Bolognesi, D. P., Helihy, W. C., Putney, S. D., and Matthews, T. J., 1990, Broadly neutralizing antibodies elicited by the hypervariable neutralizing determinant of HIV-1, *Science* **250**:1590–1593.

20. LaRosa, G. J., Weinhold, K., Profy, A. T., Langlois, A. J., Dreesman, G. R., Boswell, R. N., Shadduck, P., Bolognesi, D. P., Matthews, T. J., and Emini, E. A., 1991, Conserved sequence and structural elements in the HIV-1 principal neutralizing determinant: Further clarifications, *Science* **253**:1146.

21. Goudsmit, J., Meloen, R. H., and Brasseur, R., 1990, Map of sequential B cell epitopes of the HIV-1 transmembrane protein using human antibodies as probe, *Intervirology* **31**:327–338.

22. Horal, P., Svennerholm, B., Jeansson, S., Rymo, L., Hall, W. W., and Vahlne, A., 1991, Continuous epitopes of the human immunodeficiency virus type I (HIV-I) transmembrane glycoprotein and reactivity of human sera to synthetic peptides representing various HIV-1 isolates, *J. Virol.* **65**:2718–2723.

23. Vanini, S., Longhi, R., Lazzarin, A., Vigo, E., Siccardi, A. G., and Viale, G., 1993, Discrete

regions of HIV-1 gp41 defined by syncytia-inhibiting affinity-purified human antibodies, *AIDS* **7**:167–174.

24. Xu, J. Y., Gorny, M. K., Palker, T., Karwowska, S., and Zolla-Pazner, S., 1991, Epitope mapping of two immunodominant domains of gp41, the transmembrane protein of human immunodeficiency virus type 1, using ten human monoclonal antibodies, *J. Virol.* **65**: 4832–4838.

25. Cotropia, J. P., Ugen, K. E., Kliks, S., Broliden, K., Broliden, P.-A., Hoxie, J. A., Srikantan, V., Williams, W. V., and Weiner, D. B., 1996, A human monoclonal antibody to HIV-1 gp41 with neutralizing activity against diverse laboratory isolates, *J. Acquired Immune Defic. Syndr. Hum. Retorvirol.* **12**:221–232.

26. Cotropia, J., Ugen, K. E., Lambert, D., Ljunggrenn-Broliden, K., Kliks, S., Hoxie, J., and Weiner, D. B., Characterization of human monoclonal antibodies (HuMAb) to the HIV-1 transmembrane (TM) gp41 protein, in: *Vaccines 92: Modern Approaches to New Vaccines Including Prevention of AIDS* (F. Brown, R. M. Chanock, H. S. Ginsberg, and R. A. Lerner, eds.), Cold Spring Harbor Laboratory Press, Cold Spring Harbor, New York, pp. 157–163.

27. Kliks, S., Weiss, C., Cotropia, J., Katinger, H., and Levy, J., 1994, *Synergism of broadly neutralizing effect by two transmembrane envelope gp41 HIV-1 monoclonal antibodies*, presented at the Conference on Advances in AIDS Vaccine Development, Seventh Annual Meeting of the National Cooperative Vaccine Development Groups for AIDS (NCVDG), Reston, Virginia.

28. Muster, T., Guinea, R., and Trkola, A., 1994, Cross-neutralizing antibodies against divergent human immunodeficiency virus type 1 isolates induced by the gp41 sequence ELDKWAS, *J. Virol.* **68**:4031–4034.

29. D'Souza, M. P., Livnat, D., Bradac, J., Bridges, S., AIDS Clinical Trials Group, Antibody Selection Working Group and Investigators, 1997, Evaluation of monoclonal antibodies to human immunodeficiency virus type 1 primary isolates by neutralization assays: Performance criteria for selecting candidate antibodies for clinical trials, *J. Infect. Dis.* **175**:1056–1062.

30. Cohen, J., 1993, Jitters jeopardize AIDS vaccine trials, *Science* **262**:980–981.

31. Mascola, J. R., Louwagie, J., McCutchan, F. E., Fischer, D. L., Hegerich, P. A., Wagner, K. F., Fowler, A. K., McNeil, J. G., and Burke, D. S., 1994, Two antigenically distinct subtypes of HIV-1; viral genotype predicts neutralization serotype, *J. Infect. Dis.* **169**:48–54.

32. Matthews, T. J., 1994, The dilemma of neutralization resistance of HIV-1 field isolates and vaccine development, *AIDS Res. Hum. Retrovir.* **10**:631–632.

33. Thali, M., Furman, C., Wahren, B., Posner, M., Ho, D. D., Robinson, J., and Sodroski, J., 1992, Cooperativity of neutralizing antibodies directed against the V3 and CD4 binding regions of the human immunodeficiency virus gp120 envelope glycoprotein, *J. Acquired Immune Defic. Syndr.* **5**:591–599.

34. Tilley, S. A., Honnen, W. J., Racho, M. E., Chou, T. C., and Pinter, A., 1992, Synergistic neutralization of HIV-1 by human monoclonal antibodies against the V3 loop and the CD4 binding site of gp120, *AIDS Res. Hum. Retrovir.* **8**:461–467.

35. Vijh-Warrier, S., Pinter, A., Honnen, W. J., and Tilley, S. A., 1996, Synergistic neutralization of human immunodeficiency virus type 1 by a chimpanzee monoclonal antibody against the V2 domain of gp120 in combination with monoclonal antibodies against the V3 loop and the CD4-binding site, *J. Virol.* **70**:4466–4473.

36. Li, A., Baba, T. W., Sodroski, J., Zolla-Pazner, S., Gorny, M. K., Robinson, J., Posner, M. R., Katinger, H., Barbas III, C. F., and Burton, D., 1997, Synergistic neutralization of a chimeric SIV/HIV type 1 virus with combinations of human anti-HIV type 1 envelope monoclonal antibodies or hyperimmune globulins, *AIDS Res. Hum. Retrovir.* **13**:647–656.

37. Hioe, C. E., Xu, S., Chigurupati, P., Burda, S., Williams, C., Gorny, M. K., and Zolla-Pazner, S., 1997, Neutralization of HIV-1 primary isolates by polyclonal and monoclonal human antibodies, *Int. Immunol.* **9**:1281–1290.
38. Broliden, P. A., Moschese, V., Ljunggren, K., Rosen, J., Fundaro, C., Plebani, A., Jondal, M., Rossi, P., and Wahren, B., 1989, Diagnostic implication of specific immunoglobulin G patterns of children born to HIV-infected mothers, *AIDS* **3**:577–582.
39. Goedert, J., Mendez, H., Drummond, J., Robert-Guroff, M., Minkoff, H., Holman, S., Stevens, R., Rubinstein, A., Blattner, W., and Willoughby, A., 1989, Mother-to-infant transmission of human immunodeficiency virus type 1: Association with prematurity or low anti-gp120, *Lancet* **2**:1351–1354.
40. Rossi, P., Moschese, V., Broliden, P., Fundaro, C., Quinti, I., Plebani, A., Giaquinto, C., Tovo, P., Ljunggren, K., and Rosen, J., 1989, Presence of maternal antibodies to human immunodeficiency virus 1 envelope glycoprotein gp120 epitopes correlates with the uninfected status of children born to seropositive mothers, *Proc. Natl. Acad. Sci. USA* **86**:8055–8058.
41. Devash, Y., Calvelli, T., Wood, D., Reagan, K., and Rubinstein, A., 1990, Vertical transmission of human immunodeficiency virus is correlated with the absence of high-affinity/avidity maternal antibodies to the gp120 principal neutralizing domain, *Proc. Natl. Acad. Sci. USA* **87**:3445–3449.
42. Goedert, J. J., and Dublin, S., 1994, Perinatal transmission of HIV type 1: Associations with maternal anti-HIV serological reactivity, *AIDS Res. Hum. Retrovir.* **10**:1125–1134.
43. Ugen, K. E., Goedert, J. J., Boyer, J., Refaeli, Y., Frank, I., Williams, W. V., Willoughby, A., Landesman, S., Mendez, H., and Rubinstein, A., 1992, Vertical transmission of human immunodeficiency virus (HIV) infection. Reactivity of maternal sera with glycoprotein 120 and 41 peptides from HIV type 1, *J. Clin. Invest.* **89**:1923–1930.
44. Ugen, K. E., Srikantan, V., Goedert, J. J., Nelson, R. P., Williams, W. V., and Weiner, D. B., 1997, Vertical transmission of the human immunodeficiency virus type 1: Seroreactivity by maternal antibodies to the carboxy region of the gp41 envelope glycoprotein, *J. Infect. Dis.* **175**:63–69.
45. Loomis-Price, L. D., Cox, J. H., Mascola, J. R., VanCott, T. C., Michael, N. L., Fouts, T. R., Redfield, R. R., Robb, M. L., Wahren, B., and Sheppard, H. W., 1998, Correlation between humoral responses to human immunodeficiency virus type 1 envelope and disease progression in early-stage infection, *J. Infect. Dis.* **178**:1306–1316.
46. Boyer, J. D., Ugen, K. E., Chattergoon, M., Wang, B., Shah, A., Agadjanyan, M., Bagarazzi, M., Javadian, A., Carrano, R., and Coney, L., 1997, DNA vaccination as anti-human immunodeficiency virus immunotherapy in infected chimpanzees, *J. Infect. Dis.* **176**:1501–1509.
47. Geffin, R., Scott, G. B., Melenwicki, M., Hutto, C., Lai, S., Boots, L. J., McKenna, P., Kessler, J. A., and Conley, A. J., 1998, Association of antibody reactivity to ELDKWA, a glycoprotein 41 neutralization epitope, with disease progression in children perinatally infected with HIV type 1, *AIDS Res. Hum. Retrovir.* **14**:579–590.
48. Jackson, G. G., Rubenis, M., Knigge, M., Perkins, J. T., Paul, D. A., Despotes, J. C., and Spencer, P., 1988, Passive immunoneutralisation of human immunodeficiency virus in patients with advanced AIDS, *Lancet* **ii**:647–652.
49. Karpas, A., Hill, F., Youle, M., Cullen, V., Gray, J., Byron, N., Hayhoe, F., Tenant-Flowers, M., Howard, L., and Gilgen, D., 1988, Effects of passive immunization in patients with the acquired immunodeficiency syndrome-related complex and acquired immunodeficiency syndrome, *Proc. Natl. Acad. Sci. USA* **85**:9234–9237.
50. Jacobson, J. M., Colman, N., and Ostrow, N. A., 1993, Passive immunotherapy in the treatment of advanced HIV infection, *J. Infect. Dis.* **168**:298–305.

51. Karpas, A., 1994, Passive immunotherapy in treatment of advanced HIV infection, *J. Infect. Dis.* **170:**742–744.

52. Levy, J., Youvan, T., Lee, M. L., and Group, P. H. T. S., 1994, Passive hyperimmune plasma therapy in the treatment of acquired immunodeficiency syndrome: Results of a 12 month multicenter double-blind controlled trial, *Blood* **84:**2130–2135.

53. Vittecoq, D., Chevret, S., and Morand-Joubert, L., 1995, Passive immunotherapy in AIDS: A double-blind randomized study based on transfusion of plasma rich in anti-HIV-1 antibodies vs transfusion of seronegative plasma, *Proc. Natl. Acad. Sci. USA* **92:**1195–1199.

54. Van Rompay, K. K., Berardi, C. J., Dillard-Telm, S., P., T. R., Canfield, D. R., Valverde, C. R., Montefiori, D. C., Cole, K. S., Montelaro, R. C., and Miller, C. J., 1998, Passive immunization of newborn rhesus macaques prevents oral simian immunodeficiency virus infection, *J. Infect. Dis.* **177:**1247–1259.

55. Emini, E., 1989, Neutralization of *in vivo* HIV-1 infectivity mediated by *in vitro* neutralizing antibody, in: *V International Conference on AIDS,* Montreal, Canada, p. 538.

56. Emini, E. A., Nara, P. L., Schleif, W. A., Lewis, J. A., Davide, J. P., Lee, D. R., Kessler, J., Conley, S., Matsushita, S., and Putney, S. D., 1990, Antibody-mediated *in vitro* neutralization of human immunodeficiency virus type 1 abolishes infectivity for chimpanzees, *J. Virol.* **64:**3674–3678.

57. Prince, A. M., Reesink, H., and Pascual, D., 1991, Prevention of HIV infection by passive immunization with HIV immunoglobulin, *AIDS Res. Hum. Retrovir.* **7:**971–973.

58. Emini, A. E., Schleif, W. A., Nunberg, J. H., Conley, A. J., Eda, Y., Tokiyoshe, S., Putney, S. D., Matsushita, S., Cobb, K. E., and Jett, C. M., 1992, Prevention of HIV-1 infection in chimpanzees by gp120 V3 domain-specific monoclonal antibody, *Nature* **355:**728–730.

59. Conley, A. J., Kessler, J. A., Boots, L., McKenna, P. M., Schleif, W., Emini, E., Mark, G., Katinger, H., Cobb, E. K., and Lunceford, S. M., 1996, The consequence of passive administration of an anti-human immunodeficiency virus type 1 neutralizing antibody before challenge of chimpanzees with a primary isolate, *J. Virol.* **70:**6751–6758.

60. Hinkula, J., Bratt, G., Gilljam, G., Nordlund, S., Broliden, P.-A., Holmberg, V., Olausson-Hansson, E., Albert, J., Sandströrm, E., and Wahren, B., 1994, Immunological and virological interactions in patients receiving passive immunotherapy with HIV-1 neutralizing monoclonal antibodies, *J. Acquired Immune Defic. Syndr.* **7:**940–951.

61. Günthard, H., Gowland, P. L., Schüpback, J., Fund, M. S. C., Böni, J., Liou, R.-S., Chang, N. T., Grob, P., Graepel, P., Braun, D. G., and Lüthy, R., 1994, A phase I/IIA clinical study with a chimeric mouse–human monoclonal antibody to the V3 loop of human immunodeficiency virus type 1 gp120, *J. Infect. Dis.* **170:**1384–1393.

62. Katinger, H., Purtscher, M., Muster, T., Steindl, F., Dopper, S., Vetter, N., Armbruster, C., and Gelbmann, H., 1995, A small phase I clinical trial with the human anti-Hiv-1 mAb 2F5, in: *Dixieme Colloque des Cent Gardes,* pp. 291–297.

63. Cavacini, L. A., Samore, M. H., Gambertoglio, J., Jackson, B., Duval, M., Wisnewski, A., Hammer, S., Koziel, C., Trapnell, C., and Posner, M. R., 1998, Phase I study of a human monoclonal antibody directed against the CD4-binding site of HIV type 1 glycoprotein 120, *AIDS Res. Hum. Retrovir.* **14:**545–550.

64. Safrit, J. T., Fung, M., Andrews, C., Braun, D., Sun, W., Chang, T. W., and Koup, R. A., 1993, hu-PBL-SCID mice can be protected from HIV-1 infection by passive transfer of monoclonal antibody to the principal neutralizing determinant of envelope gp120, *AIDS* **7:**15–21.

65. Gauduin, M.-C., Jeffrey, S., Weir, R., Fung, M. S. C., and Koup, R. A., 1995, Pre- and postexposure protection against human immunodeficiency virus type 1 infection mediated by a monoclonal antibody, *J. Infect. Dis.* **171:**1203–1209.

66. Parren, P. W. H. I., Ditzel, H. J., Gulizia, R. J., Binley, J. M., Barbas III, C. F., Burton, D. R.,

and Mosier, D. E., 1995, Protection against HIV-1 infection in hu-BPBL-SCID mice by passive immunization with a neutralizing human monoclonal antibody against the gp120 CD4-binding site, *AIDS* **9**:F1–F6.

67. Andrus, L., Prince, A. M., Bernal, I., McCormack, P., Lee, D.-H., Gorny, M. K., and Zolla-Pazner, S., 1998, Passive immunization with a human immunodeficiency virus type 1-neutralizing monoclonal antibody in hu-PBL-SCID mice: Isolation of a neutralization escape variant, *J. Infect. Dis.* **177**:889–897.

68. Poignard, P., Gulizia, R. J., Picchio, G., Wang, M., Parren, P. W. H., Mosier, D. E., and Burton, D. R., 1998, *Study of the impact of neutralizing antibodies on the course of established infection with HIV-1 primary isolates in the hu-PBL-SCID mouse model,* presented at the XIIth World AIDS Conference, Geneva, Switzerland.

69. Koup, R. A., Hesselton, R. M., Safrit, J. T., Somasundaran, M., and Sullivan, J. L., 1994, Quantitative assessment of human immunodeficiency virus type 1 replication in human xenografts of acutely-infected hu-PBL-SCID mice, *AIDS Res. Hum. Retrovir.* **10**:279–284.

70. Hesselton, R. M., Koup, R. A., Cromwell, M. A., Graham, B. S., Johns, M., and Sullivan, J. L., 1993, Human peripheral blood xenografts in the SCID mouse: Characterization of immunologic reconstitution, *J. Infect. Dis.* **168**:630–640.

71. Robinson, W. E., Kawamura, T., Gorny, M. K., Lake, D., Xu, J.-Y., Matsumoto, Y., Sugan, T., Masuho, Y., Mitchell, W. M., and Hersh, E. M., 1990, Human monoclonal antibodies to the human immunodeficiency virus type 1 (HIV-1) transmembrane glycoprotein gp41 enhance HIV-1 infection *in vitro*, *Proc. Natl. Acad. Sci. USA* **87**:3185–3189.

72. Robinson, W. E., Kawamura, T., Lake, D., Masuho, Y., Mitchell, W. M., and Hersh, E. M., 1990, Antibodies to the primary immunodominant domain of human immunodeficiency virus type 1 (HIV-1) glycoprotein gp41 enhance HIV-1 infection *in vitro*, *J. Virol.* **64**:5301–5305.

73. Eaton, A. M., Ugen, K. E., Weiner, D. B., Wildes, T., and Levy, J. A., 1994, An anti-gp41 human monoclonal antibody that enhances HIV-1 infection in the absence of complement, *AIDS Res. Hum. Retrovir.* **10**:13–18.

74. Bugge, T. H., Lindhardt, B. O., Hansen, L. L., Kusk, P., Hulgaard, E., Holmback, K., Klasse, P. J., Zeuthen, J., and Ulrich, K., 1990, Analysis of a highly immunodominant epitope in the human immunodeficiency virus type 1 transmembrane glycoprotein gp41, defined by a human monoclonal antibody, *J. Virol.* **64**:4123–4129.

75. Cavacini, L. A., Emes, C. L., Wisnewski, A. V., Power, J., Lewis, G., Montefiori, D., and Posner, M. R., 1998, Functional and molecular characterization of human monoclonal antibody reactive with the immunodominant region of HIV type 1 glycoprotein 41, *AIDS Res. Hum. Retrovir.* **14**:1271–1280.

76. Burton, D. R., Pyati, J., Koduri, R., Sharp, S. J., Thornton, G. B., Parren, P. W. H. I., Sawyer, L. S. W., Hendry, M., Dunlop, N., and Nara, P. L., 1994, Efficient neutralization of primary isolates of HIV-1 by a recombinant human monoclonal antibody, *Science* **266**:1024–1027.

77. Roben, P., Moore, J. P., Thali, M., Sodroski, J., Barbas III, C. F., and Burton, D. R., 1994, Recognition properties of a panel of human recombinant Fab fragments to the CD4 binding site of gp120 that show differing abilities to neutralize human immunodeficiency virus type 1, *J. Virol.* **68**:4821–4828.

78. Trkola, A., Purtscher, M., Muster, T., Ballaun, C., Buchacher, A., Sullivan, N., Srinivasan, K., Sodroski, J., Moore, J. P., and Katinger, H., 1996, Human monoclonal antibody 2G12 defines a distinctive neutralization epitope on the gp120 glycoprotein of human immunodeficiency virus type 1, *J. Virol.* **70**:1100–1108.

79. Buchacher, A., Predl, R., Strutzenberger, K., Steinfellner, W., Trkola, A., Purtscher, M., Gruber, G., Tauer, C., Steindl, F., and Jungbauer, A., 1994, Generation of human monoclo-

nal antibodies against HIV-1 proteins; Electrofusion and Epstein–Barr virus transformation for peripheral blood lymphocyte immortalization, *AIDS Res. Hum. Retrovir.* **10:** 359–369.

80. Conley, A. J., Kessler II, J. A., Boots, L. J., Tung, J.-S., Arnold, B. A., Keller, P. M., Shaw, A. R., and Emini, E. A., 1994, Neutralization of divergent human immunodeficiency virus type 1 variants and primary isolates by IAM-41-2F5, an anti-gp41 human monoclonal antibody, *Proc. Natl. Acad. Sci. USA* **91:**3348–3352.

11

Human Immunodeficiency Virus Type 1 Accessory Genes

Targets for Therapy

SAGAR KUDCHODKAR, T. NAGASHUNMUGAM, and VELPANDI AYYAVOO

1. INTRODUCTION

Human immunodeficiency virus type 1 (HIV-1) is the etiological agent for acquired immune deficiency syndrome (AIDS), one of the world's foremost health problems. The Joint United Nations Programme on HIV/AIDS estimated that 30.6 million people were infected with HIV as of December 1997. HIV infects and selectively eliminates the T helper lymphocytes of the body's immune system via interaction with the cell surface CD4 molecule as well as various chemokine receptors.[1-3] The helper T cells act as an important regulator of the body's immune response to an infectious agent through the secretion of cellular factors that stimulate antibody formation by B cells (humoral immunity) and the destruction of infected cells by cytotoxic T cells (cellular immunity). The elimination of the T helper cells and the subsequent loss of immunity makes the body unable to fight off other infectious agents and most AIDS patients eventually succumb to opportunistic pathogens. To date, there is no known cure or vaccine for AIDS.

HIV belongs to the lentivirus subgroup of the Retroviridae virus family.

SAGAR KUDCHODKAR and T. NAGASHUNMUGAM • Stellar-Chance Laboratories, Department of Pathology and Laboratory Medicine, University of Pennsylvania, Philadelphia, Pennsylvania 19104.   VELPANDI AYYAVOO • Department of Infectious Diseases and Microbiology, University of Pittsburgh, Pittsburgh, Pennsylvania 15261.

Human Retroviral Infections, edited by Kenneth E. Ugen *et al.*
Kluwer Academic / Plenum Publishers, New York, 2000.

Members of this family of viruses all contain RNA as their genetic material as opposed to the DNA found in most common viruses. The RNA is converted into cDNA in an infected cell by a reverse transcriptase protein also encoded by these viruses. Reverse transcriptase does not possess the proofreading ability of DNA polymerase and thus lentiviruses are highly prone to genetic mutability through errors in the replication of the RNA.[4,5] In general, all retroviruses contain three long open reading frames coding for the structural and enzymatic genes *gag, pol,* and *env.* In addition to its basic structural and enzymatic genes, HIV-1 also contains sequences encoding as many as six more genes, called "accessory genes," that provide a variety of regulatory roles that enhance the ability of the virus to replicate (Fig. 1). Presence of these regulatory genes adds an additional level of complexity to the HIV life cycle, and they likely contribute to the ability of HIV to utilize its environment to its own advantage to cause pathogenic effects.

The accessory genes of HIV-1 are *tat, rev, vif, vpu, vpr,* and *nef.* Some strains of HIV-1 contain the additional gene *tev,* and related lentiviruses HIV-2 and simian immunodeficiency virus (SIV) contain the gene *vpx* in place of *vpu.* In general, scientists define the function of these individual genes in terms of their importance to the virus life cycle. Although the phylogenetic conservation of many of the regulatory genes in the lentivirus subgroup implies that they all play important roles for the virus, some are clearly more important in that productive infection cannot proceed without them.[6,7] Virion formation requires both the *tat* gene, which mediates a transactivator function on the HIV long terminal repeat, and *rev,* which

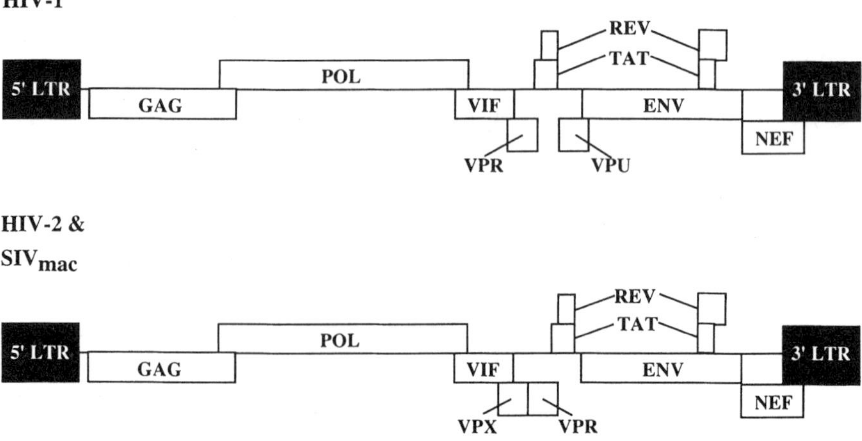

FIGURE 1. Genomic organization of HIV-1, HIV-2, and SIV.

mediates the change in gene expression from multiply spliced regulatory genes to unspliced or singly spliced structural genes. They are both regulatory genes that are absolutely essential for virus replication. Researchers deemed the other four genes "accessory genes" because initially they observed that the genes could be removed individually from the virus without affecting viral replication *in vitro*. Today, however, many studies, especially *in vivo* analysis, show naturally occurring deletion in most of these genes results in long-term nonprogression and replication-defective virus which leads to clinical latency and nontransmitter donors.[8–10] The new studies implicate the accessory genes in enhanced virion production as well as other contributions to the pathogenesis of HIV infection, and they lend support to the notion that the accessory genes are vital components of HIV.

There are no therapeutic drugs or strategies that are 100% effective for dealing with HIV and AIDS. The Food and Drug Administration has given approval for at least nine drugs that have shown some clinical efficacy in controlling viral growth. Six of the drugs target the reverse transcriptase protein of HIV, while the other three are inhibitors of the protease protein.[11] The Food and Drug Administration has not approved any drug that specifically targets the actions of the regulatory and accessory genes of HIV-1. The accessory genes remain attractive targets for future antiviral drugs due to the emerging evidence as to their importance in the viral life cycle. The success of drug treatment in AIDS patients is variable among patients. Successful containment of the virus by drug treatment involves many factors including the strain of virus infecting the patient and proximity of the start of drug treatment to the start of the infection. Drug treatments for HIV may prove futile due to the high genetic mutability of HIV and the development of drug-resistant strains.

The unreliability of drug therapies and the concern over the development of drug-resistant strains has led many scientists to pursue an anti-HIV vaccine that can confer immunity to a broad range of HIV strains. Traditional antiviral vaccination techniques have focused on the administration of live attenuated, whole killed, and nonlive preparations of the virus to the host. These three methods of vaccination have proven effective in generating both humoral and cellular immune responses against numerous viruses such as smallpox, measles, and influenza. A traditional vaccine cannot be developed against HIV because of concerns that it may revert to a virulent form. To circumvent these problems, scientists seek to develop a DNA-based vaccine against HIV. In DNA vaccination, a gene expression cassette encoding a particular gene or genes from a virus is injected into an organism. Once inside the organism, the genes are expressed, and their expression may lead to specific immune activation against the injected gene products.[12,13] DNA vaccines targeting the HIV-1 structural *env* and *gag–pol* genes are currently in clinical trials. Scientists are investigating DNA vaccines against the HIV-1 accessory and regulatory genes, but the fact that the translated products of these genes are immunogenic *in vivo* complicates their development.[14,15] The accessory genes are less susceptible to mutagenesis, making them a desirable, stable target for vaccine development.

The accessory genes of HIV contribute highly toward the virulence of HIV *in vivo*. The accessory genes and their translated products represent an important area of research in the field of lentivirus biology and in the endeavor of finding a cure or vaccine for HIV. The purpose of this chapter is to summarize what is known about the accessory genes of HIV-1. We will discuss the structure and function of these genes, and we will highlight advances made in antiviral therapies and vaccine development strategies that target the accessory genes. The conservation of the accessory genes in most lentiviruses suggests an important role for them in lentiviral infection, and through diligent study of these viral genes, a potential therapy against HIV may be found.

2. VIF

2.1. Structure and Function

The *vif* gene encodes for a highly basic, 23-kDa protein from a singly spliced, *rev*-dependent 5-kb transcript.[16–18] The *vif* gene is expressed late in the viral cycle at the same time as the genes for the structural proteins and after expression of various regulatory genes.[18,19] The notation "vif" stands for viral infectivity factor, a name assigned after initial observations that a *vif*-defective virus was 1000 times less effective in infecting cells than wild-type virus.[20] The deletion of *vif* has a significant effect on viral infectivity and moderate effects on pathogenicity.[21] The gene is highly conserved among HIV-1 isolates and is present in all but one known lentivirus, equine infectious anemia virus.[22] The exact role and mechanism of Vif protein in the virus life cycle is not well characterized. It is known that intact *vif* gene is essential for cell-free and cell–cell transmission of virus in certain types of cells that are permissive for infection.[23] Studies have identified two functions for Vif in HIV pathogenesis: the transport of the virus particles to the cell nucleus early in infection and the formation of stable virions during the late stages of the viral life cycle.

Many studies have been done to elucidate the structure of Vif. The Vif protein consists of approximately 192 amino acids. It contains two conserved cysteine residues at positions 114 and 133 that are necessary for Vif function and transcomplementation.[24] The cysteines of Vif do not have a structural role in terms of protein folding, but they do contribute to the function of Vif.[25] Earlier studies found that the C-terminus of Vif was essential in the stable association of Vif with membranes and for Vif function.[26] Phosphorylation of Vif is necessary for the targeting of Vif to specific cellular compartments. Our studies on stable cell lines expressing the *vif* gene showed that Vif is localized in the cytoplasm (Fig. 2).

The first function attributed to Vif takes place during the initial infection of a cell. It involves the transport of the virion and its contents from the cell surface to the nucleus of the cell. The Vif-mediated transport of virions to the cell nucleus is supported by evidence that shows Vif is found in a soluble form in the cytoplasm and in a cytoskeleton-associated form.[27] Vif also apparently stabilizes viral DNA intermediates after the virus has entered the cell.[11] The viral contents need to reach the nucleus, where they can hijack the native cellular mechanisms into working toward producing new viral particles. Studies indicate that Vif is present in viral particles.[11] This evidence suggests a role for Vif early in the viral life cycle.

The other function attributed to Vif manifests itself late in the viral life cycle and involves the assembly, budding, or postrelease maturation of new virions from infected cells. Viruses produced from *vif*-mutated or *vif*-deleted HIV strains produce virus particles with improperly packed nucleoprotein cores.[28] Studies show that up to 90% of Vif associates with the cell membrane and that it colocalizes with the virally encoded Gag proteins primarily through the last 22 amino acids of its C-terminal end.[29,30] The

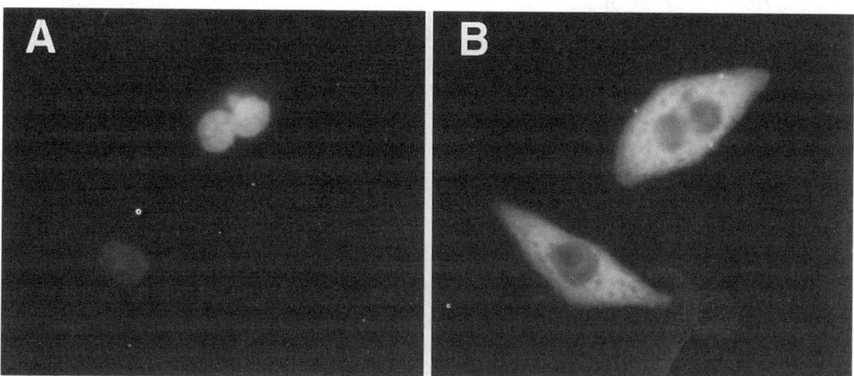

FIGURE 2. Subcellular localization of Vif in stably transfected RD cells. (A) Nuclear staining; (B) Vif staining. RD cells were transfected with pcVif, *vif* expression plasmid and selected in the presence of G418. Positive cells were stained with polyclonal anti-Vif antisera and detected with anti-rabbit secondary antibody conjugated with fluorescein isothiocyanate.

association of the Gag precursor with the inner leaflet of the plasma membrane is an initiating force in the assembly of the HIV-1 nucleoprotein core.[31,32] The association of Vif with the cell membrane can aid Vif in the role of viral assembly and morphogenesis. Viruses produced by cells infected with a *vif*-deleted virus are less infectious than wild-type virus, and the infectivity cannot be rescued by expression of Vif in the target cell, further supporting a role for Vif late in the viral life cycle.[33–35]

The consequences of infection by *vif*-deleted mutant viruses were observed to be different in various cell types. Cells can be grouped into the categories permissive, restrictive, or semi-permissive based on their ability to carry on viral replication in the absence of *vif*. The basis of these infection phenotypes is not known. A study of the protein contents of purified viral particles was unable to find any Vif- or cell-type-dependent quantitative or qualitative differences in the structural proteins of produced virions or virus-producing cells to explain this phenomena.[36] Permissive cells include HeLa, COS (monkey kidney), and M8166 (T cell line). Infection by *vif*-deleted viruses does not affect viral replication in these cells.[37] Viral replication requires the presence of *vif* in the restrictive cell group. Macrophages, PBLs, CEMx174, and H9 (rare T lymphoid cell line) cells fall into this category.[38] The viral products of restrictive cells are less infectious than wild-type virus. The virions produced by the restrictive cells are morphologically different than virions from permissive cells, and they contain altered amounts of various viral proteins.[33] Semipermissive cells include SupT1 and CEM, and a *vif*-deleted virus replicates at intermediate levels in these cells. In the laboratory the production of high-titer stocks of *vif*-deleted viruses has been difficult due to its inability to propagate in various cell lines.[39]

Earlier analyses of *in vivo vif* genetic variation have shown that most *vif* sequences are intact reading frames and the presence of intact *vif* does not have a correlation with disease status. Initial studies found that 58% of *vif* proviral sequences from 61 peripheral blood mononuclear cell (PBMC) samples from HIV-positive donors were conserved, with most sequence conservation found in the 5′ end.[40] Another study found that 38 of 46 *vif* clones derived from primary isolates of five asymptomatic and five patients with AIDS carried open *vif* reading frames and that the *vif* genes from isolates from the same patient and isolates from other patients showed very similar sequences.[41] Both studies suggest a role for Vif in natural HIV-1 infection. A study of long-term survivors (people who are HIV-1-positive for over 10 years, but have shown no clinical signs of AIDS) found that the *vif* gene in these patients contained no gross deletions or insertions and that they contained intact open reading frames.[42]

In order to determine the nature and the sequence variation of the *vif* gene *in vivo*, we cloned and analyzed *vif* variants present in uncultured

PBMC from HIV-1-positive subjects.[14] Our analysis of 20 different *vif* sequences from two subjects (10 from each subject) revealed that *vif* is highly conserved (approximately 90%) within a particular patient at a given time (Fig. 3). We found that the 5′ end (amino acids 20–85) is actually more variable and the 3′ end is more conserved in *vif*. Our findings that the 3′ end is more conserved than initially thought are supported by previous mutagenesis studies that implicate the carboxy-terminus of Vif (amino acids 171–192) plays a role in the stable association of Vif with membranes.

2.2. Vaccine and Drug Studies

An effective vaccine against Vif may be an important part of a multicomponent therapy to prevent or retard the pathogenesis of HIV in an individual. HIV-positive individuals have been shown to have antibodies against recombinant Vif protein. The presence of antibodies against Vif suggests that the protein is expressed and immunogenic during the course of HIV infection.[43–45] The immune response generated by an anti-Vif vaccine may be able to affect the early actions of Vif and inhibit viral pathogenesis.

Due to Vif's ability to affect viral replication *in trans*, an attenuated genetic vaccine design needs to be applied. In one such study design, we studied the sequence variation and immunogenic potential present in *vif* genes derived from HIV-1-infected subjects. We selected prototypic genetic variants, and studied the ability of these clones to induce humoral and cellular immune responses in animals. The selected *vif* genetic variants were also functionally characterized through transcomplementation assays utilizing cells infected with a *vif*-defective HIV-1 clone. It was demonstrated that attenuated, nonfunctional *vif* clones can induce immune responses capable of destroying native pathogen (Fig. 4). These clones could be beneficial components in a future genetic vaccine for HIV-1. This strategy may also be used to develop a safe and yet immunologically effective DNA vaccine for any potential pathogen.[14]

No antiviral therapies have been developed against Vif. Potential targets for future anti-Vif therapies include drugs that can block the association of Vif with the cytoskeleton or cellular membrane. The association of Vif with these regions of the cell appears to be important for Vif function and thus targeting them with drugs may attenuate the function of Vif *in vivo*. Therapies or drugs that block the phosphorylation of the important residues or that can inhibit the binding of the conserved cysteine residues may also have some clinical relevance. The role that Vif appears to play in both viral infectivity and virion formation suggests that anti-Vif therapies can control HIV infection.

```
1                                                             51                                                            100
N24  ---------- ---------- ------y--- r-----e--- ---q------ --------e- ----v----- ---t------ ---------- ------h---
N27  ---------- ---------- ------y--- r-----e--- ---q------ --------e- ----v----- ---t------ ---------- ------h---
N26  ---------- ---------- ------y--- r-----e--- ---q------ --------e- ----v----- ---t------ ---------- ----------
N15  ---------- ---------- ------y--- r-----e--- ---q---r-- --------e- ---------- ---t------ ---------- ----------
N17  ---------- ---------- ------y--- r-----e--- ---q------ --------e- ---------- ---t------ ---------- ----------
N23  ---------- ---------- ------y--- r-----e--- ---q------ --------e- ---------- ---t------ ---------- ------h---
N29  ---------- ---------- ------y--- r----e--n- ----hr---- --------e- ---------- ---t------ ---------- ----------
N13  ------i--- ---------- ------y--- *-----e--- *--q------ --------e- ------s--- ---t------ ---------- ------h---
N22  ---------- --.------- ----ty---- r-q--e-n-- ----h----- --------e- ----a-p--- ---t------ ---------- ----------
T39  ---------- ---------a ---------- -----kk--- ---------- ---------- ---------- ---a------ ---------- ------h---
T44  ---------- ---------a ---------- -----kk--- -------q-- ---------- ---------- ---a------ ---------- ----------
T43  ---------- ---------a ---------- -----kk--- ---------- ---------- ---------- ---a------ ---------- ----------
T37  ----e----- -e-------a ---------- -----kk--- ---------- ---------- ----v----- ---a------ ---------- ----------
T40  ---------- ------t--a ---------- -----kk--- ---------- ---------- ----v----- ---a------ ---------- ----------
T35  ---------- ---------a --------i- f----kk--- --------n- ---------- ----vt-p-- ---g-----y -a-------- ----------
T4   ---------- ---------a ---------- ------t--s ---g------ -c-------- ----v----- s--a--*--- v--r------ ----------
T3   ---------- ---------- ---------- -----kk--- ---------- ---ta----- -g---k-av- s-qa--gv-- r-hp------ ----------
T38  ---------- ---------a ---------- ---n-kk--- ---------v q--ta----- -g--qkia-- s-da------ ---------- ----------
T42  ---------- ---------a ---------- -----kk--n -----r---- ---------- ---------- ---a------ ---r------ ----------
Con  MENRWQVMIV WQVDRMRIRT WNSLVKHHMY VSKKAR-WFY RHHYESPHPK VSSEVHIPLG DARLETTTYW GLH-GERDWH LGQGVSIEWR KRRYSTQVDP

101                                                           151                                                          194
N24  ---------- ---------- ------h--- -----r---s --------i- ---a------ ---------- ---------- ------.--- ---*
N27  ---------- ---------- ------h--- -----r---s --------i- ---------- ---------- ---------- ------.--- ---*
N26  ----h----- ---------- ------h--- -----r---s --------i- ---a------ ---------- ---------- ------.--- ---*
N15  ---------- ---------- ------h--- -----r---s --------i- ---------- ---------- ---------- ------.--- ---*
N17  ---------- ---------- ------h--- -----r---s --------i- ---------- ---------- ---------- ------.--- ---*
N23  ----h----- ---------- ------h--- -----r---s --------i- ---------- ---------- ---------- ------.--- ---*
N29  ---------- ---------- ------h--- -----r---s --------i- ---------- ---------- ---------- ------.--- ---*
N13  ---------- ---------- ------h--- -----r---s --------i- ---a------ ---------- ---------- ------.--- ---*
N22  ---------- ---------- ------h--- -----r---s --------i- ---------- ---------- ---------- ------.--- ---*
T39  ---------- ---------- ---------- ---------- ---------- ---------- ---------- ---------- ------.--- ---*
T44  ---------- ---------- ---------- ---------- ---------- ---------- ---------- ---------- ------.--- ---*
T43  ---------- --g------- ---------- ---------- ---------- -------g-- ---------- ---------- ------.--- ---*
T37  ---------- ---------- ---------- ---------- ---------- ---------- ---------- ---------- ------.--- ---*
T40  ------t--- ---------- ---------- ---------- ---------- ---------- ---------- ---------- ------.--- ---*
T35  ---------- ---------- ---------- ---------- ---------- ---------- ---------- ---------- ------.--- ---*
T4   ---------- ---------- ---------- ---------- ---------- ---------- ---------- ---------- ------.--- ---*
T3   --v------- ---------- ---------- ---------- ---------- ---------- ---------- ---------- ------.--- ---*
T38  ---------- ---------- ---------- ---------- ---------- ---------- ---------- ---------- -----r.--- ---*
T42  ------t--- ---------- ---------- ---------- ---------- ---------- ---------- ---------- --te-aiq*- dt..
Con  DLADQLIHLY YFDCFSESAI RKAILGYRVS PRCEYQAGHN KVGSLQYIAL AALITPKKIK PPLPSVRKLT EDRWNKPQKT KGHRGS-HIM NGH-
```

FIGURE 3. Comparison of *vif* deduced amino acid sequences isolated from patient samples. T-** represent *vif* clones from patients with AIDS; N-** represent *vif* clones derived from an asymptomatic patient.

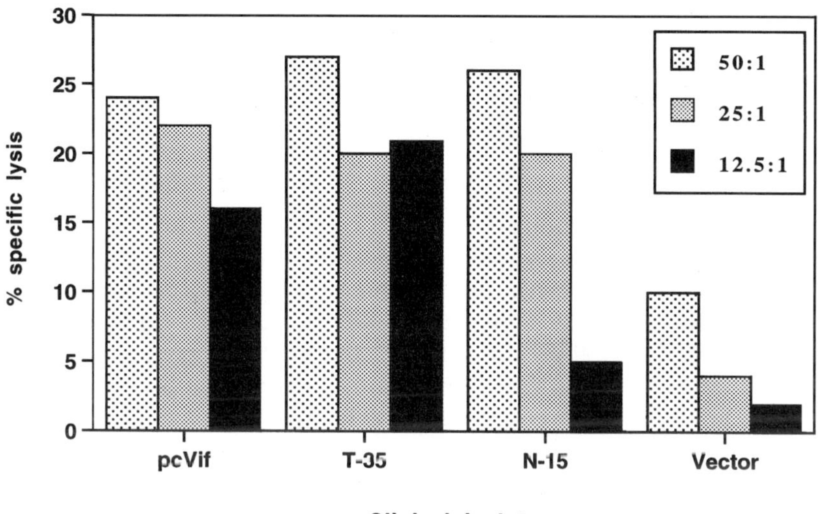

FIGURE 4. Cytotoxic T cell lysis of HeLa CD4$^+$/Dd$^+$ cells infected with clinical HIV-1 isolates by splenocytes from mice immunized with *vif* expression cassettes. HeLa CD4$^+$/Dd$^+$ cells (10^6) were infected with a cell-free HIV-1 clinical isolate followed by 1 week of incubation to allow the cells to infect and express viral proteins. One week postinfection, the target cells were labeled with ^{51}Cr for 1–2 h and used to incubate the stimulated splenocytes for 6 h. Specific lysis (%) was calculated according to the formula [(experimental release − minimum release)/maximum release − minimum release)] × 100.

3. VPU

3.1. Structure and Function

Viral protein U (Vpu) is a 16-kDa transmembrane, late viral protein found only in HIV-1 and not HIV-2 or SIV.[46] Vpu is associated with the internal cell membrane in HIV-infected cells. Vpu is synthesized from an intermediate-sized mRNA precursor at the same time as the envelope precursor gp160.[47] The two functions associated with Vpu are degradation of the CD4 receptor molecule in the endoplasmic reticulum (ER) and enhancement of virion release from the cell. The two functions occur in distinct cellular compartments, the former in the ER and the latter on the cytoplasmic side of the plasma membrane.

Vpu is a nonstructural, integral membrane protein 81 amino acids in length which associates with the membrane in oligomeric form.[48] Binding of Vpu with the cell membrane involves its hydrophobic, 27-amino acid N-terminal region.[48,49] The cytoplasmic tail of Vpu is 54 amino acids in length and highly hydrophilic.[50,51] It contains two amphipathic α-helical regions that flank a highly conserved acidic 12 amino acid segment.[52] Phosphorylation by the enzyme casein kinase-2 occurs at two serine residues at positions 52 and 56 in the acidic region, and the phosphorylation of these residues imparts functionality to Vpu.[53,54]

The first major function attributed to Vpu is the degradation of CD4 in the ER. Vpu is not a proteolytic enzyme, but it does appear to activate certain proteolytic pathways in the ER that lead to the degradation of the CD4 molecules. The CD4 molecules produced by the T cells form a tight complex with the gp160 precursor within the ER.[55–57] The interaction disrupts the proteolytic cleavage of gp160 into its component subunits gp120 and gp41, a process that precludes proper viral formation.[58] The removal of CD4 from the cell surface also attenuates the proper function of the T cell and thus enhances the effects of the virus in destroying the immune system. New studies indicate that Vpu may also contribute to the downregulation of major histocompatibility complex I (MHC-1) molecules by interfering with an early synthesis step.[59] MHC-1 molecules on antigen-presenting cells present peptide segments from immunogens to the cytotoxic T cells, so the downregulation of MHC-1 molecules aids in the escape from cytotoxic T cell (CTL) lysis of infected cells. By affecting the CD4 molecules, Vpu plays a major role in HIV pathogenesis.

Although the precise mechanism by which Vpu causes the degradation of CD4 is not known, evidence suggests that the interaction of Vpu with CD4 triggers proteolytic cellular mechanisms. The Vpu-responsive element was mapped to a six-amino acid region of the CD4 tail through studies with chimeric envelope glycoproteins.[60,61] Studies indicate that CD4 phosphorylation is not as important for Vpu-mediated proteolysis as the phosphorylation of the Ser52 and Ser56.[51] New data suggest that Vpu directs CD4 and possibly MHC-1 to ER-associated protein degradation (ERAD).[62] ERAD is a normal cell process by which misfolded and improperly processed proteins are eliminated by proteases in the ER before their release.[63] Vpu-mediated degradation of CD4 requires the action of proteosomes, does not involve the ER chaperone protein calnexin, and involves the cytosolic proteasome-ubiquitin-conjugating pathway.[62] Trafficking the naturally expressed CD4 molecule to degradative cellular processes is one way by which Vpu enhances viral infectivity.

The other major function of Vpu is in the efficient release of virus from infected cells. Unlike its role in CD4 degradation, this function appears to be compartmental nonspecific. The N-terminal transmembrane portion of

Vpu is responsible for this function. Experiments with *vpu*-deleted mutant virus show an inefficient release of these viruses into the cell culture medium by infected cells.[64,65] Viruses produced in cells infected by the *vpu*-deleted viruses are produced in internal membrane compartments and are tethered to the plasma membrane. These studies implicate Vpu in directing the new viral components to the plasma membrane for proper assembly. The effects of Vpu in virus release appear to be cell type dependent, as COS-7 cells were able to release viral particles normally both in the presence and absence of Vpu.[66] The results of this study suggest Vpu activity may mimic certain native cell factors and functions, and that may explain why lentiviruses like HIV-2 and SIV can release virus properly without Vpu. One consequence of *vpu* deletion is the premature death of infected cells brought on by the increased ratio of intracellular to extracellular proteins.[66]

The mechanism by which Vpu causes efficient viral particle release is not known, but it is believed to involve the formation of ion channels. Vpu is related to the M2 protein of influenza, which functions to produce ion channels.[11] Ion channels provide a path in the plasma membrane through which new virions can be efficiently released into the cell media.

3.2. Vaccine and Drug Studies

No anti-Vpu vaccine has been developed, but its reported role in HIV pathogenesis makes it an attractive target for vaccine development. Studies on *vpu* nucleotide sequences from long-term survivors found no gross deletions or mutations, suggesting that *vpu* is highly conserved through the course of HIV infection.[42] The implications of this study are that Vpu is a stable protein against which a vaccine could be developed. No evidence for the presence of antibodies to Vpu has been found. The lack of such evidence can be traced to the fact that Vpu is an internally expressed protein, and also suggests that it may be hard to develop humoral responses against it. Our lab has been developing a *vpu*-DNA cassette to begin vaccine studies. We have created a few constructs, and tested them *in vivo* in mice, but so far we have not seen any responses.

There are no antiviral therapies against Vpu, although the potential for one exists. The interaction of Vpu with the cytoplasmic tail of CD4 is a possible target. This interaction is essential for Vpu activity, and attacking it with therapies that inhibit this binding can theoretically inhibit T cell destruction and viral formation. Such a therapy may block the phosphorylation of the important residues on Vpu and/or CD4 that precludes activity. The formation of ion channels by the transmembrane domain is another potential target for anti-Vpu therapy.

4. NEF

4.1. Structure and Function

The negative factor (*nef*) gene encodes a 25- to 27-kDa myristylated cytoplasmic protein. Studies indicate that the *nef* gene is unique to primate lentiviruses.[67,68] The Nef protein was so named because early studies found that it suppressed the viral transcription from the long terminal repeat which downregulated viral gene expression.[69–71] It is now known that Nef is not required for infection, but it plays a prominent role in increasing viral replication and pathogenesis.[72] Studies have failed to characterize its role in HIV infection. Primate studies using SIV show that *nef* deletion attenuates the action of the virus and *nef*-deleted SIV fails to produce AIDS or death in the primates as opposed to wild-type SIV.[73] Additionally, HIV-1 positive, long-term nonprogressors show deletion in the *nef* gene and the viruses in these individuals are M-tropic, non-syncitium-inducing as opposed to the T-tropic viruses usually associated with advanced progression of HIV.[74,75] These studies indicate that *nef* plays a significant role in natural infection. Newer evidence on the less pathogenic HIV-2 indicates the presence of a truncated *nef* gene that may explain the reduced virulence of this HIV-1 cousin.[76] The two major actions attributed to Nef are the downregulation of cell surface CD4 molecules and the enhanced infectivity of cell-free virus. Evidence also links *nef* expression to the downregulation of cell surface MHC molecules and the altering of T cell-activation pathways.

The expression of *nef* and its resultant product have been well studied compared to the other accessory genes. Nef is 206 amino acids in length. It is associated with the virion at levels of 5–10 copies per molecule, and is cleaved by viral proteases expressed before gene expression in the target cell.[77–79] The *nef* gene is one of the earliest genes expressed in HIV infection, as indicated by studies showing that up to 90% of HIV-1-specific mRNAs expressed within the first 6 h of infection are from *nef*.[80] The amino acid sequence of Nef from various HIV isolates shows a high degree of heterogeneity, with the most variable regions being near the amino- and the carboxy-terminal ends.[81–84] A study of sequences of *nef* genes from HIV-infected individuals found several conserved sequences that coded for an invariant myristylation signal, an acidic region, a region with four PxxP repeat sequences, and a potential site for protein kinase C phosphorylation.[85] Residues 113–147 and the C-terminal end contain the prominent T cell epitopes of Nef.[86,87] The residues 73–97 are the major targets for CTL.[88]

Various researchers have shown that the translated Nef protein from different cell types have molecular weights of 25–27 kDa, with the 27-kDa form being the most stable.[89,90] One study found that the 25-kDa protein was more basic than the 27-kDa protein, a fact that may have importance in the folding of the proteins.[89] After translation, the 27-kDa Nef protein is myristylated, with a saturated fatty acid being added to its amino-terminal end.[11] The 25-kDa protein lacks the myristylation site.[90] Nef localizes primarily in the cytoplasm, but the 27-kDa product can associate with the cell membrane, suggesting a site for membrane association in the N-terminal end.[91] In the cytoplasm, Nef interacts with the actin molecules of the cytoskeleton via its N-terminal end in a process that is myristylation dependent.[92] There is evidence that Nef from some isolates localizes in the nucleus of target cells.[93,94] Studies on Nef structure show that Nef has a well-folded and compact tertiary structure.[95] The Nef protein contains sequences that are homologous to sequences in GTP-binding G-proteins, but there is little evidence that Nef actually binds with GTP or has GTPase activity.[96,97] The functions of Nef emanate from its interaction with various cellular factors known as Nef-associating kinases.[98–104]

The best defined function of Nef, demonstrated in various cell lines and with different isolates, is a role in the downmodulation of CD4 cell-surface molecules on T cells.[105–107] The mechanism of action invovles the rapid internalization of the CD4 molecules that leads to the accumulation of the molecules in early endosomes.[108] Phosphorylation of Nef is necessary for the activation of this mechanism, and involves the N-terminal region of Nef.[109–111] Nef expression does not affect CD4 synthesis, but does reduce its life span.[112] Interaction of Nef with CD4 occurs with the 20 proximal residues from the cytoplasmic domain of CD4, a region that contains dileucine sorting signals.[108] Recent evidence implicates Nef in the incorporation of CD4 into clathrin-coated pits that then endocytose the receptor into the cell where it is subjected to lysosomal activity.[113,114] It is not known why degradation of CD4 function is shared by both Nef and Vpu, but the actions of both proteins contribute to reducing the potency of the T cell and preventing the superinfectivity of an infected cell.

The mechanisms by which Nef enhances viral replication and infectivity are not fully known. As mentioned above, early studies described Nef as a negative regulator of viral pathogenesis, but these initial studies have been refuted by newer studies that show Nef dramatically increases viral replication. One study proposes a role for Nef in stimulating the synthesis of proviral DNA.[115] Our results from constructing chimeric viruses by exchanging the *nef* gene from highly replicating proviral (pZr6) to the low-replicating strain (pHxB2) and vice versa showed that an intact *nef* gene is essential for increased viral replication (Fig. 5). Studies on SIV, as mentioned above, clearly indicate that *nef* is essential for viral replication. Interaction with cellular factors, some of which are described above, is believed to play an important role.

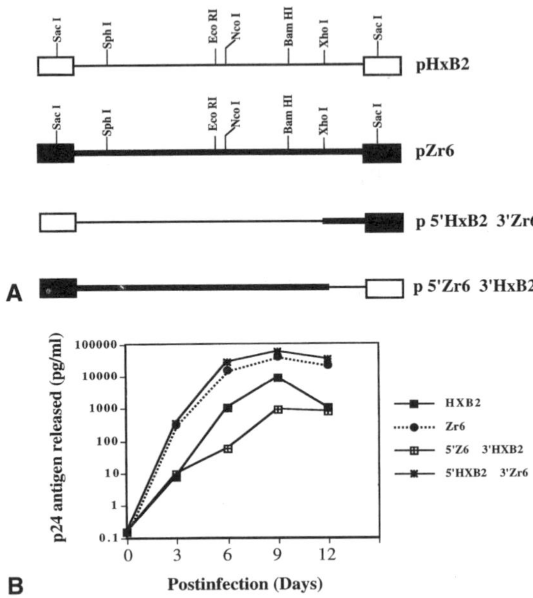

FIGURE 5. (A) Structure of chimeric proviral DNAs with exchanges involving the 3′-end of the genome. HIV proviral DNAs were digested with *Xho*I and the fragments containing the 5′-Z6 and 3′-HXB2 *Xho*I were ligated to generate chimeric proviral DNA. The chimeric constructs were verified by restriction enzyme mapping. (B) Infectivity and replication kinetics of chimeric viruses (Nef exchange) derived from pZ6 and pHxB2. Phytohemagglutinin-stimulated peripheral blood lymphocytes (PBLs) were infected with 100 ng p24 equivalent of viruses, and the kinetics of replication was monitored for 21 days.

The other aspect of Nef function at the cellular level is the alteration of normal T cell function. Evidence for this comes from experiments that show Nef expression leads to the blocking of the interleukin-2 (IL-2) and NF-κB induction.[116–118] Nef has also been shown recently to help infected cells escape degradation by cytotoxic T lymphocytes by the downregulation of MHC on the cell surface.[119] Again, the mechanisms by which Nef mediates these events are not known, but they likely involve interaction between Nef and cellular factors.

4.2. Vaccine and Drug Studies

Of all the accessory proteins, Nef is the best target for vaccine development. In the context of HIV-1 infection, Nef is one of the early proteins produced in abundance, and a majority (two thirds) of HIV-1-positive patients generate Nef-specific CTL responses, indicating that Nef is a potent immunogen.[120] Assays of sera from HIV-positive patients reveal the presence of anti-Nef antibodies.[121–123] Another study found that individuals with persistent levels of anti-Nef antibodies exhibited fewer of the markers of advanced HIV-1 disease than those whose antibody level varied.[124] Proliferative responses against recombinant Nef protein have been detected in inoculated mice.[125] Because of Nef's role early on in viral infection, a vaccine against this gene product administered alone or in conjunction with other vaccines or therapies has the potential of conferring immunity against HIV infection.

Another aspect of vaccine design that involves *nef* is vaccination using *nef*-deleted viruses. Studies using monkeys injected with *nef*-deleted SIV showed that it could protect inoculated monkeys from viral challenge with a complete SIV molecule. The *nef*-deleted virus replicated poorly in the monkeys and was nonpathogenic.[126] These data combined with other data mentioned above suggest that HIV virus lacking the *nef* gene may be sufficiently attenuated to be used as a vaccine. So far no studies have been done in humans to confirm this. Other tests that show HIV has the ability to replicate without *nef* means more studies will have to be done before a viable *nef*-deleted vaccine can be developed.

No antiviral therapies against Nef have been developed, but the interaction of Nef with its associated kinases is a prime target for rational drug design studies. Blocking Nef's ability to interact with CD4 or other factors leading to CD4 destruction may help to keep the immune system strong enough to fight off further infection. Antiviral therapies or drugs that can block Nef's role in T cell function alteration and its role in increasing viral replication may prove to be important components in an anti-HIV therapy.

5. VPR

5.1. Structure and Function

The presence of an open reading frame encoding viral protein R (Vpr) and the fact that a protein of the predicted molecular weight could be immunoprecipitated with serum from HIV-positive individuals were first described in 1987.[127] The *vpr* gene encodes a protein of 96 amino acids with a predicted molecular weight of 15 kDa and is relatively conserved by lentiviruses.[128–131] The accessory gene *vpr*, while dispensable for viral replication in T cell lines and activated PBMCs,[132–136] is required for efficient replication in primary monocytes/macrophages.

A recent study has characterized the Vpr protein as an oligomer.[137] Although Vpr and Vpx are not part of the Gag structural polyprotein, their incorporation requires an anchor to associate with the assembling capsid structures. It has been reported that Vpr is primarily localized in the nucleus when expressed in the absence of other HIV-1 proteins.[138–141] The C-terminal portion of the Gag precursor corresponding to the p6 protein appears to constitute such an anchor through an unknown mechanism. In addition, p6 is essential for the incorporation of both Vpr and Vpx into virus particles.[142–144] A putative α-helical domain near the amino terminus plays an important role in the packaging of Vpr into virions and in maintaining protein stability.[145,146]

The role of Vpr in AIDS pathogenesis is not well understood. In one study, macaques infected with SIV mac239 containing a mutation in the *vpr* initiation codon methionine progressed to AIDS at a slower rate than those infected with wild-type virus.[147] In another, a *vpr* mutant of SIVmac239 retained full pathogenecity, but in conjunction with a mutation in *vpx*, replicated at low levels and was nonpathogenic when inoculated into macaques.[148] Several possible roles have been suggested for Vpr in HIV-1 replication. Vpr modestly transactivates HIV-1 long terminal repeat[149] and thus may upregulate viral gene expression in newly infected cells before the appearance of Tat. It has been found to enhance the nuclear migration of the preintegration complex in newly infected nondividing cells.[150]

Interestingly, Vpr exhibits many dramatic effects on host cellular events in the absence of other HIV-1 proteins. Vpr induces cell proliferation and cellular differentiation in many cell lineages.[151,152] Importantly, it inhibits cell proliferation in T cells and macrophages in a dose-dependent manner (Fig. 6) and it blocks the cell cycle at the $G_{2/M}$ phase (Fig. 7). This finding has been associated with a change in the phosphorylation state of CDC2 kinase.[153–156] Furthermore, Vpr expression appears to inhibit the establishment of chronic infection in cells.[157–159] Vpr accelerates HIV-1 replication in some T-lymphoid cell lines and in primary macrophages where the effects of Vpr are more pronounced.[160–163] It has also been reported that Vpr has transcellular activity.[164] Both Vpr purified from the plasma of HIV-1-seropositive individuals and purified recombinant Vpr were capable of inducing latent cells into high-level viral producers when added to culture media at low concentrations.[164] Mechanistically, it is conceivable that this transcellular activity is mediated by the same mechanisms which modify cellular growth and differentiation.

Although no classical nuclear localization signal has been clearly identified for Vpr, it has been suggested that Vpr may gain access to the nucleus by specific interactions with nuclear proteins.[165,166] In this regard, proteins that interact with Vpr in host cells have been molecularly cloned.[167,168] Interestingly, one of these Vpr targets, designated Human Vpr-interacting protein (hVIP-1), appears to be translocated to the nucleus following its inter-

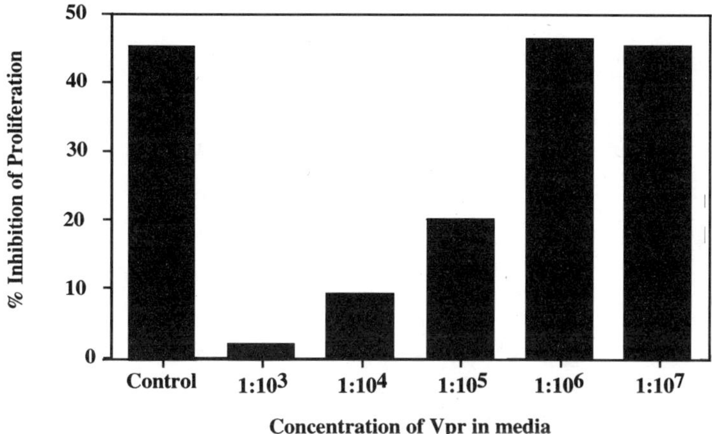

FIGURE 6. Effect of Vpr on cell proliferation. Human peripheral blood lymphocytes (PBLs) were stimulated with phytohemagglutinin (5 µg/ml) and cultured in the presence of Vpr protein for 48 h and assayed for cell proliferation by thymidine incorporation.

action with Vpr or triggering by glucocorticoid receptor ligands, supporting a possible role for the glucocorticoid receptor pathway in Vpr function.[165]

When we investigated the effects of Vpr on host cell activation, we observed that Vpr influences cellular proliferation and, more importantly, it modulates T cell receptor (TCR)-triggered apoptosis in a manner similar to glucocorticoids. The regulation of apoptosis was found to be due to the ability of Vpr to suppress NF-κB activity through the induction of IκB transcription.[169] The regulation of NF-κB by HIV-1 Vpr suggests a molecular basis for cellular effects of Vpr and supports the likelihood that some aspects of viral pathogenesis are the consequences of cellular dysregulation by this novel gene product. The results indicate that Vpr suppresses immune cells in a similar manner to that of the well-studied immunosuppressive glucocorticoid drugs. Independent of this, another study showed that Vpr, either alone or in the context of viral infection, is capable of inducing apoptosis *in vitro*.[170]

5.2. Vaccine and Drug Studies

No vaccine has been developed against Vpr. Since Vpr induces immunosuppressive effects *in vitro*, using the *vpr* gene in a vaccine construct has to be further evaluated in detail. Mapping the different functional regions and attenuating the functional domains need to be explored in depth before a

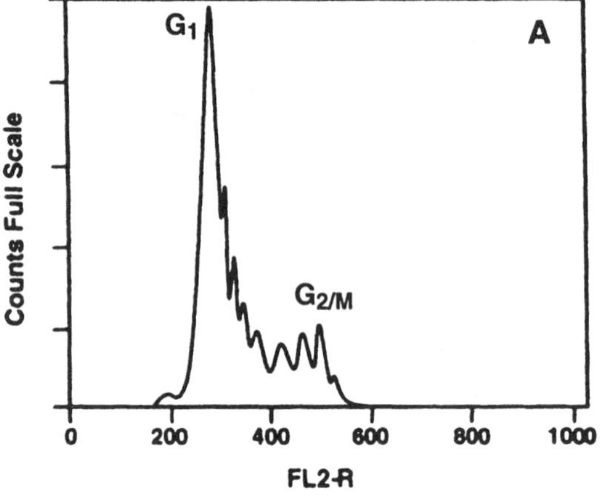

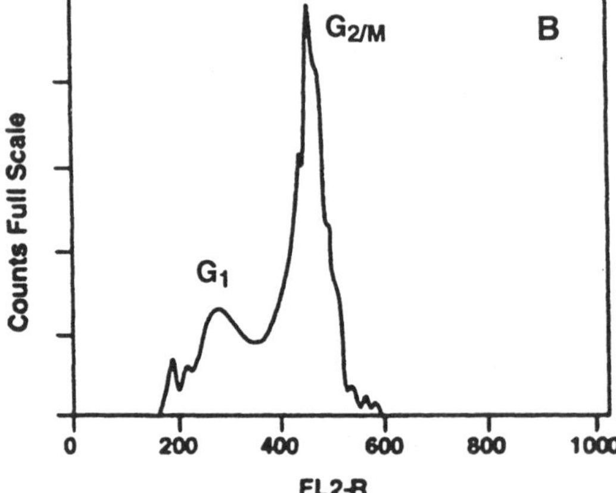

FIGURE 7. Vpr causes $G_{2/M}$ accumulation of cells. Cells (U937) were incubated with recombinant Vpr (40 ng/ml) containing supernatant or control supernatant and the cells were stained with propidium iodide (PI) and analyzed by flow cytometry. Positions of G_1 and $G_{2/M}$ are marked. A, control cells; B, Vpr treated cells.

DNA-based vaccine can be considered. As with Vpu, Vpr is expressed internally in cells and consequently the development of a vaccine would have to involve the generation of a strong cellular immune response. Because of its important role in suppressing the immune system, a vaccine that targets Vpr may by itself or in conjunction with other therapies help control HIV infection.

Developing antiviral drugs against Vpr will be worthwhile. Such a drug may inhibit the interaction of Vpr with hVip-1 and other cellular factors, thus mitigating the action of Vpr. Already available antiimmunosuppressive drugs will shed some light on this possibility.

6. CONCLUSION

Studies have proven that *vif, vpu, nef,* and *vpr* genes of HIV-1 are not accessory genes as originally believed, but rather essential components in efficient and normal HIV-1 pathogenesis. Although the virus can replicate in the absence of these genes, their presence greatly improves the quality of infection. Vif is involved in the transport of virions into the nucleus and in stabilizing viral DNA intermediates. The Vpu protein is involved in the degradation of CD4 molecules in the ER and in the enhancement of virions released from the cells. Nef downregulates CD4 molecules on the cell surface and enhances viral replication. Vpr transports the viral preintegration complex into the nucleus and freezes cell growth at the G2/M phase. Many of the functions of the accessory proteins overlap with one another, but the reason for this is still not clear.

The immunogenicity and low functional mutagenicity combine to make the accessory gene products attractive elements in the design of future antiviral immune therapeutics. A vaccine developed specifically against the accessory gene products will have to use an attenuated or mutated form of the protein since the wild-type proteins interfere with normal cell mechanisms. Studies on the interaction between the accessory proteins with various cellular factors are revealing new targets for antiviral therapies. The use of X-ray crystallography and rational drug design techniques will greatly improve the process of finding an antiviral therapy directed against the accessory gene products of HIV. Successfully mitigating the actions of the accessory genes of HIV could prove to be the key to stop this deadly virus.

REFERENCES

1. Levy, J. A., 1994, *HIV and the Pathogenesis of AIDS,* ASM Press, Washington, D.C., pp. 35–52.
2. Feng, Y., Broder, C. C., Kennedy, P. E., and Berger, E. A., 1996, HIV-1 entry cofactor: Functional cDNA cloning of a seven-transmembrane, G protein-coupled receptor, *Science* **272:**872–877.
3. Lu, Z., Berson, J. F., Chen, Y., Turner, J. D., Zhang, T., Sharron, M., Jenks, M. H., Wang, Z., Kim, J., Rucker, J., Hoxie, J. A., Peiper, S. C., and Doms, R. W., 1997, Evolution of HIV-1 coreceptor usage through interactions with distinct CCR5 and CXCR4 domains, *Proc. Natl. Acad. Sci. USA* **94:**6426–6431.
4. Preston, B. D., 1997, Technical comments: Reverse transcriptase fidelity and HIV-1 variation, *Science* **275:**228–229.
5. Prasad, V. R., Drosopoulos, W. C., and Wainberg, M. A., 1997, Technical comments: Reverse transcriptase fidelity and HIV-1 variation, *Science* **275:**230–231.
6. Adachi, A., Ono, N., Sakai, H., Ogawa, K., Shibata, R., Kiyomasu, M., Masuike, H., and

Ueda, S., 1991, Generation and characterization of the human immunodeficiency virus type 1 mutants, *Arch. Virol.* **117**:45–58.

7. Akari, H., Sakuragi, J., Takebe, Y., Tomonaga, K., Kawamura, M., Fukasawa, M., Miura, T., Shinjo, T., and Hayami, M., 1992, Biological characterization of the human immunodeficiency virus type 1 and type 2 mutants in human peripheral blood mononuclear cells, *Arch. Virol.* **123**:157–167.

8. Cheynier, R., Langlade-Demoyen, P., Marescot, M.-R., Blanche, S., Blondin, G., Wain-Hobson, S., Griscelli, C., Vilmer, E., and Plata, F., 1992, Cytotoxic T lymphocyte responses in the peripheral blood of children born to human immunodeficiency virus-1-infected mothers, *Eur. J. Immunol.* **22**:2211–2217.

9. Rowland-Jones, S., Nixon, D. F., Aldhous, M. C., Gotch, F., Ariyoshi, K., Hallam, N., Kroll, J. S., Froebel, K., and McMichael, A., 1993, HIV-specific cytotoxic T-cell activity in an HIV-exposed but uninfected infant, *Lancet* **341**:860–861.

10. Aldhous, M. C., Watret, K. C., Mok, J. Y. Q., Bird, A. G., and Froebel, K. S., 1994, Cytotoxic T lymphocyte activity and CD8 subpopulations in children at risk of HIV infection, *Clin. Exp. Immunol.* **97**:61–67.

11. Miller, R. H., and Sarver, N., 1997, HIV accessory proteins as therapeutic targets, *Nature Med.* **3**:389–394.

12. Wang, B., Ugen, K. E., Srikantan, V., Agadjanyan, M. G., Dang, K., Rafaeli, Y., Sato, A. I., Boyer, J., Williams, W. V., and Weiner, D., 1993, Gene inoculation generates immune responses against human immunodeficiency virus type 1, *Proc. Natl. Acad. Sci. USA* **90**:4156–4160.

13. Chattergoon, M., Boyer, J., and Weiner, D. B., 1997, Genetic immunization: A new era in vaccines and immune therapeutics, *FASEB J.* **11**:753–763.

14. Ayyavoo, V., Nagashunmugam, T., Boyer, J., Mahalingam, S., Fernandes, L. S., Le, P., Lin, J., Nguyen, C., Chattergoon, M., Goedert, J. J., Friedman, H., and Weiner, D. B., 1997, Development of genetic vaccine for pathogenic genes: Construction of attenuated vif DNA immunization cassettes, *AIDS* **11**:1433–1444.

15. Arya, S. K., and Gallo, R. C., 1986, Three novel genes of human T-lymphotropic virus type III: Immune reactivity of their products with sera from acquired immune deficiency syndrome patients, *Proc. Natl. Acad. Sci. USA* **83**:2209–2213.

16. Fisher, A. G., Ensoli, B., Ivanoff, L., Chamberlain, M., Petteway, S., Ratner, L., Gallo, R. C., and Wong-Staal, F., 1987, The *sor* gene of HIV-1 is required for efficient virus transmission *in vitro*, *Science* **237**:888–893.

17. Garrett, E. D., Tiley, L., and Cullen, B., 1991, *rev* activates expression of the human immunodeficiency virus type 1 *vif* and *vpr* gene products, *J. Virol.* **65**:1653–1657.

18. Schwartz, S., Felber, B., and Pavlakis, G., 1991, Expressions of human immunodeficiency virus type 1 *vif* and *vpr* mRNAs is *rev*-dependent and regulated by splicing, *Virology* **183**:677–686.

19. Cullen, B. R., and Greene, W. C., 1989, Regulatory pathways governing HIV-1 replication, *Cell* **58**:423–426.

20. Strebel, K., Daugherty, D., Clouse, K., Cohen, D., Folks, T., and Martin, M. A., 1987, The HIV "A" (*sor*) gene product is essential for virus infectivity, *Nature* **328**:728–730.

21. Aldrovandi, G. M., and Zack, J. A., 1996, Replication and pathogenicity of human immunodeficiency virus type 1 accessory gene mutants in SCID-hu mice, *J. Virol.* **70**:1505–1511.

22. Oberste, M. S., and Gonda, M. A., 1992, Conservation of amino-acid sequence motifs in lentivirus vif proteins, *Virus Genes* **6**:95–102.

23. Sakai, H., Shibata, R., Sakuragi, J.-I., Sakuragi, S., Kawamura, M., and Adachi, A., 1993, Cell-dependent requirement of human immunodeficiency virus type 1 vif protein for maturation of virus particles, *J. Virol.* **67**:1663–1666.

24. Ma, X.-Y., Sova, P., Chao, W., and Volsky, D. J., 1994, Cysteine residues in the vif protein of human immunodeficiency virus type 1 are essential for viral infectivity, *J. Virol.* **68:**1714–1720.

25. Sova, P., Chao, W., and Volsky, D. J., 1997, The redox state of cysteines in human immunodeficiency virus type 1 vif in infected cells and in virions, *Biochem. Biophys. Res. Commun.* **240:**257–260.

26. Goncalves, J., Jallepalli, P., and Gabuzda, D. H., 1994, Subcellular localization of the Vif protein human immunodeficiency virus type 1, *J. Virol.* **68:**704–712.

27. Karczewski, M. K., and Strebel, K., 1996, Cytoskeleton association and virion incorporation of the human immunodeficiency virus type 1 Vif protein, *J. Virol.* **70:**494–507.

28. Hoglund, S., Ohaen, A., Lawrence, K., and Gabuzda, D., 1994, Role of vif during packing of the core of HIV-1, *Virology* **201:**349–355.

29. Simon, J. H., Fouchier, R. A., Southerling, T. E., Guerra, C. B., Grant, C. K., and Malim, M. H., 1997, The Vif and Gag proteins of human immunodeficiency virus type 1 colocalize in infected human T cells, *J. Virol.* **71:**5259–5267.

30. Bouyac, M., Courcoul, M., Bertoia, G., Baudat, Y., Gabuzda, D., Blanc, D., Chazal, N., Boulanger, P., Sire, J., Vigne, R., and Spire, B., 1997, Human immunodeficiency virus type 1 Vif protein binds to the P155Gag precursor, *J. Virol.* **71:**9358–9365.

31. Göttlinger, H. G., Sodroski, J. G., and Haseltine, W. A., 1989, Role of capsid precursor processing and myristylation in morphogenesis and infectivity of human immunodeficiency virus type 1, *Proc. Natl. Sci. Acad. USA* **86:**5781–5785.

32. Wang, C.-T., and Barklis, E., 1993, Assembly, processing and infectivity of human immunodeficiency virus type 1 Gag mutants, *J. Virol.* **67:**4264–4273.

33. Borman, A. M., Quillent, C., Charneau, P., Dauguet, C., and Clavel, F., 1995, Human immunodeficiency virus type 1 Vif mutant particles from restrictive cells: Role of Vif in correct particle assembly and infectivity, *J. Virol.* **69:**2058–2067.

34. Courcoul, M., Patience, C., Rey, F., Blanc, D., Harmache, A., Sire, J., Vigne, R., and Spire, B., 1995, Peripheral blood mononuclear cells produce normal amounts of defective Vif human immunodeficiency virus type 1 particle which are restricted for the preretrotranscription steps, *J. Virol.* **69:**2068–2074.

35. Simon, J. H., and Malim, M. H., 1996, The human immunodeficiency virus type 1 Vif protein modulates the postpenetration stability of viral nucleoprotein complexes, *J. Virol.* **70:**5297–5305.

36. Fouchier, R. A., Simon, J. H., Jaffe, A. B., and Malim, M. H., 1996, Human immunodeficiency virus type 1 Vif does not influence expression or virion incorporation of *gag-*, *pol-*, and *env*-encoded proteins, *J. Virol.* **70:**8263–8269.

37. Fan, L., and Peden, K., 1992, Cell-free transmission of Vif mutants of HIV-1, *Virology* **190:**19–29.

38. Von Schwedler, U., Song, J., Aiken, C., and Trono, D., 1993, Vif is crucial for human immunodeficiency virus type 1 proviral DNA synthesis in infected cells, *J. Virol.* **67:**4945–4955.

39. Trono, D., 1995, HIV accessory proteins: Leading roles for the supporting cast, *Cell* **82:**189–192.

40. Wieland, U., Hartmann, J., Suhr, H., Salzberger, B., Eggers, H. J., and Kuhn, J. E., 1994, *In vivo* genetic variability of the HIV-1 *vif* gene, *Virology* **203:**43–51.

41. Sova, P., van Ranst, M., Gupta, P., Balachandran, R., Chao, W., Itescu, S., McKinley, G., and Volsky, D. J., 1995, The conservation of intact human immunodeficiency virus type 1 (HIV-1) *vif* gene in primary virus isolates from symptomatic and asymptomatic individuals and *in vivo*, *J. Virol.* **69:**2557–2564.

42. Zhang, L., Huang, Y., Yuan, H., Tuttleton, S., and Ho, D. D., 1997, Genetic characteriza-

tion of *vif, vpr,* and *vpu* sequences from long-term survivors of human immunodeficiency virus type 1 infection, *Virology* **228:**340–349.

43. Kan, N. C., Franchini, G., Wong-Staal, F., DuBois, G. C., Robey, G., Lautenberger, J. A., and Papas, T. S., 1986, Identification of HTLV-II/LAV *sor* gene product and detection of antibodies in human sera, *Science* **231:**1553–1555.

44. Schwander, S., Braun, R. W., Kuhn, J. E., Hufert, F. T., Kern, P., Dietrich, M., and Wieland, U., 1992, Prevalence of antibodies to recombinant virion infectivity factor in the sera of prospectively studied patients with HIV-1 infection, *J. Med. Virol.* **36:**142–146.

45. O'Neil, C., Lee, D., Clewley, G., Johnson, M. A., and Emery, V. C., 1997, Prevalence of anti-vif antibodies in HIV-1 infected individuals assessed using recombinant baculovirus expressed Vif protein, *J. Med. Virol.* **51:**139–144.

46. Subbramanian, R. A., and Cohen, E. A., 1994, Molecular biology of the human immunodeficiency virus accessory proteins, *J. Virol.* **68:**6831–6835.

47. Schwartz, S., Felber, B. K., Fenyo, E.-M., and Pavlakis, G. N., 1990, Env and Vpu proteins of human immunodeficiency virus type 1 are produced from multiple bicistronic mRNAs, *J. Virol.* **64:**5448–5456.

48. Maldarelli, F., Chen, M.-Y., Willey, R. L., and Strebel, K., 1993, Human immunodeficiency virus type 1 Vpu protein is an oligomeric type 1 integral membrane, *J. Virol.* **67:**5056–5061.

49. Strebel, K., Klimkait, T., Maldarelli, F., and Martin, M. A., 1989, Molecular and biochemical analyses of human immunodeficiency virus type 1 *vpu* protein, *J. Virol.* **63:**3784–3791.

50. Cohen, E. A., Terwilliger, E. F., Sodroski, J. G., and Haseltine, W. A., 1988, Identification of a protein encoded by the *vpu* gene of HIV-1, *Nature* **334:**532–534.

51. Schubert, U., and Strebel, K., 1994, Differential activities of the human immunodeficiency virus-encoded Vpu protein are regulated by phosphorylation and occurs in different compartments, *J. Virol.* **68:**2260–2271.

52. Chen, M.-T., Malderelli, F., Karczewski, M. K., Willey, R. L., and Strebel, K., 1993, Human immunodeficiency virus type 1 Vpu protein induces degradation of CD4 *in vitro*: The cytoplasmic domain of CD4 contributes to *vpu* sensitivity, *J. Virol.* **67:**3877–3884.

53. Schubert, U., Henklein, P., Boldyreff, B., Wingender, E., Strebel, K., and Porstmann, T., 1994, The human immunodeficiency virus type 1 encoded Vpu protein is phosphorylated by casein kinase-2 (CK-2) at positions ser52 and ser56 within a predicted α-helix-turn-α-helix motif, *J. Mol. Biol.* **236:**16–25.

54. Paul, M., and Jabbar, M. A., 1997, Phosphorylation of both phosphoacceptor sites in the HIV-1 Vpu cytoplasmic domain is essential for Vpu-mediated ER degradation of CD4, *Virology* **232:**207–216.

55. Jabbar, M. A., and Nayak, D. P., 1990, Intracellular interaction of human immunodeficiency virus type 1 (ARV-2) envelope glycoprotein gp160 with CD4 blocks the movement and maturation of CD4 to the plasma membrane, *J. Virol.* **64:**6297–6304.

56. Willey, R. L., Maldarelli, F., Martin, M. A., and Strebel, K., 1992, Human immunodeficiency virus type 1 vpu protein regulates formation of intracellular gp160–CD4 complexes, *J. Virol.* **66:**226–234.

57. Raja, N. U., Vincent, M. J., and Jabbar, M. A., 1993, Analysis of endoproteolytic cleavage and intracellular transport of human immunodeficiency virus type 1 envelope glycoproteins using mutant CD4 molecules bearing the transmembrane ER retention signal, *J. Gen. Virol.* **74:**2085–2097.

58. Willey, R. L., Bonifacino, J. S., Potts, B. J., Martin, M. A., and Klausner, R. D., 1988, Biosynthesis, cleavage, and degradation of the human immunodeficiency virus 1 envelope glycoprotein gp160, *Proc. Natl. Acad. Sci. USA* **85:**9580–9584.

59. Kerkau, T., Bacik, I., Bennink, J. R., Yewdell, J. W., Hunig, T., Schimpl, A., and Schubert, U., 1997, The human immunodeficiency virus type 1 (HIV-1) Vpu protein interferes with

an early step in the biosynthesis of major histocompatibility complex (MHC) class I molecules, *J. Exp. Med.* **185:**1295–1305.

60. Vincent, M. J., Raja, N. U., and Jabbar, M. A., 1993, The human immunodeficiency virus type 1 Vpu protein induces degradation of chimeric envelope glycoproteins bearing the cytoplasmic and anchor domains of CD4: Role of cytoplasmic domain in Vpu-induced degradation in the endoplasmic reticulum, *J. Virol.* **67:**5538–5549.

61. Lenburg, M. E., and Landau, N. R., 1993, Vpu-induced degradation of CD4: Requirement for specific amino acid residues in the cytoplasmic domain of CD4, *J. Virol.* **67:**6542–6550.

62. Schubert, U., Anton, L. C., Bacik, I., Cox, J. H., Bour, S., Bennink, J. R., Orlowski, M., Strebel, K., and Yewdell, J. W., 1998, CD4 glycoprotein degradation induced by human immunodeficiency virus type 1 Vpu protein requires the function of proteasomes and the ubiquitin-conjugating pathway, *J. Virol.* **72:**2280–2288.

63. Bonifacino, J. S., and Klausner, R. D., 1994, Degradation of proteins in the endoplasmic reticulum, in: *Cellular Proteolytic Systems* (A. Ciechanover and A. L. Schwartz, eds.), Vol. 15, Wiley-Liss, New York, pp. 137–160.

64. Klimkait, T., Strebel, K., Hoggan, M. D., Martin, M. A., and Orenstein, J. M., 1990, The human immunodeficiency virus type 1-specific protein vpu is required for efficient virus maturation and release, *J. Virol.* **64:**621–629.

65. Terwilliger, E. F., Cohen, E. A., Lu, Y., Sodroski, J. G., and Haseltine, W. A., 1989, Functional role of human immunodeficiency virus type 1 *vpu*, *Proc. Natl. Acad. Sci. USA* **86:** 5163–5167.

66. Gottlinger, H. G., Dorfman, T., Cohen, E. A., and Haseltine, W. A., 1993, Vpu protein of human immunodeficiency virus type 1 enhances the release of capsids produced by *gag* gene constructs of widely divergent retroviruses, *Proc. Natl. Acad. Sci. USA* **90:**7381–7385.

67. Colombini, S., Arya, S. K., Reitz, M. S., Jagodzinski, L., Beaver, B., and Wong-Staal, F., 1989, Structure of simian immunodeficiency virus regulatory genes, *Proc. Natl. Acad. Sci. USA* **86:**4813–4817.

68. Shibata, R., Miura, T., Hayami, M., Ogawa, K., Sakai, H., Kiyomasu, T., Ishimoto, A., and Adachi, A., 1990, Mutational analysis of the human immunodeficiency virus type 2 (HIV-2) genome in relation to HIV-1 and simian immunodeficiency virus (SIV), *J. Virol.* **64:**742–747.

69. Ahmad, N., Maitra, R. K., and Venkatesan, S., 1989, Rev-induced modulation of Nef protein underlies temporal regulation of human immunodeficiency virus replication, *Proc. Natl. Acad. Sci. USA* **86:**6111–6115.

70. Cheng-Mayer, C., Iannello, P., Shaw, K., Luciw, P. A., and Levy, J. A., 1989, Differential effects of *nef* on HIV replication: Implications for viral pathogenesis in the host, *Science* **246:**1629–1632.

71. Niederman, T. M. J., Thielan, B. J., and Ratner, L., 1989, Human immunodeficiency virus type 1 negative factor is a transcriptional silencer, *Proc. Natl. Acad. Sci. USA* **86:**1128–1132.

72. Miller, M. D., Warmerdam, M. T., Gaston, I., Greene, W. C., and Feinberg, M. B., 1994, The human immunodeficiency virus-1 *nef* gene product: A positive factor for viral infection and replication in primary lymphocytes and macrophages, *J. Exp. Med.* **179:**101–113.

73. Kestler, H. W., Ringler, D. J., Mori, K., Panicali, D. L., Sehgal, P. K., Daniel, M. D., and Desrosiers, R. C., 1991, Importance of the *nef* gene for maintenance of high virus loads and for development of AIDS, *Cell* **65:**651–662.

74. Deacon, N. J., Tsykin, A., Solomon, A., Smith, K., Ludford-Menting, M., Hooker, D. J., McPhee, A., Greenway, A. L., Ellett, A., Chatfield, C., *et al.*, 1995, Genomic structure of an attenuated quasi species of HIV-1 from a blood transfusion donor and recipients, *Science* **270:**988–991.

75. Kirchhoff, F., Greenough, T. C., Brettler, D. B., Sullivan, J. L., and Desrosiers, R. C., 1995,

Brief report: Absence of intact *nef* sequences in a long term survivor with nonprogressive HIV-1 infection, *N. Engl. J. Med.* **332**:228–232.

76. Switzer, W. M., Wiktor, S., Soriano, V., Silva-Grace, A., Mansinho, K., Coulibaly, I. M., Ekpini, E., Greenberg, A. E., Folks, T. M., and Heneine, W., 1998, Evidence of Nef truncation in human immunodeficiency virus type 2 infection, *J. Infect. Dis.* **177**:65–71.

77. Welker, R., Kottler, H., Kalbitzer, H. R., and Kräusslich, H. G., 1996, Human immunodeficiency virus type 1 Nef protein is incorporated into virus particles and specifically cleaved by the viral proteinase, *Virology* **219**:228–236.

78. Pandori, M. W., Fitch, N. J., Craig, H. M., Richman, D. D., Spina, C. A., and Guatelli, J. C., 1996, Producer cell modification of human immunodeficiency virus type 1: Nef is a virion protein, *J. Virol.* **70**:4283–4290.

79. Bukovsky, A. A., Dorfman, T., Weimann, A., and Gottlinger, H. G., 1997, Nef association with human immunodeficiency virus type 1 virions and cleavage by the viral protease, *J. Virol.* **71**:1013–1018.

80. Klotman, M. E., Kim, S. Y., Buchbinider, A., DeRossi, A., Baltimore, D., and Wong-Staal, F., 1992, Kinetics of expression of multiply spliced RNA in early human immunodeficiency type 1 infection of lymphocytes and monocytes, *Proc. Natl. Acad. Sci. USA* **89**:1148–1152.

81. Ratner, L., Starcich, B., Josephs, S. F., Hahn, B. H., Reddy, E. P., Livak, K. J., Petteway, S. R., Pearson, M. L., Haseltine, W. A., Arya, S. K., and Wong-Staal, F., 1985, Polymorphism of the 3′ open reading frame of the virus associated with the acquired immune deficiency syndrome, human T cell lymphotropic virus type III, *Nucleic Acids Res.* **13**:8219–8229.

82. Ratner, L., Fisher, A., Jagodzinski, L. L., Mitsuya, H., Liou, R.-S., Gallo, R. C., and Wong-Staal, F., 1987, Complete nucleotide sequences of functional clones of the AIDS virus, *AIDS Res. Hum. Retrovir.* **3**:57–69.

83. Myers, G., Berzofsky, J. A., Rabson, A. B., Smith, T. F., and Wong-Staal, F., eds., 1990, *Human Retroviruses and AIDS*, Los Alamos National Laboratory, Los Alamos, New Mexico.

84. Delassus, S., Cheynier, R., and Wain-Hobson, S., 1991, Evolution of human immunodeficiency type 1 *nef* and long terminal repeat sequences over 4 years *in vivo* and *in vitro*, *J. Virol.* **65**:225–231.

85. Shugars, D. C., Smith, M. S., Glueck, D. H., Nantermet, P. V., Seillier-Moisewisch, F., and Swanstorm, R., 1993, Analysis of human immunodeficiency virus type 1 *nef* gene sequences present *in vivo*, *J. Virol.* **67**:4639–4650.

86. Culmann, B., Gomard, E., Kieny, M.-P., Guy, B., Dreyfus, F., Saimot, A.-G., Sereni, D., and Levy, J.-P., 1989, An antigenic peptide of the HIV-1 Nef protein recognized by cytotoxic T lymphocytes of seropositive individuals in association with different HLA-B molecules, *Eur. J. Immunol.* **19**:2383–2386.

87. Bahraoui, E., Yagello, M., Billaud, J.-N., Sabatier, J.-M., Guy, B., Muchmore, E., Girard, M., and Gluckman, J. C., 1990, Immunogenicity of the human immunodeficiency virus (HIV) recombinant *nef* gene product: Mapping of T-cell and B-cell epitopes in immunized chimpanzees, *AIDS Res. Hum. Retrovir.* **6**:1087–1098.

88. Koenig, S., Fuerst, T. R., Wood, L. V., Woods, R. M., Suzich, J. A., Jones, G. M., de la Cruz, V. F., Davey, R. T., Venkatesan, S., Moss, B., Biddison, W. E., and Fauci, A. S., 1990, Mapping the fine specificity of a cytolytic T cell response to HIV-1, *J. Immunol.* **145**:127–135.

89. Zweig, M., Samuel, K. P., Showalter, S. D., Bladen, S. V., DuBois, G. C., Lautenberger, J. A., Hodge, D. R., and Papas, T. S., 1990, Heterogeneity of Nef proteins in cells infected with human immunodeficiency virus type 1, *Virology* **179**:504–507.

90. Kaminchik, J., Bashan, N., Itach, A., Sarver, Gorecki, M., and Panel, A., 1991, Genetic characterization of human immunodeficiency virus type 1 *nef* gene products translated *in vitro* and expressed in mammalian cells, *J. Virol.* **65**:583–588.

91. Bachelerie, F., Alcami, J., Hazan, U., Israel, B., Goud, F., Arenzana-Seisdedos, F., and

Virelizier, J.-I., 1990, Constitutive expression of human immunodeficiency virus (HIV) Nef protein in human astrocytes does not influence basal or induced HIV long terminal repeat activity, *J. Virol.* **64:**3059–3062.

92. Fackler, O. T., Kienzle, N., Kremmer, E., Boese, A., Schramm, B., Klimkait, T., Kucherer, C., and Mueller-Lantzsch, N., 1997, Association of human immunodeficiency virus Nef protein with actin is myristylation dependent and influences in subcellular localization, *Eur. J. Biochem.* **247:**843–851.

93. Kohleisen, B., Neumann, M., Herrmann, R., Brack-Werner, R., Krohn, K. J., Ovod, V., Ranki, A., and Erfle, V., 1992, Cellular localization of Nef expressed in persistently HIV-1 infected low-producer astrocytes, *AIDS* **6:**1427–1436.

94. Murti, K. G., Brown, P. S., Ratner, L., and Garcia, J. V., 1993, Highly localized tracks of human immunodeficiency virus type 1 Nef in the nucleus of cells of a human CD4+ T-cell line, *Proc. Natl. Acad. Sci. USA* **90:**11895–11899.

95. Franken, P., Arold, S., Padilla, A., Bodeus, M., Hoh, F., Strub, M. P., Boyer, M., Jullien, M., Benarous, R., and Dumas, C., 1997, HIV-1 Nef protein: Purification, crystallizations, and preliminary X-ray diffraction studies, *Protein Sci.* **6:**2681–2683.

96. Backer, J. M., Mendola, C. E., Fairhurst, J. L., and Kovedsi, I., 1991, The HIV-1 *nef* protein does not have guanine nucleotide binding, GTPase, or autophosphorylation activities, *AIDS Res. Hum. Retrovir.* **7:**1015–1020.

97. Harris, M., and Coates, K., 1993, Identification of cellular proteins that bind to the human immunodeficiency virus type 1 *nef* gene product *in vitro*: A role for myristoylation, *J. Gen. Virol.* **74:**1581–1589.

98. Luo, T., Livingston, R. A., and Garcia, J. V., 1997, Infectivity enhancement by human immunodeficiency virus type 1 Nef is independent of its association with cellular serine/threonine kinase, *J. Virol.* **71:**9524–9530.

99. Sawai, E. T., Baur, A., Struble, H., Peterlin, B. M., Levy, J. A., and Cheng-Mayer, C., 1994, Human immunodeficiency virus type 1 Nef associates with a cellular serine kinase in T lymphocytes, *Proc. Natl. Acad. Sci. USA* **91:**1539–1543.

100. Nunn, M. F., and Marsh, J. W., 1996, Human immunodeficiency virus type 1 Nef associates with a member of the p21-activated kinase family, *J. Virol.* **70:**6157–6161.

101. Lu, X., Wu, X., Plemenitas, A., Yu, H., Sawai, E. T., Abo, A., and Peterlin, B. M., 1996, CDC42 and Rac1 are implicated in the activation of the Nef-associated kinase and replication of HIV-1, *Curr. Biol.* **6:**1677–1684.

102. Saksela, K., Cheng, G., and Baltimore, D., 1995, Proline-rich (PxxP) motifs in HIV-1 Nef bind to SH3 domains of a subset of Src kinases and are required for the enhanced growth of Nef+ viruses but not for down regulation of CD4, *EMBO J.* **14:**484–491.

103. Arold, S., Franken, P., Strub, M. P., Hoh, F., Benichou, S., Benarous, R., and Dumas, C., 1997, The crystal structure of HIV-1 Nef protein bound to the Fyn kinase SH3 domain suggests a role for this complex in altered T cell receptor signaling, *Structure* **5:**1361–1372.

104. Rossi, F., Evstafieva, A., Pedrali-Noy, G., Gallina, A., and Milanesi, G., 1997, HsN3 proteasomal subunit as a target for human immunodeficiency virus type 1 Nef protein, *Virology* **237:**33–45.

105. Garcia, J. V., and Miller, A. D., 1991, Serine phosphorylation-independent downregulation of cell surface CD4 by nef, *Nature* **350:**508–511.

106. Anderson, S., Shugars, D. C., Swanstrom, R., and Garcia, J. V., 1993, Nef from primary isolates of human immunodeficiency virus type 1 suppresses surface CD4 expression in human and mouse T cells, *J. Virol.* **67:**4923–4931.

107. Mariani, R., and Skowronski, J., 1993, CD4 down-regulation by *nef* alleles isolated from human immunodeficiency virus type-1 infected individuals, *Proc. Natl. Acad. Sci. USA* **90:** 5549–5553.

108. Aiken, C., Konner, J., Landau, N. R., Lenburg, M. E., and Trono, D., 1994, Nef induces CD4 endocytosis: Requirement for a critical dileucine motif in the membrane-proximal CD4 cytoplasmic domain, *Cell* **76**:853–864.

109. Laurent, A. G., Hovanessian, A. G., Riviere, Y., Krust, B., Regnault, A., Montagnier, L., Findeli, A., Kieny, M. P., and Guy, B., 1990, Production of a non-functional nef protein in human immunodeficiency virus type 1 infected CEM cells, *J. Gen. Virol.* **71**:2273–2281.

110. Greenway, A. L., McPhee, D. A., Grgacic, E., Hewish, D., Lucantoni, A., Macreadie, I., and Azad, A., 1994, Nef 27, but not the Nef 25 isoform of human immunodeficiency virus type 1 pNL43 down-regulates surface CD4 and IL-2R expression in peripheral blood mononuclear cells and transformed T cell, *Virology* **198**:245–256.

111. Bandres, J. C., Shaw, A. S., and Ratner, L., 1995, HIV-1 nef protein downregulation of CD4 surface expression: Relevance of lck binding, *Virology* **207**:338–341.

112. Rhee, S. S., and Marsh, J. W., 1994, HIV-1 nef activity in murine T cells: CD4 modulation and positive enhancement, *J. Immunol.* **152**:5128–5134.

113. Foti, M., Mangasarian, A., Piguet, V., Lew, D. P., Krause, K. H., Trono, D., and Carpentier, J. L., 1997, Nef-mediated clathrin-coated pit formation, *J. Cell Biol.* **139**:37–47.

114. Greenberg, M. E., Bronson, S., Lock, M., Neumann, M., Pavlakis, G. N., and Skowronski, J., 1997, Co-localization of HIV-1 Nef with the AP-2 adaptor protein complex correlates with Nef-induced CD4 down-regulation, *EMBO* **16**:6964–6976.

115. Schwartz, O., Marechal, V., Danos, O., and Heard, J. M., 1995, Human immunodeficiency virus type-1 Nef increases the efficiency of reverse transcription in the infected cell, *J. Virol.* **69**:4053–4059.

116. Luria, S., Chambers, I., and Berg, P., 1991, Expression of the type 1 human immunodeficiency virus Nef protein in T cells prevents antigen receptor-mediated induction of interleukin 2 mRNA, *Proc. Natl. Acad. Sci. USA* **88**:5326–5330.

117. Niederman, T. M. J., Garcia, J. V., Hastings, W. R., Luria, S., and Ratner, L., 1992, Human immunodeficiency virus type 1 Nef protein inhibits NF-kB induction in human T cells, *J. Virol.* **66**:6213–6219.

118. Skowronski, J., Parks, D., and Mariani, R., 1993, Altered T-cell activation and development in transgenic mice expressing the HIV-1 *nef* gene, *EMBO J.* **12**:703–713.

119. Collins, K. L., Chen, B. K., Kalams, S. A., Walker, B. D., and Baltimore, D., 1998, HIV-1 Nef protein protects infected primary cells against killing by cytotoxic T lymphocytes, *Nature* **191**:397–401.

120. Culman, B., Gomard, E., Kieny, M.-P., Guy, B., Dreyfus, F., Saimont, A.-G., Sereni, D., Sicard, D., and Levy, J. P., 1991, Six epitopes reacting with human cytotoxic CD8+ T cells in the central region of the HIV-1 Nef protein, *J. Immunol.* **146**:1560–1565.

121. Sabatier, J. M., Clerget-Raslain, B., Fontan, G., Fenouillet, E., Rochat, H., Granier, C., Gluckman, J. C., Van Rietschoten, J., Montagnier, L., and Bahraoui, E., 1989, Use of synthetic peptides for the detection of antibodies against the nef regulation protein in sera of HIV-infected patients, *AIDS* **3**:215–220.

122. Bahraoui, E., Benjouad, A., Sabatier, J. M., Allain, J. P., Laurian, Y., Montagnier, L., and Gluckman, J.-C., 1990, Relevance of anti-nef antibody detection as an early serologic marker of human immunodeficiency virus infection, *Blood* **76**:257–264.

123. Kienzle, N., Broke, M., Harthus, H.-P., Enders, M., Erfle, V., Buck, M., and Muller-Lantzsch, N., 1991, Immunological study of the nef protein from HIV-1 polyclonal and monoclonal antibodies, *Arch. Virol.* **124**:123–132.

124. Reiss, P., deRonde, A., Lange, J., deWolf, F., Dekker, J., Debouck, C., and Goudsmit, J., 1989, Antibody response to the viral negative factor (nef) in HIV-1 infection: A correlate of levels of HIV-1 expression, *AIDS* **3**:227–235.

125. Winter, N., Lagranderie, M., Rauzier, J., Timm, J., Leclerc, C., Guy, B., Kieny, M. P., Gheorghiu, M., and Gicquel, B., 1991, Expression of heterologous genes in *Mycobacterium bovis* BCG: Induction of a cellular response against HIV-1 Nef protein, *Gene* **109**:47–54.

126. Daniel, M. D., Kirchhoff, F., Czajak, S. C., Sehgal, P. K., and Desrosiers, R. C., 1992, Protective effects of a live attenuated SIV vaccine with a deletion in the *nef* gene, *Science* **258**:1938–1941.

127. Wong-Staal, F., Chanda, P. K., and Gbrayeb, J., 1987, Human immunodeficiency virus type 1: The eighth gene, *AIDS Res. Hum. Retrovir.* **3**:33–39.

128. Myers, G., Korber, B., Barzofsky, J. A., Smith, R. F., and Pavlakis, G. N., 1992, *Human Retroviruses and AIDS*, Theoretical Biology and Biophysics Group T-10, Los Alamos National Laboratory, Los Alamos, New Mexico.

129. Arrigo, S. J., and Chen, I. S. Y., 1991, Rev is necessary for translation but not cytoplasmic accumulation of HIV-I *vif, vpr,* and *env/vpu* 2 RNAs, *Genes Dev.* **5**:808–819.

130. Cohen, E. A., Dehni, G., Sodroski, J. G., and Haseltine, W. A., 1990, Human immunodeficiency virus *vpr* product is a virion-associated regulatory protein, *J. Virol.* **64**:3097–3099.

131. Garrett, E. D., Tiley, L. S., and Cullen, B. R., 1991, Rev activates expression of the human immunodeficiency virus type 1 *vif* and *vpr* gene products, *J. Virol.* **65**:1653–1657.

132. Adachi, A., Ono, N., Sakai, H., Ogawa, K., Shibata, R., Kiyomasu, T., Masuike, H., and Ueda, S., 1991, Generation and characterization of the human immunodeficiency virus type 1 mutants, *Arch. Virol.* **117**:45–58.

133. Akari, H., Sakuragi, J., Takebe, Y., Tomonaga, K., Kawamura, M., Fukasawa, M., Miura, T., Shinjo, T., and Hayami, M., 1992, Biological characterization of the human immunodeficiency virus type 1 and type 2 mutants in human peripheral blood mononuclear cells, *Arch. Virol.* **123**:157–167.

134. Cohen, E. A., Terwilliger, E. F., Jalinoos, Y., Proulx, J., Sodroski, J. G., and Haseltine, W. A., 1990, Identification of HIV-1 *vpr* product and function, *J. Acquired Immune Defic. Syndr.* **3**:11–18.

135. Dedera, D., Hu, W., Vander Heyden, N., and Ratner, L., 1989, Viral protein R of human immunodeficiency virus types 1 and 2 is dispensable for replication and cytopathogenicity in lymphoid cells, *J. Virol.* **63**:3205–3208.

136. Balliet, J. W., Kolson, D. L., Eiger, G., Kim, F. M., Mcgann K. A., Srinivasan, A., and Collman, R. G., 1994, Distinct effects in primary macrophages and lymphocytes of human immunodeficiency virus type 1 accessory genes *vpr, vpu* and *nef:* Mutational analysis of a primary HIV-1 isolate, *Virology* **200**:623–631.

137. Zhao, L.-J., Wang, L., Mukherjee, S., and Narayan, O., 1994, Biochemical mechanism of HIV-1 Vpr function: Oligomerization mediated by the N-terminal domain, *J. Biol. Chem.* **269**:32131–32137.

138. Lu, Y. L., Spearman, P., and Ratner, L., 1993, Human immunodeficiency virus type 1 viral protein R localization in infected cells and virions, *J. Virol.* **67**:6542–6550.

139. Mahalingam, S., Collman, R. G., Patel, M. P., Monken, C. E., and Srinivasan, A., 1995, Functional analysis of HIV-1 Vpr: Identification of determinants essential for subcellular localization, *Virology* **212**:331–339.

140. Yao, X. J., Subramanian, R., Rougeau, N., Boisvert, F., Bergeron, D., and Cohen, E. A., 1995, Mutagenic analysis of human immunodeficiency virus type 1 Vpr: Role of a predicted N-terminal alpha helical structure in Vpr nuclear localization and virion incorporation, *J. Virol.* **69**:7032–7044.

141. DiMarzio, P., Choe, S., Ebright, M., Knoblough, R., and Landau, N. R., 1995, Mutational analysis of cell cycle arrest, nuclear localization, and virion packaging of the human immunodeficiency virus type 1 Vpr, *J. Virol.* **69**:7909–7916.

142. Kondo, E., Mammano, F., Cohen, E. A., and Gottlinger, H. G., 1995, The p6gag domain of human immunodeficiency virus type 1 is sufficient for the incorporation of Vpr into heterologous virus particles, *J. Virol.* **69:**2759–2764.

143. Paxton, W., Connor, R. I., and Landau, N. R., 1993, Incorporation of vpr into human immunodeficiency virus type 1 virions: Requirement for the p6 region of gag and mutational analysis, *J. Virol.* **67:**7229–7237.

144. Wu, X., Conway, J. A., Kim, J., and Kappes, J. C., 1994, Localization of the Vpx packing signal within the C-terminus of the human immunodeficiency virus type 2 *gag* precursor protein, *J. Virol.* **68:**6161–6169.

145. Mahalingam, S., Kahn, S. A., Jabbar, M. A., Monken, C. E., Collman, R. G., and Srinivasan, A., 1995, Identification of residues in the N-terminal acidic domain of HIV-1 Vpr essential for virion incorporation, *Virology* **207:**297–302.

146. Mahalingam, S., Kahn, S. A., Murali, R., Jabbar, M. A., Monken, C. E., Collman, R. G., and Srinivasan, A., 1995, Mutagenesis of the putative alpha helical domain of HIV-1 Vpr: Effect on stability and virion incorporation, *Proc. Natl. Acad. Sci. USA* **92:**3794–3798.

147. Lang, S. M., Weeger, M., Stahl-Hennig, C., Coulibaly, C., Hunsmann, G., Muller, J., Muller-Hermelink, H., Fuchs, D., Wachter, H., Daniel, M., Desrosiers, R. C., and Fleckenstein, B., 1993, Importance of Vpr infection of rhesus monkeys with simian immunodeficiency virus, *J. Virol.* **67:**902–912.

148. Gibbs, J. S., Lackner, A. A., Lang, S. M., Simon, M. A., Sehgal, P. K., Daniel, M. D., and Desrosiers, R. C., 1995, Progression to AIDS in the absence of a gene for Vpr and Vpx, *J. Virol.* **69:**2378–2383.

149. Cohen, E. A., Dehni, G., Sodroski, J. G., and Haseltine, W. A., 1990, Human immunodeficiency virus *vpr* product is a virion-associated regulatory protein, *J. Virol.* **64:**3097–3099.

150. Heinzinger, N. K., Bukrinsky, M. I., Haggerty, S. A., Ragland, A. M., Kewalramani, V., Lee, M. A., Gendelman, H. E., Ratner, L., Stevensson, M., and Emerman, M., 1994, The Vpr protein of human immunodeficiency virus type 1 influences nuclear localization of viral nucleic acids in non-dividing host cells, *Proc. Natl. Acad. Sci. USA* **91:**7311–7315.

151. Levy, D. N., Fernandes, L. S., Williams, W. V., and Weiner, D. B., 1993, Induction of cell differentiation by human immunodeficiency virus 1 vpr, *Cell* **72:**541–550.

152. Mahalingam, S., MacDonald, B., Ugan, K. E., Ayyavoo, V., Agadjayan, M. G., Williams, W. V., and Weiner, D. B., 1997, *In vitro* and *in vivo* tumor growth suppression by HIV-1 Vpr, *DNA Cell Biol.* **16:**137–143.

153. Rogel, M. E., Wu, L. I., and Emerman, M., 1995, The human immunodeficiency virus type 1 *vpr* gene prevents cell proliferation during chronic infection, *J. Virol.* **69:**882–888.

154. He, J., Choe, S., Walker, R., DiMarzio, P., Morgan, D. O., and Landau, N. R., 1995, The human immunodeficiency virus type 1 viral protein R (Vpr) arrests cells in the G2 phase of the cell cycle by inhibiting p34cdc2 activity, *J. Virol.* **69:**6705–6711.

155. Jowett, J. B. M., Planelles, V., Poon, B., Shah, N. L., Chen, M. L., and Chen, I. S. Y., 1995, The human immunodeficiency virus type 1 *vpr* gene arrests infected T cells in the G2+M phase of the cell cycle, *J. Virol.* **69:**6304–6313.

156. Re, F., Braaten, D., Franke, E. K., and Luban, J., 1995, The human immunodeficiency virus type 1 Vpr arrests the cell cycle in G2 by inhibiting the activation of p34cdc2-cyclin B, *J. Virol.* **69:**6859–6864.

157. Rogel, M. E., Wu, L. I., and Emerman, M., 1995, The human immunodeficiency virus type 1 *vpr* gene prevents cell proliferation during chronic infection, *J. Virol.* **69:**882–888.

158. Mustafa, F., and Robinson, H. L., 1993, Context-dependent role of human immunodeficiency virus type 1 auxiliary genes in the establishment of chronic virus producers, *J. Virol.* **67:**6909–6915.

159. Levy, D. N., Refaeli, Y., and Weiner, D. B., 1995, The Vpr regulatory gene of human

immunodeficiency virus, in: *Transacting Functions of Human Retroviruses* (I. S. Y. Chen, H. Koprowski, A. Srinivasan, and P. K. Vogt, eds.), Springer-Verlag, New York, pp. 209–236.

160. Macreadie, I. G., Castelli, L. A., Hewish, D. R., Kirkpatrick, A., Ward, A. C., and Azad, A. A., 1995, A domain of human immunodeficiency virus type 1 Vpr containing repeated H(S/F)RIG amino acid motifs causes cell growth arrest and structural defects, *Prov. Natl. Acad. Sci. USA* **92:**2770–2774.

161. Balliet, J. W., Kolson, D. L., Eiger, G., Kim, F. M., Mcgann, K. A., Srinivasan, A., and Collman, R. G., 1994, Distinct effects in primary macrophages and lymphocytes of human immunodeficiency virus type 1 accessory genes *vpr, vpu* and *nef:* Mutational analysis of a primary HIV-1 isolate, *Virology* **200:**623–631.

162. Balotta, C., Lusso, P., Crowley, R., Gallo, R. C., and Franchini, G., 1993, Antisense phosphorothioate oligodeoxynucleotides targeted to the *vpr* gene inhibit human immunodeficiency virus type 1 replication in primary human macrophages, *J. Virol.* **67:**4409–4414.

163. Connor, R. I., Chen, B. K., Choe, S., and Landau, N. R., 1995, Vpr is required for efficient replication of human immunodeficiency virus type-1 in mononuclear phagocytes, *Virology* **206:**935–944.

164. Levy, D. N., Refaeli, Y., and Weiner, D. B., 1995, Extracellular Vpr protein increase cellular permissiveness of human immunodeficiency virus replication and reactivates virus from latency, *J. Virol.* **69:**1243–1252.

165. Refaeli, Y., Levy, D. N., and Weiner, D. B., 1995, The glucocorticoid receptor type II complex is a target of the HIV 1 Vpr gene product, *Proc. Natl. Acad. Sci. USA* **92:**3621–3625.

166. Zhao, L.-J., Mukherjee, S., and Narayan, O., 1994, Biochemical mechanism of HIV-1 Vpr function: Specific interaction with a cellular protein, *J. Biol. Chem.* **269:**15577–15582.

167. BouHamdan, M., Benichou, S., Rey, F., Navarro, J. M., Agostini, I., Spire, B., Camonis, J., Sluppaug, G., Vigne, R., Benarous, R., and Sire, J., 1996, Human immunodeficiency virus type 1 Vpr protein binds to the uracil DNA glycosylase (UNG) DNA repair enzyme, *J. Virol.* **70:**697–704.

168. Mahalingam, S., Ayyavoo, V., Patel, M., Kao, G. D., Muschel, R. J., and Weiner, D. B., 1998, HIV-1 Vpr interacts with a 38 kDa cellular factor important for G2/M phase transition of the mammalian cell cycle, *Proc. Natl. Acad. Sci. USA* **95:**3419–3424.

169. Ayyavoo, V., Mahboubi, A., Mahalingam, S., Ramalingam, R., Kudchodkar, S., Williams, W. V., Green, D. R., and Weiner, D. B., 1997, HIV-1 Vpr suppresses immune activation and apoptosis through regulation of nuclear factor *k*b, *Nature Med.* **3:**1117–1123.

170. Stewart, S. A., Poon, B., Jowett, J. B., and Chen, I. S., 1997, Human immunodeficiency virus type 1 Vpr induces apoptosis following cell cycle arrest, *J. Virol.* **71:**5579–5592.

12

A New Generation of Antiviral Therapeutics Designed to Prevent the Use of Chemokine Receptors for Entry by HIV-1

BENJAMIN J. DORANZ and ROBERT W. DOMS

1. INTRODUCTION

Human immunodeficiency virus type 1 (HIV-1) is characterized by significant viremia that is maintained by massive viral replication, with between 1 and 10 billion virus particles being generated each day in an infected adult in the absence of therapy.[1,2] When coupled with the high error rate of the viral reverse transcriptase, these replication characteristics allow HIV-1 to evade pharmaceutical and immune challenges that might otherwise stem the virus. Nevertheless, combination chemotherapy utilizing two inhibitors of reverse transcriptase in conjunction with a protease inhibitor have led to significant sustained suppression of circulating virus in some individuals (see Gazzard[3] for review). However, combination chemotherapy is not always effective since multidrug-resistant virus strains may arise, and the cost, side effects, and complicated drug regimen preclude use of the triple-drug cocktail for many individuals, especially in less-developed countries. Thus,

BENJAMIN J. DORANZ and ROBERT W. DOMS • Department of Pathology and Laboratory Medicine, University of Pennsylvania, Philadelphia, Pennsylvania 19104.

Human Retroviral Infections, edited by Kenneth E. Ugen *et al.*
Kluwer Academic / Plenum Publishers, New York, 2000.

improved versions of existing antiretrovirals, as well as the development of novel therapeutic approaches, are called for.

The discovery that members of the chemokine receptor family are absolutely required for viral entry and transmission opens a new realm of possibilities for inhibiting the virus at a novel step in its replication life cycle. Prior to their identification as HIV-1 coreceptors, the chemokine receptors were pursued as targets for immunologically oriented therapeutics. These early efforts are now allowing progress in targeting the chemokine receptors to be made at an unprecedented pace. For example, the fortuitous development of HIV-1 inhibitors has already yielded small-molecule inhibitors of the coreceptors as well as preliminary clinical trials of some of these compounds, in effect yielding answers to HIV-1 clinical therapies before the questions had even been raised. Given the requirement of coreceptors for both transmission and pathogenesis of infection as well as the readily targeted family of proteins to which the chemokine receptors belong, the speed at which the coreceptor field is progressing can be expected to be overshadowed only by the productivity of these discoveries in the near future.

2. CHEMOKINE RECEPTORS AS HIV-1 CORECEPTORS

The use of CD4 as the primary cellular receptor for HIV-1 binding and entry was discovered within 2 years of the identification of HIV-1 as the etiological agent of acquired immune deficiency syndrome (AIDS). CD4 binds directly to the viral envelope (Env) protein, antibodies directed to CD4 block virus entry, and the introduction of CD4 into most human cell lines allows viral entry and replication.[4–6] These early studies noted, however, that human CD4 alone was not sufficient to render nonprimate cell types susceptible to virus infection.[4] Later studies showed that the block to virus infection was at the level of viral entry, and that nonprimate cell types lack one or more viral coreceptors that are required in conjunction with CD4 to render cells permissive for viral entry.[7,8] Further indication that cell surface molecules in addition to CD4 are necessary for HIV-1 to enter cells came from studies which demonstrated that HIV-1 strains can be divided into two groups based on cellular tropism (see Table I for summary): T cell-tropic (T-tropic) viruses that replicate in T cell lines and primary T cells, but not in macrophages, and macrophage-tropic (M-tropic) viruses that replicate in primary macrophages and primary T cells, but not in T cell lines (for review see Miedema *et al.*[9]). A number of dual-tropic virus strains have also been described that can infect T cell lines while retaining the ability to infect macrophages.[10] This ability to differentially infect CD4-positive cell types

TABLE I
M- and T-Tropic Virus Characteristics

Characteristic	M-tropic	T-tropic
Major coreceptor	CCR5	CXCR4
Cells infected	1° T cells, macrophages	1° T cells, T cell lines
Syncytia in MT-2 cells	Non-syncytia-inducing	Syncytia-inducing
Replication rates	Slow/low	Rapid/high
Transmission	Via all routes	Rare
Emergence	Earliest detected	Later in disease
Typical V3 loop charge	3–5 basic residues	7–9 basic residues
Prototypical HIV-1 strains	JRFL, BaL	HXB, BH8, NL43

largely maps to the Env of each isolate, with the third variable loop (V3) region of Env being particularly influential in determining viral tropism.[11–13]

Understanding the molecular basis of viral tropism is of interest because changes in tropism *in vivo* can be predictive of disease progression. The M-tropic strains are preferentially transmitted by sexual contact, direct blood transfer, and vertical transmission, and are the first types of virus detected early after infection.[14] The T-tropic strains arise later as disease progresses and their emergence is correlated with a steep decline in T cell counts and more rapid progression to AIDS (for review see Miedema *et al.*[9]). Despite this correlation with accelerated immune depletion, T-tropic isolates have not been identified in all individuals who progress to AIDS. However, studies addressing this point were conducted before the identification of the HIV-1 coreceptors and the realization that dual-tropic strains, which have properties of both T- and M-tropic isolates, are far more prevalent than had previously been suggested.[15,16] With the identification of the viral coreceptors and the development of new therapeutics with the potential to selectively block specific strains of HIV-1 by inhibiting use of specific coreceptors, the distinction among T-, M-, and dual-tropic strains and their relationship to AIDS pathogenesis assume greater importance.

The first HIV-1 coreceptor was identified by E. Berger's lab at the NIH using a novel cell–cell based fusion assay in which a HeLa cDNA library was used to identify a gene which rendered murine 3T3-CD4 cells permissive for membrane fusion mediated by a T-tropic Env.[17] The gene encoded a seven-transmembrane protein and was shown to allow fusion and entry by T-tropic, but not M-tropic, virus strains. This molecule, initially termed fusin, was later renamed CXCR4 when it was shown to be the fourth member of a family of G-protein-coupled receptors called CXC chemokine receptors.[18,19] The homology of fusin to the chemokine receptor family did not escape the attention of the HIV-1 research community, and only 2 months after the publication of CXCR4 as a coreceptor for T-tropic viruses, five independent groups identified CCR5, a member of the closely related CC chemokine receptor family, as the major coreceptor for M-tropic strains.[20–24] The rapid identification of CCR5 was aided not only by the homology of CXCR4 to chemokine receptors, but also by the observation made only months earlier that the chemokines RANTES, MIP-1β, and MIP-1α were inhibitory factors secreted by $CD8^+$ T cells that blocked replication of M-tropic strains.[25] The contemporaneous discovery of CCR5, a chemokine receptor that could bind the chemokines RANTES, MIP-1β, and MIP-1α paved the way for its rapid identification as a coreceptor that allows fusion and entry of M-tropic strains of HIV-1.[26] In conjunction with CD4, the ability of CXCR4 and CCR5 to mediate Env binding, Env fusion, and viral entry classifies the chemokine receptors as true coreceptors that are required for HIV-1 entry.[27–29]

3. CHEMOKINE RECEPTORS AS PATHOGENIC DETERMINANTS

Shortly after the identification of CXCR4 and CCR5 as coreceptors, a highly prevalent CCR5 polymorphism (allele frequency 10% in Whites) termed Δ32CCR5 was identified.[30,31] This polymorphism, a frameshift 32-bp deletion in the middle of CCR5 that precludes full-length translation and transport to the cell surface, helps explain the molecular basis of resistance demonstrated by some highly exposed individuals who nonetheless remain uninfected. Homozygotes for this mutation are nearly completely resistant to HIV-1 infection, with a handful of exceptions,[32–34] and heterozygotes for this mutation demonstrate a 2- to 4-year delay in progression to AIDS.[35–37] More than anything else, though, this discovery demonstrates that CCR5 is the primary coreceptor used by HIV-1 for transmission and that it plays a major role in determining the rate of disease progression once an individual is infected. Moreover, the lack of any apparent deficiency in individuals who are homozygous for this mutation makes CCR5 a nearly ideal candidate for pharmaceutical targeting directed to block HIV-1 at the level of entry. Additional mutations in CCR5 have also been reported, but these are much less prevalent and none are predicted to have any obvious protective consequences. One CCR5 polymorphism, however, that results in a severely truncated cytoplasmic tail is predicted to lack signaling properties.[38] Although signaling of CCR5 is clearly not required for virus fusion and entry,[39–42] signaling may play a role in postentry events such as integration into quiescent cells such as macrophages. Any possible protective effects of this mutation would therefore be of great interest in determining alternative mechanisms of inhibiting CCR5 coreceptor function.

The identification of the Δ32CCR5 polymorphism helped trigger a search for additional polymorphisms in either chemokines or chemokine receptors that might influence viral transmission and pathogenesis. A single-base pair mutation in CCR2 that changes a valine to an isoleucine at position 64 in the CCR2 protein (V64I-CCR2) has been shown to correlate with a delay in disease progression.[43] However, CCR2 is utilized by only a handful of HIV-1 strains for infection, and a complete explanation for its protective effects awaits resolution. Unlike Δ32CCR5, V64I-CCR2 reaches the cell surface, functions as both a coreceptor and as a signal-transducing chemokine receptor, and does not have any gross effects on cell function or other chemokine receptors that might explain its protective effects (unpublished data). Subtle effects, however, such as its ability to cross-regulate the surface expression of CCR5 and CXCR4, offer a viable explanation that might account for its effects on HIV-1 pathogenesis. Other alternatives, such as linkage to a mutation in the nearby CCR5 gene, or its contribution to the replication of a small percentage of highly pathogenic strains of HIV-1 *in vivo*, cannot yet be excluded, but appear less likely due to its subtle effects on progression, but not transmission. The V64I-CCR2 polymorphism may be most significant, however, in demonstrating that protective effects from mutations in other genes, and ultimately by alternative clinical interventions, may contribute to the suppression of HIV-1 pathogenesis through mechanisms other than direct coreceptor blockade.

Thus far, all HIV-1 strains identified use CCR5, CXCR4, or both as coreceptors. However, nearly a dozen additional chemokine receptors, orphan receptors, and viral chemokine-receptor homologs have been identified as capable of mediating fusion with one or more HIV-1, HIV-2, or simian immunodeficiency virus (SIV) Envs *in vitro* (a current tabulation of coreceptor usage is being kept in the Los Alamos National Laboratory's annual database of human retroviruses and AIDS). The significance of these alternative coreceptors for viral pathogenesis *in vivo* is uncertain, but it is possible that use of receptors other than CCR5 or CXCR4 may help explain some of the variable outcomes following HIV-1 infection. For example, it has been suggested that use of CCR3 may be associated with neurotropism.[44] Thus, as new coreceptors are identified, it will be important to determine if they can support virus infection in relevant target cells *in vivo*.

4. CHEMOKINE INHIBITION OF HIV-1

The discovery that the HIV-1 coreceptors have natural ligands and signaling functions suggested immediate therapeutic possibilities (a list of chemokine-related inhibitory strategies is presented in Table II and discussed throughout the text). The chemokine receptors are normally in-

TABLE II
Chemokine Receptor Inhibitors and Strategies

Strategy	Inhibitor
Chemokine ligands	MIP-1α, MIP-1β, RANTES, SDF
Viral chemokines	vMIPII
Chemokine variants	BB-10010
Chemokine derivatives	AOP-RANTES, Met-RANTES
HIV suppression by chemokines	MCP1
Receptor downregulation	SDF/CXCR4
Receptor cross-regulation	CCR2
Gene therapy delivery	Reverse pseudotyped vectors
Intracellular receptor interference	Antisense, intrakine
Selective cell targeting	Zidovudine, didanosine
Small-molecule inhibitors	T22, AMD3100, ALX40-4C

volved in the chemoattractive localization of cells by transducing a signal upon binding a chemokine ligand (for review see Premack and Schall[45]). As such, nearly all cells of the hematopoietic lineage express chemokine receptors of some sort, although chemokine receptors are also found on a surprisingly large number of other primary cell types and cell lines. Binding of a chemokine ligand to the extracellular loops of its receptor not only induces a signal within the cell, but can also prevent HIV-1 utilization of the receptor.[20,21,23,24] Chemokines directed to fusogenic chemokine receptors such as CXCR4, CCR5, CCR3, and CCR8 inhibit use of their cognate receptors by HIV-1, although the degree of inhibition can vary tremendously depending on the Env protein.[46] The mechanism of inhibition by chemokine ligands largely involves direct steric hindrance of the Env protein, but downregulation of the receptor caused by ligand-induced signaling may also contribute to entry inhibition.[47] In addition, indirect signaling effects by other chemokines, such as MCP1 or MDC, may also play a role in regulating the availability of the chemokine receptor to Env.[48,49]

5. THERAPEUTICS DIRECTED AT CHEMOKINE RECEPTORS

The remarkable resistance to virus infection exhibited by individuals who lack CCR5, coupled with the lack of side effects associated with loss of CCR5 function, indicates that CCR5 and the other HIV-1 coreceptors may represent important new targets for antiretroviral therapy. In addition, even partial inhibition of chemokine receptor use by HIV-1 may have beneficial effects, as evidenced by Δ32CCR5 heterozygotes. Nevertheless, coreceptor inhibitors carry imposing concerns: mutations in HIV-1 Env are likely to allow usage of other coreceptors or other parts of the same coreceptor, chemokine receptor activation may transduce immune signals that enhance rather than inhibit HIV-1 replication, and chemokine receptor expression levels may not be a limiting factor for entry into all cell types. Further, chemokine receptor inhibitors might create an artificial pressure that pushes a virus to evolve into using another coreceptor and possibly becoming more virulent. Whether these considerations prove to be true obstacles in the development of effective antiretroviral agents that target chemokine receptors remains to be determined.

Of the three natural chemokines that target CCR5, RANTES is generally the most potent in inhibiting HIV-1 entry.[46] With this in mind, several groups have tested N-terminally modified derivatives of RANTES for inhibitory activity. Identifying antagonists of CCR5 is especially important since agonists of CCR5 may enhance HIV-1 infection in some cell populations[50] and can potentially stimulate other infected cell populations.[51,52] Thus far, the most potent CCR5 antagonist is AOP-RANTES, a RANTES molecule with a modified N-terminus that has a greater affinity for CCR5 and that is approximately 10-fold more potent than RANTES in inhibiting HIV-1.[53] AOP-RANTES is not known to induce signaling through CCR5, and unlike RANTES, it effectively prevents HIV-1 infection of macrophages, possibly because of its higher affinity for the receptor. A fourth naturally occurring chemokine, vMIPII, encoded by the herpesvirus HHV8, targets CCR5 and additional chemokine receptors with antagonist properties and can inhibit use of several receptors by HIV-1.[54,55] While vMIPII is not as potent as AOP-RANTES, it demonstrates that molecules can be developed to target multiple viral coreceptors.

Due to their role in inflammation, advances in targeting chemokine receptors were made even before their identification as viral coreceptors. A fully functional variant of MIP1α (also called stem cell inhibitor, SCI, or LD78) called BB-10010, was developed for aiding immune therapies involved in acute myeloid leukemia, stem cell transplantation, and cancer chemotherapy. BB-10010 has been shown to recruit hematopoietic stem cells into the peripheral blood and enhances leukocyte progenitor repopulation following chemotherapy in murine models.[56–58] Prior to the identification of HIV-1 coreceptors, BB-10010 was found to be safe in phase I clinical trials involving normal and chemotherapy-treated (e.g., immune system-depleted) patients.[59,60] While the application of this drug toward treating HIV-1 infection is still being evaluated, preliminary indications suggest that the formulation of this chemokine, as is the case of most labile proteins, will limit its dosage *in vivo* to well below that required to inhibit HIV-1.[60] Nevertheless, this initial study lays the groundwork for follow-up safety trials of chemically synthesized drugs that may possess CCR5-agonist properties.

In retrospect, currently employed anti-HIV-1 drugs may already be affecting HIV-1 coreceptor use inadvertently. The nucleoside inhibitors zidovudine (ZDV) and didanosine (ddI) have previously been shown to affect M- and T-tropic viral populations differently.[61–63] This may now be explained by the selective activation of these prodrugs by specific cell types.[64] Thus, ZDV selectively inhibits M-tropic strains of HIV-1, possibly by the selective activation of ZDV in activated T cells that express high levels of CCR5. In contrast, ddI selectively inhibits T-tropic strains of HIV-1, possibly by the selective activation of ddI in resting T cells that express higher levels of CXCR4 than CCR5.

Various gene therapy approaches have also been proposed to target the chemokine receptors. One of these methods expresses an endoplasmic reticulum-retained version of a chemokine, termed an intrakine, in order to bind and prevent expression of functional chemokine receptors.[65,66] Preliminary work with chemokine receptors has also allowed the construction of novel reverse-pseudotype viral vectors that are capable of targeting HIV-1-infected cells.[67–69] While these approaches add an interesting depth to the arsenal of tools that the chemokine receptors have provided, their utility for the therapeutic delivery of genes is likely to be overshadowed by their use as marking tools for tracking HIV-1 infection *ex vivo* and for studies using animal models.

6. SMALL-MOLECULE INHIBITORS OF CHEMOKINE RECEPTORS

The most recent progress in targeting chemokine receptors for therapeutic intervention has been the identification of small molecules that target chemokine receptors directly and prevent HIV-1 entry.[70–73] These initial molecules will serve as lead compounds rather than pharmaceutical endpoints in the effort to identify drugs that target the coreceptors. Each of the three molecules recently identified as inhibitors of CXCR4 were previously published and characterized for their ability to inhibit T-tropic strains of HIV-1 prior to the identification of the HIV-1 coreceptors. With the identification of the viral coreceptors, all three have now been shown to inhibit T-tropic strains of HIV-1 that use CXCR4, but not M-tropic strains that use CCR5. They have also proven to block SDF activation of CXCR4, to block the antibody 12G5 from binding to CXCR4, and to block dual-tropic viruses only when CXCR4 is present without CCR5 (see Table III for comparison of salient inhibitor features). These criteria of selectivity have become the standard for determining whether a compound blocks a corecep-

TABLE III
Properties of Small-Molecule Chemokine Receptor Inhibitors

	T22	AMD3100	ALX40-4C
Composition	18 amino acids	Bicyclam	Nine D-Arg residues
Structure	NMR solved	Chemically synthesized	None
Structural motifs	Tyr–Arg–Lys	Aromatic linker	None
Blocks T-tropics	Yes	Yes	Yes
Blocks SDF	Yes	Yes	Yes
Blocks 12G5	Yes	Yes	Yes
Blocks CCR5	No	No	No
Resistant strains	Unknown	Yes	Yes
Clinical use	None	SCID-hu micd	Phase I/II trials

tor directly or, for example, binds to Env at a location (e.g., V3) that blocks its coreceptor usage. While these studies will surely be followed quickly with more advanced compounds encompassing a broader range of coreceptors and with more potent activity, each of the three initially identified small-molecule inhibitors directed against CXCR4 provides a unique insight into the directions subsequent studies will follow.

T22 is an 18-amino acid peptide derived from the hemolymph of the American horseshoe crab, *Limulus polyphemus*.[74] Its fortuitous identification as an inhibitor of HIV-1 escaped a mechanistic explanation until its ability to block T-tropic, but not M-tropic strains was evaluated and its ability to block SDF activation of CXCR4 was demonstrated.[71] Prior to this discovery, studies with T22 focused on mapping a critical three-amino acid motif of the molecule, Tyr–Arg–Lys, that plays a vital role in mediating its antiviral effects.[75] Alteration of this motif, or of the two Cys bonds that hold it in place, reveals that the arrangement of amino acids and three-dimensional structure rather than its mere charge ratio contribute to antiviral activity. Such mutations and structural mapping studies can now be revisited in terms of mapping the structures in T22 that bind directly to CXCR4.[76] These structures may be critical for designing molecules that can bind to other chemokine receptors, as well as for modeling how SDF and Env bind to CXCR4.

AMD3100 (also known as JM3100, JM2987, and SID791) is the first true chemical derivative identified that blocks HIV-1 Env coreceptor usage by directly binding to CXCR4.[72,73] Derivatives of this family of molecules, the bicyclams, have been synthesized and tested for their inhibitor properties and will ultimately yield structural information about how a chemical product can bind to a chemokine receptor.[77,78] The regions of CXCR4 with which it interacts will be especially interesting given that even with its small size it is capable of inhibiting Env, SDF, and 12G5. Importantly, however, strains of HIV-1 resistant to bicyclams can be derived *in vitro* after passage in standard cell lines,[73,79] suggesting that the utility of small-molecule inhibitors of the chemokine receptors may face a battle against the flexibility of most Envs in their usage of the coreceptors. AMD3100 has been tested in preliminary *in vivo* pharmacokinetic assays[77] and has also been used to inhibit HIV-1 replication in SCID-hu mouse models,[80] proving at least in theory that such models can be exploited for preliminary safety and efficacy trials. The bicyclam family of molecules will surely be pursued as a class of lead compounds for the screening and development of compounds directed against the chemokine receptors.

ALX40-4C is a small, positively charged molecule composed of nine modified Argine residues that is capable of inhibiting the interaction of CXCR4 with Env, SDF, and the monoclonal antibody 12G5.[70] Previous work with ALX40-4C focused on mapping the viral determinants of its inhibitory properties.[81,82] Although this compound offers little structural insight for designing better inhibitors, the binding site of ALX40-4C appears to involve the first and second extracellular loops of CXCR4.[70] In combination with other structure–function studies, this observation suggests that inhibitors directed to the second extracellular loop of CXCR4 may be particularly effective. More importantly, however, ALX40-4C proves to be the first small-molecule chemokine receptor inhibitor to be tested in clinical trials. Such trials were initiated by the Canadian company that developed the compound, Allelix Pharmaceuticals, several years prior to the identification of the HIV-1 coreceptors. Two clinical trials were conducted, the first a phase I safety trial[83] and the second a phase I/II safety/efficacy trial.[84] While the detailed results of these trials are being reevaluated based on the current knowledge of the mechanism of action, the results were promising in the sense that ALX40-4C proved safe even at maximal doses.[84] The safety of an antagonist targeted to a chemokine receptor is critical, especially since elimination of some chemokines, such as SDF, can be lethal.[85] That such pathways do not appear disrupted to any consequential extent favors the continuation of such trials in a more expanded fashion.

While ALX40-4C proved safe, its effectiveness could not be demonstrated.[84,86] In retrospect, however, shifts in a subpopulation, such as T-tropic strains, could easily be overlooked in the context of the larger viral burden. M-tropic or dual-tropic strains that use CCR5 or other alternate coreceptors would not be expected to be inhibited by a compound that selectively targets CXCR4. The relevant measurement for such a compound must therefore be the quantitation of T-tropic strains of HIV-1 that use only CXCR4 as a coreceptor. Dual-tropic strains that could persist *in vivo* without CXCR4, but that can still replicate in CXCR4-containing cells *in vitro* must be strictly accounted for in such a study. Retrospective analysis of this valuable clinical study is being performed, but the methodology used for future clinical trials of chemokine receptor inhibitors must surely consider the design and effects of this earliest clinical trial.

7. FUTURE DIRECTIONS

In the short period of time since the identification of the HIV-1 coreceptors, tremendous insight has been gained into HIV-1 tropism, transmission, and pathogenesis. In terms of understanding the mechanisms that underlie viral tropism at the level of entry into the cell, the coreceptors have fulfilled a decade-long search for the host determinants that are responsible for triggering fusion between the viral envelope and cell membrane. In terms of host protection factors, the coreceptors explain some of the epidemiological data and anecdotal evidence that has surfaced over the last decade concerning exposed uninfected individuals. In terms of clinical relevance, the coreceptors have provided molecular insight into transmission, pathogenesis, and disease progression. Finally, in terms of therapeutic intervention, the coreceptors offer an important opportunity to combat a highly variable virus by targeting invariable host proteins that are critically important for virus entry. It is fortuitous that the HIV-1 coreceptors are members of the seven-transmembrane G-protein-coupled receptor family, since this is possibly the single most heavily funded and well studied class of proteins targeted by pharmaceutical companies. The hectic pace of work over the past 2 years has witnessed the identification of the coreceptors, investigation into the nature of their function, the identification of natural ligands that prevent HIV-1 entry, and the identification of small-molecule lead compounds that prevent HIV-1 entry. If the coming years are anything like the past, we should expect the chemokine receptors to become major clinical targets of HIV-1 therapeutics.

ACKNOWLEDGMENTS. We thank Bill O'Brien and Allelix Pharmaceuticals for information about the use of ALX40-4C. This work was supported by NIH grants to R.W.D. and a Howard Hughes Medical Institute predoctoral fellowship to B.J.D.

REFERENCES

1. Wei, X., Ghosh, S. K., Taylor, M. E., Johnson, V. A., Emini, E. A., Deutsch, P., Lifson, J. D., Bonhoeffer, S., Nowak, M. A., Hahn, B. H., Sang, M. S., and Shaw, G. M., 1995, Viral dynamics in human immunodeficiency virus type 1 infection, *Nature* **373**:117–122.
2. Ho, D. D., Neumann, A. U., Perelson, A. S., Chen, W., Leonard, J. M., and Markowitz, M.,

1995, Rapid turnover of plasma virions and CD4 lymphocytes in HIV-1 infection, *Nature* **373**:123–126.

3. Gazzard, B., 1996, What we know so far, *AIDS* **10**(Suppl. 1):S3–S7.

4. Maddon, P. J., Dalgleish, A. G., McDougal, J. S., Clapham, P. R., Weiss, R. A., and Axel, R., 1986, The T4 gene encodes the AIDS virus receptor and is expressed in the immune system and the brain, *Cell* **47**:333–348.

5. Klatzmann, D., Champagne, E., Chamaret, S., Gruest, J., Guetard, D., Hercend, T., Gluckman, J. C., and Montagnier, L., 1984, T-lymphocyte T4 molecule behaves as the receptor for human retrovirus LAV, *Nature* **312**:767–768.

6. Dalgleish, A. G., Beverley, P. C., Clapham, P. R., Crawford, D. H., Greeves, M. F., and Weiss, R. A., 1984, The CD4 (T4) antigen is an essential component of the receptor for the AIDS retrovirus, *Nature* **312**:763–767.

7. Dragic, T., Charneau, P., Clavel, F., and Alizon, M., 1992, Complementation of murine cells for human immunodeficiency virus envelope/CD4-mediated fusion in human/murine heterokaryons, *J. Virol.* **66**:4794–4802.

8. Broder, C. C., Dimitrov, D. S., Blumenthal, R., and Berger, E. A., 1993, The block to HIV-1 envelope glycoprotein-mediated membrane fusion in animal cells expressing human CD4 can be overcome by a human cell component(s), *Virology* **193**:483–491.

9. Miedema, F., Meyaard, L., Koot, M., Klein, M. R., Roos, M. T. L., Groenink, M., Fouchier, R. A. M., Van't Wout, A. B., Tersmette, M., Schellekens, P. T. A., and Schuitemaker, H., 1994, Changing virus–host interactions in the course of HIV-1 infection, *Immunol. Rev.* **140**: 35–72.

10. Collman, R., Balliet, J. W., Gregory, S. A., Friedman, H., Kolson, D. L., Nathanson, N., and Srinivasan, A., 1992, An infectious molecular clone of an unusual macrophage-tropic and highly cytopathic strain of human immunodeficiency virus type 1, *J. Virol.* **66**:7517–7521.

11. O'Brien, W. A., Koyanagi, Y., Namazie, A., Zhao, J. Q., Diagne, A., Idler, K., Zack, J. A., and Chen, I. S. Y., 1990, HIV-1 tropism for mononuclear phagocytes can be determined by regions of gp120 outside the CD4-binding domain, *Nature* **348**:69–73.

12. Hwang, S. S., Boyle, T. J., Lyerly, H. K., and Cullen, B. R., 1991, Identification of the envelope V3 loop as the primary determinant of cell tropism in HIV-1, *Science* **253**:71–74.

13. Shioda, T., Levy, J. A., and Cheng-Meyer, C., 1991, Macrophage and T cell-line tropisms of HIV-1 are determined by specific regions of the envelope gp120 gene, *Nature* **349**:167–169.

14. Van't Wout, A. B., Kootstra, N. A., Mulder-Kampinga, G. A., Albrecht van Lent, N., Scherpbier, H. J., Veenstra, J., Boer, K., Coutinho, R. A., Miedema, F., and Schuitemaker, H., 1994, Macrophage-tropic variants initiate human immunodeficiency virus type 1 infection after sexual, parenteral, and vertical transmission, *J. Clin. Invest.* **94**:2060–2067.

15. Simmons, G., Wilkinson, D., Reeves, J. D., Dittmar, M. T., Beddows, S., Weber, J., Carnegie, G., Desselberger, U., Gray, P. W., Weiss, R. A., and Clapham, P. R., 1996, Primary, syncytium-inducing human immunodeficiency virus type 1 isolates are dual-tropic and most can use either Lestr or CCR5 as coreceptors for virus entry, *J. Virol.* **70**:8355–8360.

16. Connor, R. I., Sheridan, K. E., Ceradini, D., Choe, S., and Landau, N. R., 1997, Change in coreceptor use correlates with disease progression in HIV-1 infected individuals, *J. Exp. Med.* **185**:621–628.

17. Feng, Y., Broder, C. C., Kennedy, P. E., and Berger, E. A., 1996, HIV-1 entry cofactor: Functional cDNA cloning of a seven-transmembrane, G protein-coupled receptor, *Science* **272**:872–877.

18. Bleul, C. C., Farzan, M., Choe, H., Parolin, C., Clark-Lewis, I., Sodroski, J., and Springer, T. A., 1996, The lymphocyte chemoattractant SDF-1 is a ligand for LESTR/fusin and blocks HIV-1 entry, *Nature* **382**:829–833.

19. Oberlin, E., Amara, A., Bachelerie, F., Bessia, C., Virelizier, J. L., Arenzana-Seisdedos, F.,

Schwartz, O., Heard, J. M., Clark-Lewis, I., Legler, D. F., Loetscher, M., Baggiolini, M., and Moser, B., 1996, The CXC chemokine SDF-1 is the ligand for LESTR/fusin and prevents infection by T-cell-line-adapted HIV-1, *Nature* **382**:833–835.

20. Choe, H., Farzan, M., Sun, Y., Sullivan, N., Rollins, B., Ponath, P. D., Wu, L., Mackay, C. R., LaRosa, G., Newman, W., Gerard, N., Gerard, C., and Sodroski, J., 1996, The β-chemokine receptors CCR3 and CCR5 facilitate infection by primary HIV-1 isolates, *Cell* **85**:1135–1148.

21. Deng, H., Liu, R., Ellmeier, W., Choe, S., Unutmaz, D., Burkhart, M., DiMarzio, P., Marmon, S., Sutton, R. E., Hill, C. M., Davis, C. B., Peiper, S. C., Schall, T. J., Littman, D. R., and Landau, N. R., 1996, Identification of a major co-receptor for primary isolates of HIV-1, *Nature* **381**:661–666.

22. Doranz, B. J., Rucker, J., Yi, Y., Smyth, R. J., Samson, M., Peiper, S. C., Parmentier, M., Collman, R. G., and Doms, R. W., 1996, A dual-tropic primary HIV-1 isolate that uses fusin and the β-chemokine receptors CKR-5, CKR-3, and CKR-2b as fusion cofactors, *Cell* **85**: 1149–1158.

23. Alkhatib, G., Combadiere, C., Broder, C. C., Feng, Y., Kennedy, P. E., Murphy, P. M., and Berger, E. A., 1996, CC CKR5: A RANTES, MIP-1α, MIP-1β receptor as a fusion cofactor for macrophage-tropic HIV-1, *Science* **272**:1955–1958.

24. Dragic, T., Litwin, V., Allaway, G. P., Martin, S. R., Huang, Y., Nagashima, K. A., Cayanan, C., Maddon, P. J., Koup, R. A., Moore, J. P., and Paxton, W. A., 1996, HIV-1 entry into CD4$^+$ cells is mediated by the chemokine receptor CC-CKR-5, *Nature* **381**:667–673.

25. Cocchi, F., DeVico, A. L., Garzino-Demo, A., Arya, S. K., Gallo, R. C., and Lusso, P., 1995, Identification of RANTES, MIP-1α, and MIP-1β as the major HIV-suppressive factors produced by CD8$^+$ T cells, *Science* **270**:1811–1815.

26. Samson, M., Labbe, O., Mollereau, C., Vassart, G., and Parmentier, M., 1996, Molecular cloning and functional expression of a new human CC-chemokine receptor gene, *Biochemistry* **35**:3362–3367.

27. Trkola, A., Dragic, T., Arthos, J., Binley, J. M., Olson, W. C., Allaway, G. P., Cheng-Mayer, C., Robinson, J., Maddon, P. J., and Moore, J. P., 1996, CD4-dependent, antibody-sensitive interactions between HIV-1 and its co-receptor CCR-5, *Nature* **384**:184–187.

28. Wu, L., Gerard, N. P., Wyatt, R., Choe, H., Parolin, C., Ruffing, N., Borsetti, A., Cardoso, A. A., Desjardin, E., Newman, W., Gerard, C., and Sodroski, J., 1996, CD4-induced interaction of primary HIV-1 gp120 glycoproteins with the chemokine receptor CCR-5, *Nature* **384**: 179–183.

29. Lapham, C. K., Ouyang, J., Chandrasekhar, B., Nguyen, N. Y., Dimitrov, D. S., and Golding, H., 1996, Evidence for cell-surface association between fusin and the CD4–gp120 complex in human cell lines, *Science* **274**:602–605.

30. Samson, M., Libert, F., Doranz, B. J., Rucker, J., Liesnard, C., Farber, C. M., Saragosti, S., Lapouméroulie, C., Cognaux, J., Forceille, C., Muyldermans, G., Verhofstede, C., Burtonboy, G., Georges, M., Imai, T., Rana, S., Yi, Y., Smyth, R. J., Collman, R. G., Doms, R. W., Vassart, G., and Parmentier, M., 1996, Resistance to HIV-1 infection in Caucasian individuals bearing mutant alleles of the CCR-5 chemokine receptor gene, *Nature* **382**:722–725.

31. Liu, R., Paxton, W. A., Choe, S., Ceradini, D., Martin, S. R., Horuk, R., MacDonald, M. E., Stuhlmann, H., Koup, R. A., and Landau, N. R., 1996, Homozygous defect in HIV-1 coreceptor accounts for resistance of some multiply-exposed individuals to HIV-1 infection, *Cell* **86**:367–377.

32. Biti, R., French, R., Young, J., Bennetts, B., and Stewart, G., 1997, HIV-1 infection in an individual homozygous for the CCR5 deletion allele, *Nature Med.* **3**:252–253.

33. O'Brien, T. R., Winkler, C., Dean, M., Nelson, J. A. E., Carrington, M., Michael, N. L., and White, II, G. C., 1997, HIV-1 infection in a man homozygous for CCR5Δ32, *Lancet* **349**:1219.

34. Theodorou, I., Meyer, L., Magierowska, M., Katlama, C., Rouzioux, C., and the Seroco

Study Group, 1997, HIV-1 infection in an individual homozygous for CCR5Δ32, *Lancet* **349**:1219–1220.

35. Dean, M., Carrington, M., Winkler, C., Huttley, G. A., Smith, M. W., Allikmets, R., Goedert, J. J., Buchbinder, S. P., Vittinghoff, E., Gomperts, E., Donfield, S., Vlahov, D., Kaslow, R., Saah, A., Rinaldo, C., Detels, R., Hemophilia Growth and Development Study, Multicenter AIDS Cohort Study, Multicenter Hemophilia Cohort Study, San Francisco City Cohort, ALIVE Study, and O'Brien, S. J., 1996, Genetic resistance of HIV-1 infection and progression to AIDS by a deletion allele of the CKR5 structural gene, *Science* **273**:1856–1862.

36. Michael, N. L., Chang, G., Louie, L. G., Mascola, J. R., Dondero, D., Birx, D. L., and Sheppard, H. W., 1997, The role of viral phenotype and CCR-5 gene defects in HIV-1 transmission and disease progression, *Nature Med.* **3**:338–340.

37. Huang, Y., Paxton, W. A., Wolinsky, S. M., Neumann, A. U., Zhang, L., He, T., Kang, S., Ceradini, D., Jin, Z., Yazdanbakhsh, K., Kunstman, K., Erickson, D., Dragon, E., Landau, N. R., Phair, J., Ho, D. D., and Koup, R. A., 1996, The role of a mutant CCR5 allele in HIV-1 transmission and disease progression, *Nature Med.* **2**:1240–1243.

38. Ansari-Lari, M. A., Liu, X. M., Metzker, M. L., Rut, A. R., and Gibbs, R. A., 1997, The extent of genetic variation in the CCR5 gene, *Nature Genet.* **16**:221–222.

39. Doranz, B. J., Lu, Z., Rucker, J., Zhang, T., Sharron, M., Cen, Y., Wang, Z., Guo, H., Du, J., Accavitti, M. A., Doms, R. W., and Peiper, S. C., 1997, Two distinct CCR5 domains can mediate coreceptor usage by human immunodeficiency virus type 1, *J. Virol.* **71**:6305–6314.

40. Farzan, M., Choe, H., Martin, K. A., Sun, Y., Sidelko, M., Mackay, C. R., Gerard, N. P., Sodroski, J., and Gerard, C., 1997, HIV-1 entry and macrophage inflammatory protein-1b-mediated signaling are independent functions of the chemokine receptor CCR5, *J. Biol. Chem.* **272**:6854–6857.

41. Gosling, J., Monteclaro, F. S., Atchison, R. E., Arai, H., Tsou, C., Goldsmith, M. A., and Charo, I. F., 1997, Molecular uncoupling of C-C chemokine receptor 5-induced chemotaxis and signal transduction for HIV-1 coreceptor activity, *Proc. Natl. Acad. Sci. USA* **94**:5061–5066.

42. Alkhatib, G., Locati, M., Kennedy, P. E., Murphy, P. M., and Berger, E. A., 1997, HIV-1 coreceptor activity of CCR5 and its inhibition by chemokines: Independence from G protein signaling and importance of coreceptor downmodulation, *Virology* **234**:340–348.

43. Smith, M. W., Dean, M., Carrington, M., Winkler, C., Huttley, G. A., Lomb, D. A., Goedert, J. J., O'Brien, T. R., Jacobson, L. P., Kaslow, R., Buchbinder, S., Vittinghoff, E., Vlahov, D., Hoots, K., Hilgartner, M. W., Hemophilia Growth and Development Study (HGDS), Multicenter AIDS Cohort Study (MACS), Multicenter Hemophilia Cohort Study (MHCS), San Francisco City Cohort (SFCC), ALIVE Study, and O'Brien, S. J., 1997, Contrasting genetic influence of CCR2 and CCR5 variants on HIV-1 infection and disease progression, *Science* **277**:959–965.

44. He, J., Chen, Y., Farzan, M., Choe, H., Ohagen, A., Gartner, S., Busciglio, J., Yang, X., Hofmann, W., Newman, W., Mackay, C. R., Sodroski, J., and Gabuzda, D., 1997, CCR3 and CCR5 are co-receptors for HIV-1 infection of microglia, *Nature* **385**:645–649.

45. Premack, B. A., and Schall, T. J., 1996, Chemokine receptors: Gateways to inflammation and infection, *Nature Med.* **2**:1174–1178.

46. Trkola, A., Paxton, W. A., Monard, S. P., Hoxie, J. A., Siani, M. A., Thompson, D. A., Wu, L., Mackay, C. R., Horuk, R., and Moore, J. P., 1998, Genetic subtype-independent inhibition of human immunodeficiency virus type 1 replication by CC- and CXC-chemokines, *J. Virol.* **72**:396–404.

47. Amara, A., Le Gall, S., Schwartz, O., Salamero, J., Montes, M., Loetscher, P., Baggiolini, M., Virelzier, J.-L., and Arenzana-Seisdedos, F., 1997, HIV coreceptor downregulation as anti-

viral principle: SDF-1α-dependent internalization of the chemokine receptor CXCR4 contributes to inhibition of HIV replication, *J. Exp. Med.* **186:**139–146.

48. Frade, J. M. R., Liorente, M., Mellado, M., Alcami, J., Gutierrez-Ramos, J. C., Zaballos, A., del Real, G., and Martinez-A, C., 1997, The amino-terminal domain of the CCR2 chemokine receptor acts as coreceptor for HIV-1 infection, *J. Clin. Invest.* **100:**497–502.

49. Pal, R., Garzino-Demo, A., Markham, P. D., Burns, J., Brown, M., Gallo, R. C., and DeVico, A. L., 1997, Inhibition of HIV-1 infection by the β-chemokine MDC, *Science* **278:**695–698.

50. Schmidtmayerova, H., Sherry, B., and Bukrinsky, M., 1996, Chemokines and HIV replication, *Nature* **382:**767.

51. Weissman, D., Rabin, R. L., Arthos, J., Rubbert, A., Dybul, M., Swofford, R., Venkatesan, S., Farber, J. M., and Fauci, A. S., 1997, Macrophage-tropic HIV and SIV envelope induce a signal through the CCR5 chemokine receptor, *Nature* **389:**981–985.

52. Davis, C. B., Dikic, I., Unutmaz, D., Hill, C. M., Arthos, J., Siani, M. A., Thompson, D. A., Schlessinger, J., and Littman, D. R., 1997, Signal transduction due to HIV-1 envelope interactions with chemokine receptors CXCR4 or CCR5, *J. Exp. Med.* **186:**1793–1798.

53. Simmons, G., Clapham, P. R., Picard, L., Offord, R. E., Rosenkilde, M. M., Schwartz, T. W., Buser, R., Wells, T. N. C., and Proudfoot, A. E. I., 1997, Potent inhibition of HIV-1 infectivity in macrophages and lymphocytes by a novel CCR5 antagonist, *Science* **276:**276–279.

54. Boshoff, C., Endo, Y., Collins, P. D., Takeuchi, Y., Reeves, J. D., Schweickart, V. L., Siani, M. A., Sasaki, T., Williams, T. J., Gray, P. W., Moore, P. S., Chang, Y., and Weiss, R. A., 1997, Angiogenic and HIV inhibitory functions of KSHV-encoded chemokines, *Science* **278:**290–294.

55. Kledal, T. N., Rosenkilde, M. M., Coulin, F., Simmons, G., Johnsen, A. H., Alouani, S., Power, C. A., Lüttichau, H. R., Gerstoft, J., Clapham, P. R., Clark-Lewis, I., Wells, T. N. C., and Schwartz, T. W., 1997, A broad-spectrum chemokine antagonist encoded by Kaposi's sarcoma-associated herpesvirus, *Science* **277:**1656–1659.

56. Marshall, F., Woolford, L. B., and Lord, B. I., 1997, Continuous infusion of macrophage inflammatory protein MIP-1α enhances leucocyte recovery and haemopoietic progenitor cell mobilization after cyclophosphamide, *J. Cancer* **75:**1715–1720.

57. Lord, B. I., Woolford, L. B., Wood, L. M., Czaplewski, L. G., McCourt, M., Hunter, M. G., and Edwards, R. M., 1995, Mobilization of early hematopoetic progenitor cells with BB-10010: A genetically engineered variant of human macrophage inflammatory protein-1α, *Blood* **12:**3412–3415.

58. Owen-Lynch, P. J., Adams, J. A., Brereton, M. L., Czaplewski, L. G., Whetton, A. D., and Yin, J. A. L., 1996, The effect of the chemokine rhMIP-1α, and a non-aggregating variant BB-10010, on blast cells from patients with acute myeloid leukaemia, *Br. J. Haematol.* **95:**77–84.

59. Gordon, M. S., McCaskill-Stevens, W. J., Broxmeyer, H. E., Battiato, L. A., Harrison-Mann, B., Kovalsky, M., Rasmussen, H. S., and Sledge, G. W., 1996, A phase I trial of subcutaneous BB-10010 in breast-cancer patients receiving high-dose cyclophosphamide, *Exp. Hematol.* **24:**1104.

60. Czaplewski, L. G., 1997, *Preclinical and clinical evaluation of the therapeutic potential of hMIP-1 alpha, hMIP-1 beta, and Rantes,* Presented at the IBC Conference on HIV Coreceptors. Baltimore, Maryland.

61. Van't Wout, A. B., De Jong, M. D., Kootstra, N. A., Veenstra, J., Lange, J. M. A., Boucher, C. A. B., and Schuitemaker, H., 1996, Changes in cellular virus load and zidovudine resistance of syncytium-inducing and non-syncytium-inducing human immunodeficiency virus populations under zidovudine pressure: A clonal analysis, *J. Infect. Dis.* **174:**845–849.

62. Delforge, M. L., Liesnard, C., Debaisieux, L., Tchetcheroff, M., Farber, C. M., and Van

Vooren, J. P., 1995, *In vivo* inhibition of syncytium-inducing variants of HIV in patients treated with didanosine, *AIDS* **9**:89–90.

63. Zheng, N. N., McQueen, P. W., Hurren, L., Evans, L. A., Law, M. G., Forde, S., Barker, S., Cooper, D. A., and Delaney, S. F., 1996, Changes in biologic phenotype of human immunodeficiency virus during treatment of patients with didanosine, *J. Infect. Dis.* **173**:1092–1096.

64. Van't Wout, A. B., Ran, L. J., de Jong, M. D., Bakker, M., van Leeuwen, R., Notermans, D. W., Loeliger, A. E., de Wolf, F., Danner, S. A., Reiss, P., Boucher, C. A. B., Lange, J. M. A., and Schuitemaker, H., 1997, Selective inhibition of syncytium-inducing and non-syncytium-inducing HIV-1 variants in individuals receiving didanosine or zidovudine, respectively, *J. Clin. Infect.* **100**:2325–2332.

65. Chen, J. D., Bai, X., Yang, A. G., Cong, Y., and Chen, S. Y., 1997, Inactivation of HIV-1 chemokine co-receptor CXCR-4 by a novel intrakine strategy, *Nature Med.* **3**:1110–1116.

66. Yang, A. G., Bai, X., Huang, X. F., Yao, C., and Chen, S. Y., 1997, Phenotypic knockout of HIV type 1 chemokine coreceptor CCR-5 by intrakines as potential therapeutic approach for HIV-1 infection, *Proc. Natl. Acad. Sci. USA* **94**:11567– 11572.

67. Mebatsion, T., Finke, S., Weiland, F., and Conzelmann, K. K., 1997, A CXCR4/CD4 pseudotype rhabdovirus that selectively infects HIV-1 envelope protein-expressing cells, *Cell* **90**: 841–847.

68. Schnell, M. J., Johnson, J. E., Buonocore, L., and Rose, J. K., 1997, Construction of a novel virus that targets HIV-1-infected cells and controls HIV-1 infection, *Cell* **90**:849–857.

69. Endres, M. J., Jaffer, S., Haggarty, B., Turner, J. D., Doranz, B. J., O'Brien, P. J., Kolson, D. L., and Hoxie, J. A.,. 1997, Targeting of HIV- and SIV-infected cells by CD4-chemokine receptor pseudotypes, *Science* **278**:1462–1464.

70. Doranz, B. J., Grovit-Ferbas, K., Sharron, M. P., Mao, S., Goetz, M. B., Daar, E. S., Doms, R. W., and O'Brien, W. A., 1997, A small-molecule inhibitor directed against the chemokine receptor CXCR4 prevents its use as an HIV-1 coreceptor, *J. Exp. Med.* **186**:1395–1400.

71. Murakami, T., Nakajima, T., Koyanagi, Y., Tachibana, K., Fujii, N., Tamamura, H., Yoshida, N., Waki, M., Matsumoto, A., Yoshie, O., Kishimoto, T., Yamamoto, N., and Nagasawa, T., 1997, A small molecule CXCR4 inhibitor that blocks T cell line-tropic HIV-1 infection, *J. Exp. Med.* **186**:1389–1393.

72. Schols, D., Struyf, S., Van Damme, J., Esté, J. A., Henson, G., and De Clerq, E., 1997, Inhibition of T-tropic HIV strains by selective antagonization of the chemokine receptor CXCR4, *J. Exp. Med.* **186**:1383–1388.

73. Donzella, G. A., Schols, D., Lin, S. W., Esté, J. A., Nagashima, K. A., Maddon, P. J., Allaway, G. P., Sakmar, T. P., Henson, G., De Clerq, E., and Moore, J. P., 1998, AMD3100, a small molecule inhibitor of HIV-1 entry via the CXCR4 co-receptor, *Nature Med.* **4**:72–77.

74. Nakashima, H., Masuda, M., Murakami, T., Koyanagi, Y., Matsumoto, A., Fujii, N., and Yamamoto, N. Y., 1992, Anti-human immunodeficiency virus activity of a novel synthetic peptide, T22 ([Tyr-5,12, Lys-7] polyphemusin II): A possible inhibitor of virus-cell fusion, *Antimicrob. Agents Chemother.* **36**:1249–1255.

75. Tamamura, H., Murakami, T., Masuda, M., Otaka, A., Takada, W., Ibuka, T., Nakashima, H., Waki, M., Matsumoto, A., Yamamoto, N., and Fujii, N., 1994, Structure–activity relationships of an anti-HIV peptide, T22, *Biochem. Biophys. Res. Commun.* **205**:1729–1735.

76. Tamamura, H., Kuroda, M., Masuda, M., Otaka, A., Funakoshi, S., Nakashima, H., Yamamoto, N., Waki, M., Matsumoto, A., Lancelin, J. M., Kohda, D., Tate, S., Inagaki, F., and Fujii, N., 1993, A comparative study of the solution structures of tachyplesin I and a novel anti-HIV synthetic peptide, T22 ([Tyr^{5,12}, Lys^{7}]-polyphemusin II), determined by nuclear magnetic resonance, *Biochim. Biophys. Acta* **1163**:209–216.

77. De Clerq, E., Yamamoto, N., Pauwels, R., Balzarini, J., Witvrouw, M., De Vreese, K., Debyser, Z., Rosenwirth, B., Peichl, P., Datema, R., Thornton, D., Skerlj, R., Gaul, F., Padmanabhan,

S., Bridger, G., Henson, G., and Abrams, M., 1994, Highly potent and selective inhibition of human immunodeficiency virus by the bicyclam derivative JM3100, *Antimicrob. Agents Chemother.* **38**:668–674.

78. De Clerq, E., Yamamoto, N., Pauwels, R., Baba, M., Schols, D., Nakashima, H., Balzarini, J., Debyser, Z., Murrer, B. A., Schwartz, D., Thornton, D., Bridger, G., Fricker, S., Henson, G., Abrams, M., and Picker, D., 1992, Potent and selective inhibition of human immunodeficiency virus (HIV)-1 and HIV-2 replication by a class of bicyclams interacting with a viral uncoating event, *Proc. Natl. Acad. Sci. USA* **89**:5286–5290.

79. De Vreese, K., Kofler-Mongold, V., Leutger, C., Weber, V., Vermeire, K., Schacht, S., Anne, J., De Clerq, E., Datema, R., and Werner, G., 1996, The molecular target of bicyclams, potent inhibitors of human immunodeficiency virus replication, *J. Virol.* **70**:689–696.

80. Datema, R., Rabin, L., Hincenbergs, M., Moreno, M. B., Warren, S., Linquist, V., Rosenwirth, B., Seifert, J., and McCune, J. M., 1996, Antiviral efficacy *in vivo* of the anti-human immunodeficiency virus bicyclam SDZ SID 791 (JM 3100), an inhibitor of infectious cell entry, *Antimicrob. Agents Chemother.* **40**:750–754.

81. O'Brien, W. A., Sumner-Smith, M., Mao, S., Sadeghi, S., Zhao, J., and Chen, I. Y., 1996, Anti-human immunodeficiency virus type 1 activity of an oligocationic compound mediated via gp120 V3 interactions, *J. Virol.* **70**:2825–2831.

82. Sumner-Smith, M., Zheng, Y., Zhang, Y. P., Twist, E. M., and Climie, S. C., 1995, Anti-herpetic activities of N-α-acetyl-nona-D-arginine amide acetate, *Drugs Exp. Clin. Res.* **21**:1–6.

83. "35th Interscience Conference Unveils Results from Several Important Clinical Studies," in: TAGline [database online] Volume 2, Issue 11, November 1995.

84. "Allelix Refocuses Its Transcription Therapeutics Program," in: Businesswire [database online] San Francisco: Business Wire, January 20, 1997, 10:30:00-0500.

85. Nagasaw, T., Hirota, S., Tachibana, K., Takakura, N., Nishikawa, S., Kitamura, Y., Yoshida, N., Kikutani, H., and Kishimoto, T., 1996, Defects of B-cell lymphopoiesis and bone-marrow myelopoiesis in mice lacking the CXC chemokine PBSF/SDF-1, *Nature* **382**:635–638.

86. Dummett, B., "Allelix Supports AIDS Drug Despite Poor Test Results," in: HIV-News [electronic bulletin board], 19 December 1996, 14:45:00-0600 [cited 21 January 1997].

13

Protease Inhibitors and HIV-1 Genetic Variability in Infected Children

MAUREEN M. GOODENOW, ELENA E. PEREZ, and JOHN W. SLEASMAN

1. ANTIVIRAL DRUG THERAPIES

1.1. Introduction

Two classes of antiviral drugs, the reverse transcriptase (RT) inhibitors and the protease (PR) inhibitors, have provided the first successful impact on the human immunodeficiency virus type 1 (HIV-1) epidemic. Zidovudine (ZDV), one of a number of inhibitors targeted to the viral RT enzyme, significantly reduces pediatric infection by maternal transmission when administered to infected women during pregnancy and delivery, and to their newborns during the first 6 weeks of life.[1] Inhibitors targeted at the HIV-1 PR can suppress plasma virus by greater than 1.5 $\log_{10}$ copies per milliliter below pretreatment levels, and in some cases to levels that are undetectable (<50 RNA copies/ml plasma). Diminished virus burden is accompanied by increases in CD4 T lymphocytes, although the extent of immune cell reconstitution appears to be influenced by factors in addition

MAUREEN M. GOODENOW and JOHN W. SLEASMAN • Department of Pathology, Immunology, and Laboratory Medicine, and Division of Immunology and Infectious Diseases, Department of Pediatrics, College of Medicine, Health Science Center, University of Florida, Gainesville, Florida 32610.   ELENA E. PEREZ • Department of Pathology, Immunology, and Laboratory Medicine, and Medical Scientist Training Program, College of Medicine, Health Science Center, University of Florida, Gainesville, Florida 32610.

Human Retroviral Infections, edited by Kenneth E. Ugen *et al.*
Kluwer Academic / Plenum Publishers, New York, 2000.

to virus decline. Five PR inhibitors, namely saquinavir, ritonavir, indinavir, amprenavir, and nelfinavir, are approved by the Food and Drug Administration (FDA) for use in combination therapies to treat adult HIV-1 infection. Ritonavir, amprenavir, and nelfinavir are also approved for HIV-1-infected children, while indinavir and saquinavir are in Phase I/II studies for pediatric patients.

1.2. HIV-1 Targets for Drugs

Therapy for treatment of HIV-1 infection involves combinations of antiviral drugs aimed at RT and PR, which function at different stages of the virus life cycle (Fig. 1).[2] RT activity is crucial for synthesis of the double-stranded DNA form of HIV-1, so RT inhibitors act at an early stage of the virus life cycle following infection of target cells (Fig. 1, step 2). PR activity is essential for virion maturation, which occurs late in the virus life cycle as virions bud from the surface of infected cells (Fig. 1, step 7).[3] Virus particles are formed in the presence of PR inhibitors, but infectivity of new cells is drastically reduced if PR activity is blocked.[4]

1.3. HIV-1 Protease

PR is encoded within the *pol* region of the HIV-1 genome (Fig. 2A). The enzyme is an aspartic protease, which functions as a molecular scissor to cleave the Gag and Gag/Pol polyprotein precursors into structural proteins (matrix, capsid, nucleocapsid and p6) and enzymes (PR, RT, and integrase) (Fig. 2B). At least eight different target sequences (Fig. 2B, A–G) in the polyprotein serve as substrates for PR and are cleaved in an ordered process during maturation of virions (Fig. 2C).[5]

The functional form of HIV-1 PR is a dimeric molecule comprised of two identical 99-amino acid monomers. Three functional domains are involved in PR activity: the active-site cleft with two catalytic aspartic acid

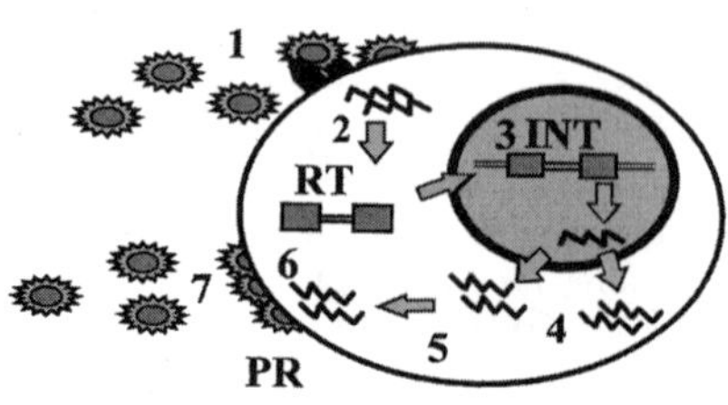

FIGURE 1. HIV-1 Life Cycle. The main steps of the HIV-1 life cycle are designated by numbers and include (1) viral attachment and entry, (2) reverse transcription, (3) integration, (4) viral RNA transcription and processing, (5) translation of viral mRNAs into polyprotein precursors of structural and enzymatic proteins, (6) assembly, (7) budding. Retroviral enzymes reverse transcriptase (RT), integrase (INT), and protease (PR) are active during specified steps in the virus life cycle.

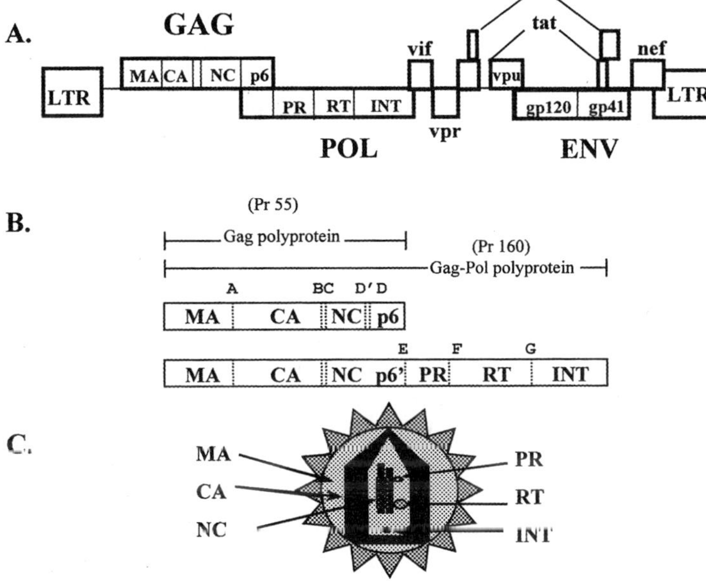

FIGURE 2. HIV-1 Genome. HIV-1 is an RNA virus comprised of two identical, 10-kb RNA molecules. (A) Genome organization of HIV-1. The three main coding regions of HIV-1, *gag*, *pol*, and *env*, and the accessory genes are represented with respect to their reading frames. Two identical long terminal repeats (LTR) flank the viral genome. (B) Processing of precursor polyproteins. The Pr55 *gag* and Pr160 *gag–pol* polyproteins are depicted with respect to the reading frame. Dotted lines represent protease cleavage sites for the generation of mature viral proteins. The structural proteins encoded by *gag* and the enzymatic proteins encoded by *pol* are shown in their relative locations within the virion. The ribosomal frameshift is located at D'. Frameshift removes the D' and D cleavage sites. The Pr55 and Pr160 polyproteins are translated at a ratio of 20:1. (C) Virion structure. MA, matrix or p17; CA, capsid or p24; NC, nucleocapsid or p7.

residues, the flaps, and the psi loop. Binding of substrate in the active site induces conformational changes which close the flaps and bring the catalytic aspartate residues into position to mediate substrate cleavage. PR inhibitors are small molecules that compete with substrates for active-site binding. Once bound, the inhibitors cannot be cleaved and consequently inactivate the enzyme by remaining in the active site.

1.4. Viral Drug Resistance

HIV-1 develops resistance to RT and to PR inhibitors as a result of errors introduced during replication by reverse transcriptase and selection for continued replication. A number of characteristics raised the possibility that

PR could be an Achilles' heel in the HIV genome. For example, the impact of drug-resistant mutations on PR activity, the small target (297 nucleotides) for mutagenesis that PR presents in the 10-kb HIV-1 genome, and the number of mutations (four or more) required for high-level resistance all pointed to a region of the genome that was predicted to be intolerant to extensive variation in the absence of selective pressures.

In view of the theoretical constraints on natural variation in PR, it was a surprise to find that PR alleles in untreated individuals harbored amino acid polymorphisms at almost half of the residues comprising PR (Fig. 3).[6–8] Amino acid variation is localized to regions that surround the functional domains. Sequences in the active-site domain and the psi loop are virtually identical among virus populations in different patients, whereas changes in the flap domains can occur, but with limited frequency.

Reduced sensitivity of PR to inhibitors involves multiple amino acid substitutions, which occur with high frequency in the functional domains of PR (Fig. 3). The types of amino acid substitutions are, for the most part, unique to PR alleles with reduced sensitivity to inhibitors and occur predominantly in viruses from individuals who have received treatment with PR

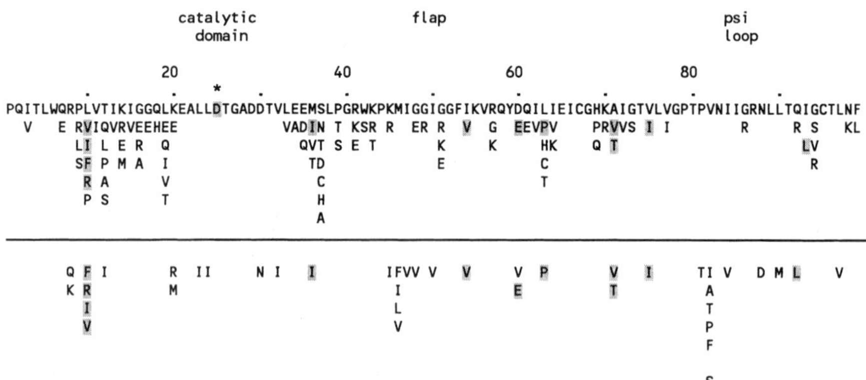

FIGURE 3. Amino acid polymorphisms in protease alleles. Protease monomer is comprised of 92 amino acids. Active enzyme is a homodimeric molecule comprised of two monomers. The reference amino acid sequence, displayed as the single-letter code, is protease from HIV_{LAI}. Catalytic domain, flaps, and psi loop are indicated. The asterisk indicates the catalytic aspartic acid at amino acid position 25. Amino acid substitutions in proteases from viruses in patients in the absence of protease inhibitors are summarized between the reference sequence and the horizontal line. Below the horizontal line are amino acids associated with reduced sensitivity or resistance to protease inhibitors identified in viruses from treated patients or in culture. Amino acid residues from resistant alleles that are found in proteases in the absence of inhibitors are shaded.

inhibitors. HIV-1 PR alleles from patients prior to PR inhibitor therapy seldom display more than one amino acid substitution that is known to contribute to resistance. Similarity among inhibitors in binding to the active-site cleft of the enzyme selects for combinations of amino acid replacements in PR that can confer resistance to the administered inhibitor and cross-resistance to other PR inhibitors as well.[9–11]

Although multiple amino acid changes in PR are required for high-level resistance to drug, continuous replication by the virus, selection for variants that are "more fit" in the presence of inhibitors, and chance all contribute to the appearance of highly resistance alleles of PR within weeks or months of therapy.[12–14] Changes to HIV-1 PR amino acids in adult patients receiving PR inhibitors occur in an ordered sequence. For example, mutations that produce amino acid substitutions in position 82 appear early in therapy and are highly associated with initial decline in inhibitor effectiveness.[10] Additional changes in amino acids at positions 20, 36, and 54 accumulate within weeks after initiation of therapy, correspond to rebounds in plasma virus levels, and produce in viruses in culture an 8- to 23-fold increased resistance to inhibitors.[10]

2. ANTIVIRAL DRUGS AND PEDIATRIC PATIENTS

2.1. Drugs for HIV-1-Infected Children

Requests by pharmaceutical companies for FDA approval of drugs for treatment of HIV-1 disease in children have lagged behind approval for antiviral drugs in the adult population. For example, ZDV therapy for HIV-1-infected children was approved about 10 years after approval for use in infected adults. FDA recommendations now facilitate, and even require, Phase I/II studies that assess drug safety and pharmacokinetics in the pediatric population. Efficacy data for drugs in children are not mandatory if efficacy has been demonstrated in adults for diseases that have similar etiology. HIV-1 disease in both children and adults results from virus infection and PR inhibitors are effective treatment in adults. Consequently, in contrast to ZDV, approval for pediatric use of ritonavir occurred within 2 years of adult approval, while amprenavir and nelfinavir were approved simultaneously for use in adults and children.

Problems associated with providing antiretroviral drugs to children include the formulation, bioavailability, and pharmacokinetics, which all differ from adults.[15] Drugs for the adult population are frequently formulated as capsules or tablets, which are not the optimal form for delivery to young children who may not be able to swallow capsules. Dosages of drugs formulated in capsule or tablet form for adults are difficult to adjust for pediatric patients, while liquid or suspension formulations, which facilitate pediatric dosage, can have poor solubility or unpleasant tastes. In addition, differences between adults and children in hepatic metabolism and renal clearance result in different pharmacokinetics in pediatric patients. For the pediatric population, PR inhibitors were formulated as liquid for ritonavir or as powder for nelfinavir. Initial suspension formulations of indinavir or saquinavir had reduced bioavailability, which has delayed availability of these PR inhibitors for young children.

2.2. Protease Inhibitors and Clinical Response

Three Phase I/II clinical studies of protease inhibitors ritonavir, indinavir, and nelfinavir have been carried out in children. In addition to evaluation of safety and pharmacokinetics, these studies indicated that protease inhibitors can control virus replication and restore CD4 T cell counts. Consequently, the Public Health Service recommends that all HIV-1-infected children should be treated with combination antiretroviral therapy, which includes protease inhibitors.[16]

2.2.1. Ritonavir

A liquid formulation of ritonavir administered as monotherapy and in combination with RT inhibitors was evaluated in a Phase I/II study involving 48 children, who ranged in age from 0.5 to 14.4 years (median age, 7.7 years).[17] Over the course of the 24-week study, 7 (15.9%) children responded with a reduction in plasma virus to undetectable levels that was

TABLE I

Response of Pediatric Cohort Following 24 Weeks of PR Inhibitor Therapy[a]

Percentage of patients	CD4 T cell reconstitution[b]	Decrease in plasma virus[c]
59	2- to 100-fold	$\geqslant 1.7 \log_{10}$
26	1- to 100-fold	$\leqslant 0.5 \log_{10}$
15	<2-fold	$\leqslant 0.5 \log_{10}$

[a]Data are summarized from a study of 44 children treated with ritonavir plus zidovudine and didanosine.[19]

[b]Children who were CDC immune stage 1 displayed median 2- to 4-fold increases in CD4 T cells, which generally restored CD4 T cells to values that were normal for age. Children who were immune stage 3 could display median increases as great at 100-fold in CD4 T cell numbers.

[c]Decreases greater than 3 logs to undetectable plasma levels were displayed by some children.

sustained for at least 8 weeks. About 60% of the total study group experienced greater than a 1.7 $\log_{10}$ decrease in virus RNA and a concomitant increase of 2- to 100-fold in CD4 T cells (Table I). Approximately 16% of the children failed to demonstrate a sustained decrease in plasma virus or more than a transient increase in CD4 T cells. Almost 25% of the children displayed an unexpected response, in that CD4 T cells increased dramatically, whereas virus RNA was suppressed only transiently and rebounded to baseline levels. Recent reports identify a similar response in the adult population.[18] One explanation for these results is that viruses which emerge during combination therapy may have diminished pathogenic potential relative to pretherapy viruses.

2.2.2. Indinavir

Clinical evaluation of indinavir alone or in combination with RT inhibitors in 54 HIV-1-infected children (ages 3.1–18.9 years) revealed that the free-base suspension was less bioavailable than the capsule formulation.[19] Even with the limitations of the initial study and the relatively high plasma virus levels (4.76 $\log_{10}$) at entry in the cohort, 16 of 35 patients achieved a sustained decrease of at least 1 $\log_{10}$ viral RNA copies/ml of plasma, while undetectable virus levels were achieved in 3 children. A sustained increase in absolute CD4 T cells occurred in all patients, with a median increase of 60 cells/mm^3 after 16 weeks of therapy. No significant difference in baseline CD4 values between children who did or did not respond at 16 weeks was detected. In a pilot study of combination indinavir in capsule form plus RT inhibitors in 12 children, median plasma HIV RNA concentrations declined by 1.3 $\log_{10}$ copies/ml after 24 weeks, while median CD4 T cell counts increased by 294 cells/mm^3 during the same period.[20] Based on these studies, Phase I/II studies of a new formulation of indinavir were initiated in 1998 within the Pediatric AIDS Clinical Trial Group (PACTG). Similar studies of combination therapy with saquinavir and RT inhibitors in children are in the planning stages through the PACTG.

2.2.3. Nelfinavir

A third large-scale Phase I/II pediatric study evaluated nelfinavir powder and tablet formulations in combination with one or two nucleoside RT inhibitors in a cohort of 62 HIV-1-infected children (ages 3 months to 13 years).[21] The two formulations displayed similar pharmacokinetics and oral bioavailability with limited toxicity. Baseline median plasma HIV RNA levels for 55 subjects were approximately 4.5 $\log_{10}$, which decreased by a median 0.7 $\log_{10}$ copies/ml. Viral RNA declined in 36% (15/55) of the children to undetectable levels that were sustained for a median time of 42 weeks. CD4

T lymphocytes increased by as much as 7–8% at week 26 of therapy, although the increase in CD4 T cells was more pronounced in patients who sustained plasma virus at undetectable levels.

2.3. Immune Reconstitution Following PR Inhibitor Therapy in Children

Combination therapy including PR inhibitors in adults can produce a positive impact on CD4 T cell homeostasis that involves initial trafficking of CD4 CD45RO memory lymphocytes from lymph nodes to peripheral blood, decreased CD4 T cell destruction, and, finally, increased production of CD4 CD45RA naive T cells, which express CD62L (l-selectin), from thymic precursors.[22] Children display a number of immunological differences from adults that could result in an expanded capacity for T cell reconstitution in response to triple combination therapy. For example, 70–90% of pediatric peripheral blood CD4 T cells express phenotypic markers of thymic-derived naive T cells. HIV-1 infection in children can lead to higher steady-state levels of virus than in infected adults, perhaps reflecting differences in the immune response or in susceptible target cells.[23] Greater levels of infection in memory than naive CD4 T cells in children could account, at least in part, for immune dysfunction.[24]

A comprehensive assessment of immune reconstitution in 44 HIV-1-infected children following triple therapy with ritonavir identified a positive impact on multiple lymphocyte lineages (Fig. 4).[25] Transient changes in CD8 T cells coupled with sustained increases in both B lymphocytes and CD4 T lymphocytes in children differed in dynamics and magnitude from adult responses. HIV-infected children, even in late-stage disease, displayed considerable potential for reconstitution of thymic-derived CD4 CD45RA

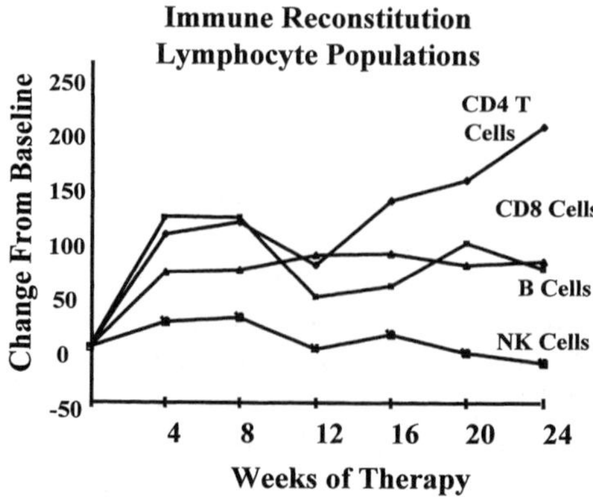

FIGURE 4. Lymphocyte reconstitution in a cohort of children treated with ritonavir. Therapy had little impact on median natural killer cell numbers in the cohort. CD8 T lymphocytes and B lymphocytes increased by about 100 cells during 24 weeks of therapy, although the kinetics of repopulation were different. CD4 T lymphocytes displayed the greatest increase in median cell numbers. An initial phase of increase during the first 12 weeks was followed by a second phase that resulted in a median increase of >200 CD4 T cells by week 24 of therapy.

naive T cells, implying that the thymus maintains its functional role in lymphocyte development.

3. GENETIC ANALYSIS OF PROTEASE IN RESPONSE TO PROTEASE INHIBITORS

3.1. Patients

To determine the interaction between response to PR inhibitor therapy and development of drug resistance in PR, we selected three children for detailed genetic analysis of HIV-1 PR before and during therapy with PR inhibitors. The children were representative of each response group to the Phase I/II studies of ritonavir or indinavir (Fig. 5). Patient A was a 14-year-old hemophiliac (Centers for Disease Control [CDC] stage A3) who responded to ritonavir therapy with sustained suppression of plasma virus below the levels of detection for almost 1 year and an increase of about 200 CD4 T cells (Fig. 5A). Patient B was a 6.7-year old, CDC stage C3 child who was infected by maternal transmission. Patient B had a 2 log increase in CD4 T cells over the course of 1 year of ritonavir therapy, but displayed only a transient decline in virus levels (Fig. 5B). Patient D was a 14-year-old, perinatally infected adolescent with advanced disease (CDC C3) who experienced no sustained increase in CD4 T cells or suppression of virus levels (Fig. 5C).

3.2. Methods

PR alleles were amplified from plasma RNA or from peripheral blood mononuclear cell DNA from samples collected at entry, after 24–32 weeks of therapy, and at two intermediate time points. Amplified products of about 1.7 kb, which extended from CA in the *gag* region through PR and into RT in the *pol* region (see Fig. 2B), were cloned so that sequences could be determined for individual PR alleles. Nucleotide sequences from five to ten clones at each time point were merged and translated in Gene Runner 3.0 for Windows© (Hastings Software Inc., 1994). Sequences were aligned using COMPARE in the DNA sequence alignment editor 2.4 (Dr. Alan Goldin, Department of Biology, Cal Tech, 1994). Phylogenetic analysis was performed using neighbor-joining and Kimura two-parameter distance matrix software in the PHYLIP package.[26] Bootstrap analysis, based on 100 bootstrap trees, was performed to determine the confidence levels of branching, and branch values were superimposed on the neighbor-joining trees. Trees were constructed in the program TREEVIEW version 1.40 and further refined in Harvard Graphics version 1.03 (Software Publishing, 1992).

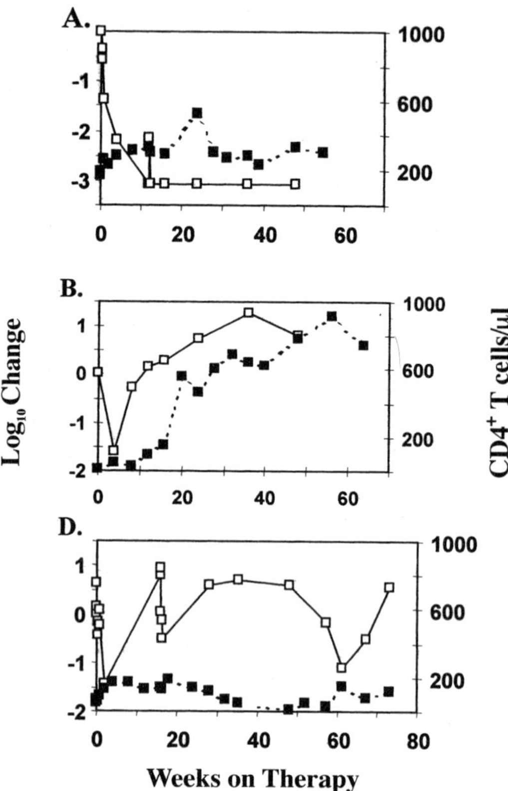

FIGURE 5. Clinical response in children during protease inhibitor therapy. Clinical response to therapy was monitored in three patients (A, upper panel; B, middle panel; and D, lower panel) by measuring $\log_{10}$ change from baseline of HIV-1 plasma RNA copies per milliliter (left *y* axis) and CD4 T cells per microliter (right *y* axis) over time (*x* axis). Symbols: open, plasma RNA; closed, absolute CD4 T cell count.

3.3. Phylogenetic Trees

Phylogenetic trees, which were constructed from all the nucleotide sequences in the dataset, established the epidemiological linkages among sequences from each patient and demonstrated the integrity of the dataset (Fig. 6). The virus populations in each child were distinct, which was expected, as none of the children who were infected by maternal transmission were siblings, nor were hemophiliac patients infected by the same blood products. The relationship among viruses from treated individuals was essentially identical among untreated children and their mothers.[7] In

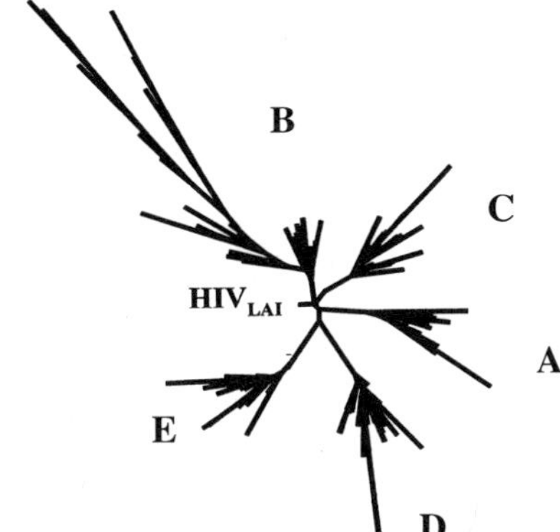

FIGURE 6. Phylogenetic tree of protease nucleotide sequences from five children treated with protease inhibitors. Letters next to branches identify patients. Patients A, B, and D are described in this chapter. The HIV$_{LAI}$ protease sequence was used as an outgroup in the generation of this tree.

addition, every PR sequence within each patient was different at the nucleotide level, indicating genetic complexity in the samples.

3.4. Genotypic Resistance in Protease

Nucleotide sequences were translated and aligned to facilitate analysis of amino acids in PR by type and position. Amino acid substitutions associated with resistance to PR inhibitors were summarized for each patient (Fig. 7). Patient A virus levels in plasma were reduced to undetectable, even though pretreatment PR alleles displayed one of three amino acids that contribute to reduced sensitivity to various inhibitors (M36I, ritonavir; L63P, indinavir; V77I, other inhibitors). Even though HIV-1 was undetectable in plasma, virus persisted in peripheral blood cells and contained a subset of PR alleles that displayed an additional resistant mutation at position 82.

Patient B, whose pretherapy virus population contained only one indinavir-resistant L63P substitution, displayed no sustained virus response to ritonavir therapy. Viral PRs developed amino acid substitutions at multiple positions, including 20, 36, 54, 71, and 82. Viruses with similar PR alleles displayed about 40-fold increased resistance to ritonavir in culture.[10]

In contrast, viruses in patient D, who displayed only transient response to therapy, contained pretreatment PR alleles that uniformly contained four amino acids associated with resistance to various inhibitors. Two changes, L10I and A71V, are associated specifically with resistance to indinavir.[27] PR alleles with some additional, resistance-associated substitutions appeared during the course of therapy. Nonetheless, virus in this patient appeared "primed" to resist indinavir.

Amino Acid Positions in Protease

	SAMPLE:	20	32	33	36	46	54	63	71	77	82	90	93
Patient A	ENTRY				RNA			RNA		RNA			
Patient A	WK 2-4				DNA			DNA			DNA		
Patient A	WK 14-34				DNA			DNA			DNA		
Patient B	ENTRY							RNA					
Patient B	WK 12-14	RNA			RNA	RNA	RNA	RNA/DNA			RNA		
Patient B	WK 20-28	RNA/DNA	RNA/DNA	RNA/DNA	RNA/DNA	RNA/DNA	RNA/DNA	RNA/DNA	RNA/DNA		RNA/DNA	RNA/DNA	
Patient D	ENTRY-WK 4	RNA							RNA	RNA			RNA
Patient D	WK 6-8	RNA/DNA						RNA	RNA	DNA			RNA/DNA
Patient D	WK 24-32	RNA/DNA	DNA					DNA	RNA	RNA	DNA		RNA/DNA

FIGURE 7. Summary of amino acid substitutions in HIV-1 protease in children receiving therapy with PR inhibitor. Amino acid substitutions in key positions are associated with, or contribute directly to, protease resistance to inhibitors. Protease sequences were determined in plasma virus RNA (dotted triangles) and in cell-associated viral DNA (black triangles) at baseline (B), and at indicated times in weeks for each of three patients.

3.5. Genetic Distance in Protease over Time

Considerable insight into response to therapy was apparent when genetic changes over time were analyzed relative to baseline (Fig. 8). PR distance declined over time in patient A, who responded to therapy. In contrast, virus from patient B exhibited a distinct increase in distance over the course of 28 weeks of therapy. Failure of therapy to suppress virus was not invariably related to increased distance relative to baseline. For example, little change appeared in viral PR in patient D, whose virus contained multiple resistance-associated amino acids at the start of therapy (Fig. 8).

Even though all proteases contained specific amino acid substitutions related to resistance, none of the PR populations exhibited a preponderance of replacement versus silent nucleotide substitutions, i.e., nonsynonymous or synonymous changes, respectively (Fig. 8). In each case, silent substitutions comprised the majority of nucleotide changes and made the greater contribution to the variability. Thus, strong selection by inhibitor for amino acid substitutions in PR was insufficient to tip the balance of nucleotide changes so that nonsynonymous substitutions exceeded silent substitutions.

3.6. Models for Evolution of Virus with Inhibitor

Based upon genetic analysis of PR resistance during therapy, we developed three models to account for the types of changes found in the patients

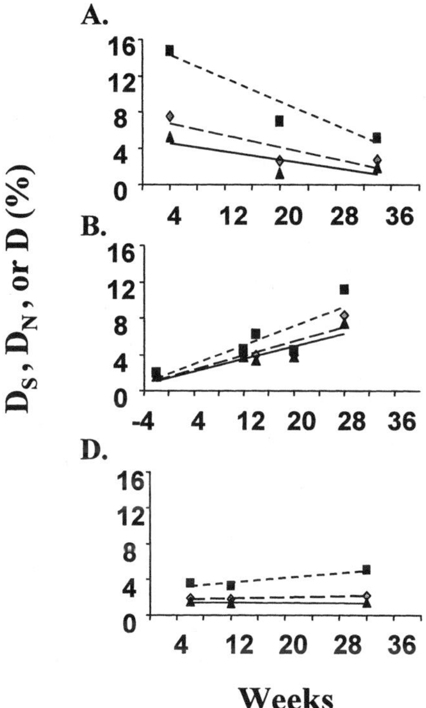

FIGURE 8. Change in PR alleles during PR inhibitor therapy. Virus populations in patients A (top panel), B (middle panel), and D (lower panel) were evaluated in PR for total distance and nonsynonymous or synonymous substitutions (x axis) over time (y axis). Symbols: filled diamonds, total distance; triangles, nonsynonymous (replacement) substitutions; filled squares: synonymous (silent) substitutions.

(Fig. 9). Model 1, the suppression model, accounts for the results in patient A, whose virus levels were reduced to undetectable in plasma. Variants of the virus that were detected in plasma at entry were suppressed. Infection persisted in cells, suggesting that there may be reservoirs of long-lived infected cells or continued viral replication within tissues inaccessible to protease inhibitors. Failure to achieve sustained suppression of virus levels or prolonged increase in CD4 T cells, as in patient D, was associated with emergence of multiple lineages of virus variants (model 2) that were directly related to the baseline population. Noncompliance, subtherapeutic drug levels, preexisting resistance, or resistance mapping outside of protease may account for this type of evolutionary pattern. Model 3 represents

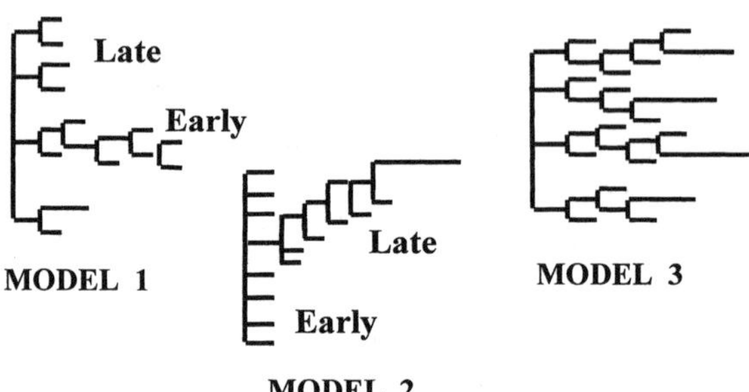

FIGURE 9. Models for emergence of virus based on protease variability. Model 1—suppression is represented by patient A, whose viral load became undetectable early in therapy. Model 2—progression reflects incompletely suppressed viral replication in the presence of inhibitor and the outgrowth of a highly genotypically resistant subpopulation, as in patient B. Model 3—multiple lineages demonstrates concurrent evolution of multiple populations without the outgrowth or selection for a specific population, as in patient D.

selective outgrowth of a minor population from the multiple variants present at entry, as in patient B.

4. VIRUS FROM OTHER PERSPECTIVES

4.1. Variability in *env* Region of the Virus Genome

Views of the virus under selective pressure of PR inhibitors could differ depending upon the region of the genome examined. For example, models for emergence of resistance based on analysis of PR could apply specifically to the sequences under direct selection by inhibitors or, alternatively, reflect the population of viruses. We evaluated variability in the virus populations by analysis of the *env* region. Several criteria were used in development of this strategy. The *env* region is unselected by PR inhibitors, thereby providing an independent assessment of the virus at the population level. V3 and other hypervariable domains in *env* are generally more variable than PR. In addition, the types of amino acid substitutions in V3 can provide insights into biological characteristics of the viruses.

4.1.1. Genetic Changes in env Relative to Baseline

Nucleotide distances in V3 during therapy were analyzed relative to baseline V3 sequences in the three patients (Fig. 10, squares). Patient A, whose virus levels in plasma were reduced to undetectable, nonetheless

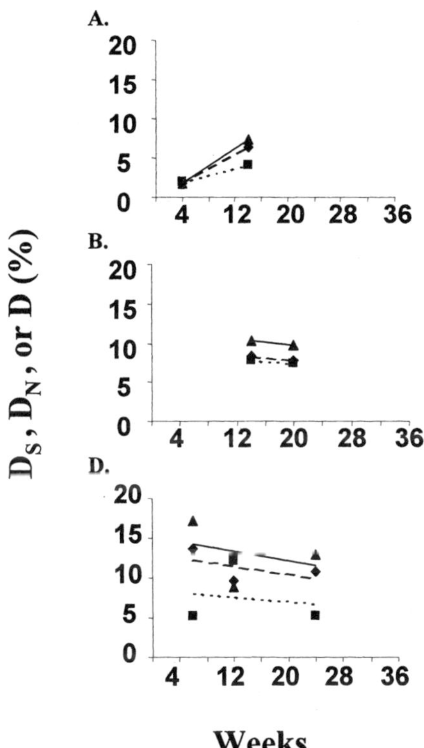

FIGURE 10. Changes in envelope V3 during therapy with PR inhibitors. Virus populations in patients A (top panel), B (middle panel), and D (lower panel) were evaluated in V3 for total distance and nonsynonymous or synonymous substitutions (*x* axis) over time (*y* axis). Symbols: filled diamonds, total distance; triangles, nonsynonymous (replacement) substitutions; filled squares: synonymous (silent) substitutions. Reduced variability in V3 in patients B and D was associated with a change from V3 sequences with high positive charge and SI genotype to V3 sequences with low positive charge and NSI genotype (data not shown).

displayed increased distance in V3 in peripheral blood cells. Patients B and D, both of whom experienced only transient decline in plasma virus, displayed reductions in distance in V3 domains during therapy. Changes in viral levels in plasma were unrelated to V3 variability over time. Moreover, changes in V3 were not necessarily related to the patterns of change which emerged in the protease regions in populations of viruses from the same patients. When the types of nucleotide changes in V3 were evaluated, replacement or nonsynonymous substitutions contributed somewhat more to distance in V3 than silent substitutions (Fig. 10, triangles). The results in V3 were in contrast to PR, where silent substitutions predominated.

Phylogenetic trees were constructed from V3 sequences over time from each of the patients and assessed relative to the models developed for protease trees. In all cases, a lineage of V3 sequences emerged during therapy, producing V3 trees that resembled model 2 (Fig. 11). Concordance between V3 and PR trees occurred only in patient B. V3 trees were distinctly different from PR trees in patients A and D.

4.1.2. Changes in V3 Amino Acids

Analysis of V3 amino acids provided unique insights into the virus populations during therapy. V3 loops comprised of amino acids with a low net charge (range 2–4) usually occur in viruses that have a non-syncytium-inducing (NSI) phenotype in culture, while V3 loops with high net charge are usually syncytium-inducing (SI).[28] When V3 amino acids were assessed in the three patients, patient A had NSI V3 sequences that were uniformly low in charge throughout therapy. In contrast, patients B and D had pre-therapy viruses that were exclusively SI or a mixture of NSI and SI based on V3 charge. During the course of therapy in both patients, V3 sequences with NSI charge characteristics emerged as the predominant genotype. NSI viruses tend to have reduced variability relative to SI viruses (S. L. Lamers and M. M. Goodenow, unpublished). Emergence of NSI viruses in patients B and D accounts for decreased genetic variability despite continued virus replication. Even though virus replication was not suppressed, viruses that

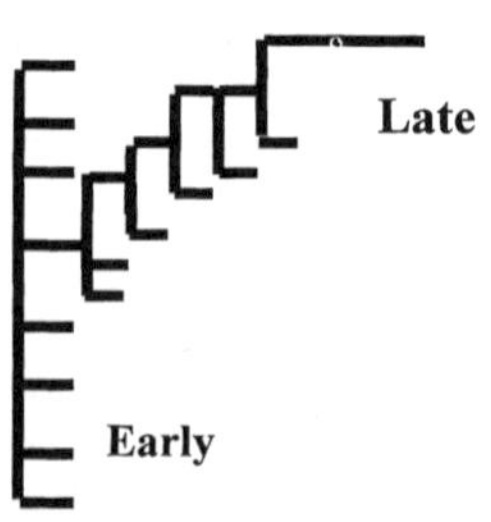

MODEL 2

FIGURE 11. Progression model (model 2) for emergence of virus based on envelope V3 variability. Virus variants progressed over time and emerged sequentially during therapy from a limited number of baseline lineages. Similar patterns of envelope sequences appeared in each patient independent of response to therapy or phylogenetic relationships identified in protease.

emerged during the course of therapy could be qualitatively different from pretreatment viruses.

4.2. Genetic Analysis and Reservoirs of Viruses

Suppression of virus to undetectable levels in plasma does not necessarily indicate that virus is eradicated in the individual. Virus sequences can persist in the peripheral blood cells and, under certain conditions in culture, infectious virus can be isolated from these cells.[29,30] It is possible that reservoirs of long-lived infected cells are sequestered in the body. One cellular reservoir could be tissue macrophages, which tend to be infected by NSI viruses. Results from our study support this model, and provide an explanation for the qualitative changes from SI to NSI that we found in viruses from two patients.

Quantitative analysis of virus during antiretroviral therapy has provided critical insights into kinetics of replication in patients.[31,32] Amino acid changes in protease that contribute directly to resistance accumulate rapidly and with some degree of progression.[10,27] Analysis of different regions of the virus genome by phylogenetic relationships and genetic distances reveals additional complexities of the virus populations. Combinations of genetic analysis plus quantitative assessment of virus burdens will expand our understanding of the virus response to multiple pressures that limit its survival.

ACKNOWLEDGMENTS. The authors are grateful to Susanna Lamers LaFleur for genetic analysis, to Brant Burkhardt for determinations of protease genotypes in plasma, and to Paul Krogsted and colleagues for sharing results from the pediatric nelfinavir studies.

The work was supported by PHS Awards AI28571 (M.M.G. and B.M. Dunn), HD32259 (M.M.G. and J.W.S.), and GM14899 (E.E.P.), and by an award from the Elizabeth Glaser Pediatric AIDS Foundation (PG-550956) (J.W.S.).

REFERENCES

1. Connor, E. M., Sperling, R. S., Gelber, R., Kiselev, P., Scott, G., O'Sullivan, M. J., VanDyke, R., Bey, M., Shearer, W., Jacobson, R. L., *et al.*, 1994, Reduction of maternal–infant transmission of human immunodeficiency virus type 1 with zidovudine treatment. Pediatric AIDS Clinical Trials Group Protocol 076 Study Group, *N. Engl. J. Med.* **331:**1173–1180.
2. Folks, T., Hart, C., Curran, J., Essex, M., Fauci, A., DeVita, V., Hellman, S., and Rosenberg, S., eds., 1997, *AIDS: Etiology, Diagnosis, Treatment, and Prevention,* 4th ed., Lippincott-Raven, Philadelphia, Chapter 2, The life cycle of HIV-1.

3. Kaplan, A. H., Manchester, M., and Swanstrom, R., 1994, The activity of the protease of human immunodeficiency virus type 1 is initiated at the membrane of infected cells before the release of viral proteins and is required for release to occur with maximum efficiency, *J. Virol.* **68:**6782–6786.

4. Kaplan, A. H., Zack, J. A., Knigge, M., Paul, D. A., Kempf, D. J., Norbeck, D. W., and Swanstrom, R., 1993, Partial inhibition of the human immunodeficiency virus type 1 protease results in aberrant virus assembly and the formation of noninfectious particles, *J. Virol.* **67:**4050–4055.

5. Erickson-Viitanen, S., Manfredi, J., Viitanen, P., Tribe, D. E., Tritch, R., Hutchison, C. A., Loeb, D. D., and Swanstrom, R., 1989, Cleavage of HIV-1 gag polyprotein synthesized *in vitro*: Sequential cleavage by the viral protease, *AIDS Res. Hum. Retrovir.* **5:**577–591.

6. Winslow, D. L., Stack, S., King, R., Scarnati, H., Bincsik, A., and Otto, M. J., 1995, Limited sequence diversity of the HIV type 1 protease gene from clinical isolates and *in vitro* susceptibility to HIV protease inhibitors, *AIDS Res. Hum. Retrovir.* **11:**107–113.

7. Barrie, K. A., Perez, E. E., Lamers, S. L., Farmerie, W. G., Dunn, B. M., Sleasman, J. W., and Goodenow, M. M., 1996, Natural variation in HIV-1 protease, Gag p7 and p6, and protease cleavage sites within gag/pol polyproteins: Amino acid substitutions in the absence of protease inhibitors in mothers and children infected by human immunodeficiency virus type 1, *Virology* **219:**407–417.

8. Lech, W. J., Wang, G., Yang, Y. L., Chee, Y., Dorman, K., McCrae, D., Lazzeroni, L. C., Erickson, J. W., Sinsheimer, J. S., and Kaplan, A. H., 1996, *In vivo* sequence diversity of the protease of human immunodeficiency virus type 1: Presence of protease inhibitor-resistant variants in untreated subjects. *J. Virol.* **70:**2038–2043.

9. Dunn, B. M., Gustchina, A., Wlodawer, A., and Kay, J., 1994, Subsite preferences of retroviral proteinases, *Meth. Enzymol.* **241:**254–278.

10. Molala, A., Korneyeva, M., Gao, Q., Vasavanonda, S., Schipper, P. J., Mo, H. M., Markowitz, M., Chernyavskiy, T., Niu, P., Lyons, N., *et al.*, 1996, Ordered accumulation of mutations in HIV protease confers resistance to ritonavir, *Nature Med.* **2:**760–766.

11. Swanstrom, R., 1994, Characterization of HIV-1 protease mutants: Random, directed, selected, *Curr. Opin. Biotechnol.* **5:**409–413.

12. Brown, A. J., and Richman, D. D., 1997, HIV-1: Gambling on the evolution of drug resistance? *Nature Med.* **3:**268–271.

13. Coffin, J. M., 1995, HIV population dynamics *in vivo*: Implications for genetic variation, pathogenesis, and therapy, *Science* **267:**483–489.

14. Bonhoeffer, S., Coffin, J. M., and Nowak, M. A., 1997, Human immunodeficiency virus drug therapy and virus load, *J. Virol.* **71:**3275–3278.

15. Mueller, B. U., 1997, Antiviral chemotherapy, *Curr. Opin. Pediatr.* **9**(2):178–183.

16. Members of the Working Group on Antiretroviral Therapy and Medical Management of HIV-Infected Children, 1998, Guidelines for the use of antiretroviral agents in pediatric HIV infection, *Morbid. Mortal. Weekly Rep.* **17**(RR-4):3–31.

17. Mueller, B., Nelson, R., Sleasman, J., and Heath-Chiozzi, M., 1998, A phase I/II study of the protease inhibitor ritonavir in children with human immunodeficiency virus infection, *Pediatrics* **1998:**101335–101343.

18. Cohen, J., 1997, The daunting challenge of keeping HIV suppressed, *Science* **277:**32–33.

19. Mueller, B., Sleasman, J., Nelson, R., Smith, S., Deutsch, P., Ju, W., Steinberg, S., Balis, F., Jaronsinski, P., Brouwers, P., *et al.*, 1998, A Phase I/II Study of the protease inhibitor indinavir in children with HIV infection, *J. Pediatr.* **102:**101–109.

20. Kline, M. W., Fletcher, V., Harris, A., Evans, K., Brundage, R., Remmel, R., Calles, N., Kirkpatrick, S., and Simon, C., 1998, A pilot study of combination therapy with indinavir, stavudine (d4T), and didanosine (ddI) in children infected with the human immunodeficiency virus, *J. Pediatr.* **132:**543–546.

21. Krogstad, P., Wiznia, A., Luzuriaga, K., Dankner, W., Nielsen, K., Gersten, M., Kerr, B., Hendricks, A., Boczany, B., Rosenberg, M., Jung, D., Spector, S. A., and Bryson, Y., 1999, Treatment of human immunodeficiency virus 1-infected infants and children with the protease inhibitor nelfinavir mesylate, *Clin. Infect. Dis.* **28:**1109–1118.

22. Autran, B., Carcelain, G., Li, T. S., Blanc, C., Mathez, D., Tubiana, R., Katlama, C., Debre, P., and Leibowitch, J., 1997, Positive effects of combined antiretroviral therapy on CD4$^+$ T cell homeostasis and function in advanced HIV disease, *Science* **277:**112–116.

23. Shearer, W. T., Quinn, T. C., LaRussa, P., Lew, J. F., Mofenson, L., Almy, S., Rich, K., Handelsman, E., Diaz, C., Pagano, M., *et al.*, 1997, Viral load and disease progression in infants infected with human immunodeficiency virus type 1. Women and Infants Transmission Study Group, *N. Engl. J. Med.* **336:**1337–1342.

24. Sleasman, J. W., Aleixo, L. F., Morton, A., Skoda-Smith, S., and Goodenow, M. M., 1996, CD4$^+$ memory T cells are the predominant population of HIV-1-infected lymphocytes in neonates and children, *AIDS* **10**(13):1477–1484.

25. Sleasman, J. W., Nelson, R. P., Goodenow, M., Wilfert, D., Hutson, A., Health-Chiozzi, M., Baseler, M., Zuckerman, J., Pizzo, P. A., and Mueller, B. U., 1999, Immunoreconstitution following ritonavir therapy in HIV-infected children with human immunodeficiency virus infection involves multiple lymphocyte lineages, *J. Pediatr.* **134:**597–606

26. Felsenstein, J., 1993, PHYLIP 3.5c [computer program], Department of Genetics, University of Washington, Seattle, Washington.

27. Condra, J. H., Holder, D. J., Schleif, W. A., Blahy, O. M., Danovich, R. M., Gabryelski, L. J., Graham, D. J., Laird, D., Quintero, J. C., Rhodes, A., *et al.*, 1996, Genetic correlates of *in vivo* viral resistance to indinavir, a human immunodeficiency virus type 1 protease inhibitor, *J. Virol.* **70:**8270–8276.

28. Milich, L., Margolin, B., and Swanstrom, R., 1993, V3 loop of the human immunodeficiency virus type 1 Env protein: Interpreting sequence variability, *J. Virol.* **67:**5623–5634.

29. Chun, T. W., Carruth, L., Finzi, D., Shen, X., DiGiuseppe, J. A., Taylor, H., Hermankova, M., Chadwick, K., Margolick, J., Quinn, T. C., *et al.*, 1997, Quantification of latent tissue reservoirs and total body viral load in HIV-1 infection, *Nature* **387:**183–188.

30. Finzi, D., Hermankova, M., Pierson, T., Carruth, L., Buck, C., Chaisson, R., Quinn, T., Chadwick, K., Margolick, J., Brookmeyer, R., *et al.*, 1998, Identification of a reservoir for HIV-1 in patients on highly active antiretroviral therapy, *Science* **278:**1295–1300.

31. Perelson, A. S., Neumann, A. U., Markowitz, M., Leonard, J. M., and Ho, D. D., 1996, HIV-1 dynamics *in vivo:* Virion clearance rate, infected cell life-span, and viral generation time, *Science* **271:**1582–1586.

32. Wei, X., Ghosh, S. K., Taylor, M. E., Johnson, V. A., Emini, E. A., Deutsch, P., Lifson, J. D., Bonhoeffer, S., Nowak, M. A., Hahn, B. H., *et al.*, 1995, Viral dynamics in human immunodeficiency virus type 1 infection, *Nature* **373:**117–122.

14

Gene Therapy and HIV-1 Infection

Experimental Approaches, Shortcomings, and Possible Solutions

RALPH DORNBURG and ROGER POMERANTZ

1. HIV-1 INFECTION AND CONVENTIONAL PHARMACEUTICAL AGENTS

Extensive studies of the life cycle and pathogenesis of the human immunodeficiency virus (HIV-1) have provided a large number of possible targets for the development of antiviral drugs. For example, specific chemical compounds have been developed to block the activity or function of viral proteins such as the reverse transcriptase or protease.[1-5]

Recent studies have demonstrated that combination chemotherapy may lead to significant reduction of viral load *in vivo*. Utilizing three agents, including two reverse transcriptase inhibitors and one protease inhibitor in treatment-naive patients, the serum HIV-1 RNA levels may be reduced to an undetectable level. How long this response will last in these patients remains an open issue.

Unfortunately, the clinical application of many drugs for the treatment of acquired immune deficiency syndrome (AIDS) has given disappointing results. The failure of such drugs is mainly due to the facts that (1) drug-resistant virus mutants emerge rapidly in HIV-1-infected patients[6] and

RALPH DORNBURG and ROGER POMERANTZ • The Dorrance H. Hamilton Laboratories, Center for Human Virology, Division of Infectious Diseases, Department of Medicine, Jefferson Medical College, Thomas Jefferson University, Philadelphia, Pennsylvania 19107.

Human Retroviral Infections, edited by Kenneth E. Ugen *et al.* Kluwer Academic / Plenum Publishers, New York, 2000.

(2) high virus loads persist in the clinically latent stage.[7-9] Even in the case of the most efficient treatment using "triple therapy" as outlined above, drug-resistant variants of the HIV-1 virus start to emerge. Thus, enormous efforts are underway in many laboratories to develop alternative therapeutics.

The primary target cells for HIV-1 are cells of the hematopoietic system, in particular CD4+ T lymphocytes and macrophages. During HIV-1 infection these cells are destroyed by the virus, leading to immunodeficiency of the infected individual. Thus, AIDS patients usually die from secondary opportunistic infections such as pneumonia or from tumors. In order to prevent the destruction of the cells of the immune system, many efforts are now underway to make such cells resistant against the HIV-1 virus. This approach has been termed "intracellular immunization."[10] In particular, the development of genetic agents which attack the virus at several points simultaneously inside the cell and/or are independent from viral mutations has gained great attention.

Such potential agents, also termed "genetic antivirals," should have four features which overcome the shortcomings of conventional treatments: First, they should be directed against a highly conserved moiety in HIV-1 which is absolutely essential for virus replication, eliminating the chance that new mutant variants can arise which can escape this attack. Second, they have to be highly effective, greatly reducing or, ideally, completely blocking the production of progeny virus. Third, they have to be nontoxic. A fourth criterion, which should not be overlooked, is that the antiviral agent has to be tolerated by the immune system. It would not make much sense to endow immune cells with an antiviral agent which elicits an immune response against itself leading to the destruction of the HIV-1-resistant cell after a short period of time.

In the past few years many strategies have been developed and proposed for clinical application to block HIV-1 replication inside the cell (see Fig. 1). Such strategies use either antiviral RNAs or proteins. They include antisense oligonucleotides, ribozymes, RNA decoys, transdominant mutant proteins, products of toxic genes, and immunogens such as single-chain antigen-binding proteins (for reviews, see refs. 11–23). Antiviral strategies that employ RNAs have the advantage that they are less likely to be immunogenic than protein-based antiviral agents. However, protein-based systems have been engineered that use inducible promoters which only become active upon HIV-1 infection.

2. ANTISENSE RNAs AND RIBOZYMES

Prokaryotes and bacteriophages express antisense RNAs which provide regulatory control over gene expression by hybridizing to specific RNA

VIRUS LIFE-CYCLE   MOLECULAR INTERCEPTION

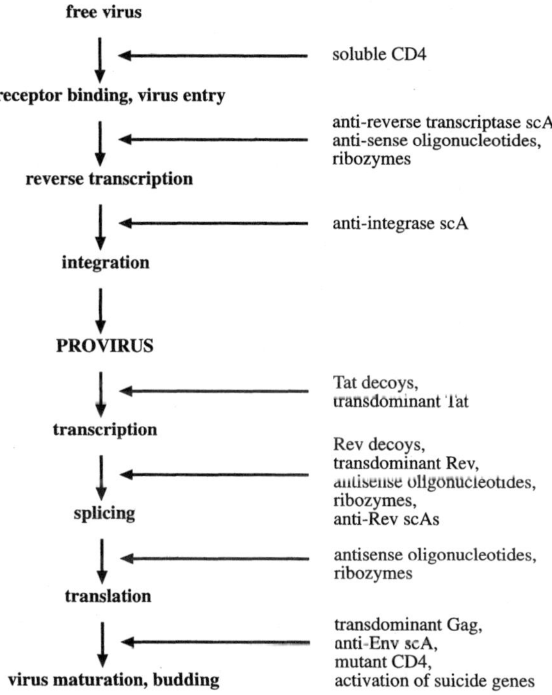

FIGURE 1. Molecular intervention into the HIV-1 life cycle. The steps of the HIV-1 life cycle are shown on the left. Molecular therapeutics which block the life cycle at various steps are shown at the right. For further explanation, see text.

sequences.[24] In animal cells, artificial antisense oligonucleotides (RNAs or single-stranded DNAs) have been successfully used to selectively prevent expression of various genes (e.g., oncogenes, differentiation genes, viral genes).[24,25] Furthermore, the presence of double-stranded RNA inside the cell can induce the production of interferon and/or other cytokines, stimulating an immune response. Indeed, it has also been reported that the expression of RNAs capable of forming a double-stranded RNA molecule with the HIV-1 RNA (antisense RNAs) can significantly reduce the expression of HIV-1 proteins and consequently the efficiency of progeny virus production.[24-29]

Ribozymes are very similar to antisense RNAs, e.g., they bind to specific RNA sequences, but they are also capable of cleaving their target at the binding site catalytically. Thus, they have the advantage that they may not need to be overexpressed in order to fulfill their function. Certain ribozymes (e.g., hairpin and hammerhead ribozymes) require only a GUC sequence. Thus, many sites in the HIV genome can be targeted. However, several questions still remain to be answered. For example, it is unclear (1) whether efficient subcellular colocalization can be obtained, in particular *in vivo*, or (2) whether the target RNA will be efficiently recognized due to its secondary and tertiary folding, or (3) whether RNA-binding proteins would prevent efficient binding. Thus, more experimentation is necessary to address these problems.[30–43]

3. RNA DECOYS

In contrast to ribozymes and antisense RNAs, RNA decoys do not attack the viral RNAs directly. RNA decoys are mutant RNAs that resemble authentic viral RNAs that have crucial functions in the viral life cycle. They mimic such RNA structures and decoy viral and/or cellular factors required for the propagation of the virus.[30,44–53] For example, HIV-1 replication largely depends on the two regulatory proteins Tat and Rev. These proteins bind to specific regions in the viral RNA, which are termed TAR and RRE, respectively. Cells that overexpress short RNAs containing TAR or RRE sequences will capture Tat or Rev proteins, preventing the binding of such proteins to their actual targets. This strategy has the advantage over antisense RNAs and ribozymes in that mutant Tat or Rev, which will not bind to the RNA decoys, will also not bind to their actual targets. Thus, the likelihood is low that mutant strains would arise which would bypass the RNA decoy trap. However, it still remains to be elucidated whether cellular factors also bind to Tat or Rev decoys, and whether overexpression of decoy RNAs would lead to the sequestering of the resulting protein–RNA complexes in the cell.

4. TRANSDOMINANT MUTANT PROTEINS

During HIV-1 replication, several regulatory proteins are essential for viral gene expression and gene regulation. Mutant forms of such proteins greatly reduce the efficiency of viral replication. Transdominant (TD) mutants are genetically modified viral proteins that still bind to their targets, but are unable to perform their actual function. They compete with the corresponding native, wild-type protein inside the cell. The competition of several TD proteins with the wild-type counterpart has been shown greatly to reduce virus replication, especially when such TD mutants are expressed from strong promoters (e.g., the cytomegalovirus [CMV] immediate early promoter.[10,17,54–62]

For example, transcription from the HIV-1 long terminal repeat (LTR) promoter is dependent on the Tat protein. Mutant Tat proteins which still bind to the nascent viral RNA, but which are unable to further trigger RNA elongation of transcription greatly reduce the production of HIV-1 RNAs and consequently the production of progeny virus. In a similar way, mutant Rev proteins interfere with regulated posttranscriptional events and also greatly reduce the efficiency of virus replication in an infected cell.

Although TD mutants have been shown to be effective *in vitro*, it still remains unclear how long cells endowed with such proteins will survive *in vivo*. There is a significant possibility that peptides of such proteins will be displayed via human leukocyte antigen (HLA), leading to the destruction of the HIV-1-resistant cell by the patient's own immune system. There is, however, the potential to express such proteins from the HIV1 LTR promoter, which only becomes activated upon HIV-1 infection. However, this would rule out that a TD Tat can be used, because TD Tat may also abolish its own expression. Even if other TD proteins are expressed from inducible promoters, it remains unclear whether such inducible promoters are really silent enough so that no protein is made (and no immune response) as long as there is no viral infection.

5. TOXIC GENES

Another approach to reduce the production of progeny virus is to endow the target cells of HIV-1 with toxic genes which become activated immediately after virus infection. The activation of the toxic gene leads to immediate cell death, and therefore no new progeny virus particles can be produced. Theoretically, this would lead to an overall reduction of the virus load in the patient. *In vitro* experiments have shown that the production of HIV-1 particles was indeed reduced if target cells were endowed with genes coding for the herpes simplex virus (HSV) thymidine kinase or a mutant form of the bacterial diphtheria toxin protein. Such genes were inserted downstream of the HIV-1 LTR promoter, which only becomes activated upon HIV-1 infection, when the viral Tat protein is expressed.[63–68]

Besides the question regarding the "silence" of the HIV-1 LTR promoter without Tat (discussed above), the main problem with this approach is the actual number of cells that carry a toxic gene present in the patient. Since HIV-1 will not only infect cells that carry the toxin gene, but also many other cells of its host, this approach may only slow down virus replication for a short period of time until all cells that carry the toxin genes undergo self-destruction upon infection.

6. CD4 AS DECOY

The CD4 molecule is the major receptor for HIV-1 for entry into T lymphocytes. Thus, in a similar way to RNA decoys, mutant CD4, which stays inside the endoplasmatic reticulum, has been shown to inactivate HIV-1 envelope maturation, preventing formation of infectious particles. In another approach, soluble CD4 has been used to block the envelope of free extracellular virus particles and to prevent binding to fresh target cells. However, the question remains whether soluble and/or mutant CD4 in the blood of the patient will also serve as a trap for natural CD4 ligands leading to the impairment of important physiological functions.[69–71]

7. SINGLE-CHAIN ANTIBODIES

Single-chain antibodies (scA) were originally developed for *Escherichia coli* expression to bypass the costly production of monoclonal antibodies in tissue culture or mice.[72,73] They comprise only the variable domains of both the heavy chain and the light chain of an antibody. These domains are expressed from a single gene, in which the coding regions for these domains are separated by a short spacer sequence coding for a peptide bridge which connects the two variable-domain peptides. The resulting single-chain antibody (scA, also termed single-chain variable fragment, scFv) can bind to its antigen with similar affinity as an Fab fragment of the authentic antibody molecule.

scFvs have been developed by our group and others to combat HIV-1 replication when expressed intracellularly.[23,70,74–80] Both preintegration (e.g., integrase, reverse transcriptase, matrix protein) and postintegration (e.g., Rev and Tat) sites of the viral life cycle have been targeted, with varying success. Further studies utilizing constructs combining multiple scFvs for potential synergistic antiviral potency are under analysis and should be important in developing robust anti-HIV-1 molecular therapeutics.

8. GENE DELIVERY OF ANTIVIRAL AGENTS

In all the above therapeutic approaches the therapeutical agent cannot be delivered directly to the cell. Instead, the corresponding genes have to be transduced to express the therapeutic agent of interest within the target cell. Genes can be delivered using a large variety of tools. Such tools range from nonviral delivery agents (liposomes or even naked DNA) to viral vectors. Since HIV-1 remains and replicates in the body of an infected person for many years, it will be essential to stably introduce therapeutic genes into the genome of target cells for either continuous expression or for availability upon demand. Thus, gene delivery tools such as naked DNA, liposomes, or adenoviruses (AV), which are highly effective for transient expression of therapeutic genes, are certainly not very useful for gene therapy of HIV-1 infection.

Adeno-associated virus (AAV), a nonpathogenic, single-stranded DNA virus of the parvovirus family, has recently gained a great deal of attention as a vector because it is not only capable of inserting its genome specifically at one site at chromosome 19 in human cells, but it is also capable of infecting nondividing cells. However, vectors derived from AAV are much less effi cient and lose their ability to target chromosome 19.[81,82] Another shortcoming of AAV is the need for it to be propagated with replication-competent AV, since AAV alone is replication defective. It also remains to be shown how efficiently AAV vectors transduce genes into human hematopoietic stem cells and/or mature T lymphocytes and macrophages. Since gene therapy of HIV-1 infection may also require multiple injections of the vector (*in vivo* gene therapy, see below), or of *ex vivo* manipulated cells, it also remains to be shown whether even small amounts of contaminating AV which is used as a helper agent to grow AAV will cause immune problems.

The most efficient tools for stable gene delivery are retroviral vectors.[83–89] Retroviruses are RNA viruses which stably integrate their genome as DNA into the genome of the host cell as part of their life cycle. This is why they have been used in the first place to develop virus-derived gene transfer tools. This is also why they are being used in almost all current human gene therapy trials, including all on-going clinical AIDS trials. They most certainly will also continue to be a major standard tool for gene delivery in the future.

All retroviral vectors used today are basically retroviral particles that contain a genome in which all viral protein-coding sequences have been replaced with the gene(s) of interest. As a result, such viruses cannot further replicate after one round of infection. Furthermore, infected cells do not express any retroviral proteins, which makes cells that carry a vector provirus (the integrated DNA form of a retrovirus) invisible to the immune system.[83–89]

Retroviral vector particles are produced by helper cells (Fig. 2). Helper cells contain plasmid constructs which express all retroviral proteins necessary for replication. After transfection of the vector genome into such helper cells, the vector genome is encapsidated into virus particles (due to

A

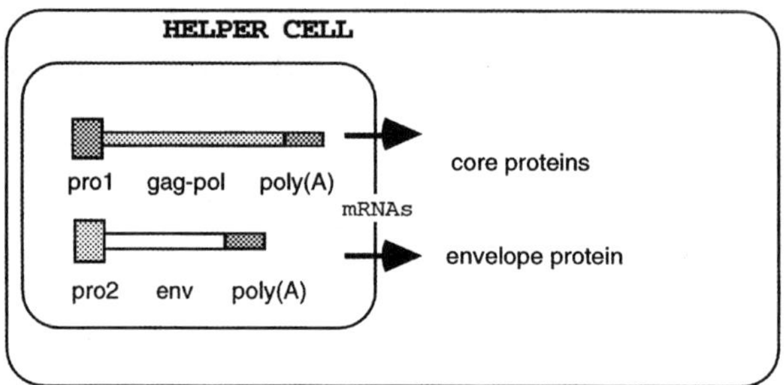

B

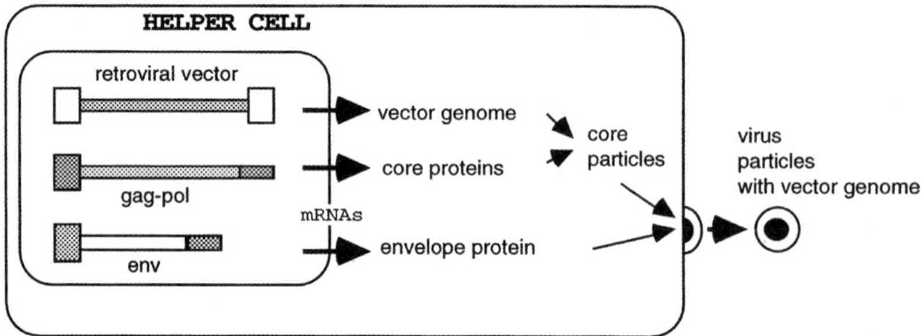

FIGURE 2. Retroviral packaging cells. (A) A standard retroviral packaging cell (also termed helper cell) contains two separate plasmid DNAs to express the retroviral core (*gag–pol*) and envelope (*env*) proteins. (B) After the transduction of a retroviral vector genome in which all viral protein-coding sequences have been replaced with therapeutic gene(s), the helper cell produces retroviral vector particles which transduce the gene of interest into a fresh target cell. pro1, pro2: promoters to express retroviral proteins; poly(A): polyadenylation signal sequence.

the presence of specific encapsidation sequences). Virus particles are released from the helper cell carrying a genome containing only the gene(s) of interest (Fig. 2). Thus, once established, retrovirus helper cells can produce gene transfer particles for very long time periods (e.g., several years). In the last decade, several retroviral vector systems derived from chicken or murine retroviruses have been developed for the transduction and expression of various genes.[83,85,89]

All retroviral vectors currently used in human gene therapy trials have been derived from amphotropic (ampho) murine leukemia virus (MLV). In contrast to ecotropic (eco) MLV, which only infects mouse cells, ampho-MLV has a very broad host range and can infect various tissues of many species including humans. This also applies to ampho-MLV-derived retroviral vectors. Such vectors have been proven to transduce efficiently many different therapeutic genes into a large variety of human tissues *ex vivo*.

However, MLV-derived vectors have several major shortcomings in regard (but not solely) to gene therapy of HIV-1 infection: First, they do not efficiently infect human hematopoietic cells due to low levels or the absence of a viral receptor on such cells.[90] Second, MLV-derived vectors do not infect quiescent cells. Third (this is a problem with all current gene delivery tools), due to the broad host range of ampho-MLV, gene therapy protocols have to be performed *ex vivo*. If injected directly into the blood stream, the chance that the vector particles would infect their actual target cells is very low. Furthermore, such vector particles may infect germ line cells (which are continuously dividing).

The cumbersome and very expensive *ex vivo* protocols may be bypassed by a cell-type-specific gene delivery system, which enables the injection of the gene delivery vehicle directly into the patient's bloodstream or tissue of interest. In the past few years, several attempts have been made to develop cell-type-specific gene delivery tools, again with retrovirus-derived vectors leading the field.

The cell-type specificity of a virus particle is determined by the nature of the retroviral envelope protein which mediates the binding of the virus to a receptor of the target cell.[91] Thus, experiments have been initiated in several laboratories to modify the envelope protein of retroviruses in order to alter the host range of the vector. One of the first attempts specifically to deliver genes into distinct target cells was performed in the laboratory of H. Varmus. Using retroviral vectors derived from avian leukosis virus (ALV), these investigators incorporated the human CD4 molecule into virions specifically to transduce genes into HIV-1-infected cells.[92] However, such particles were not infectious, for unknown reasons (J. Young, personal communication). However, recent reports indicate that MLV particles that carry CD4 can infect HIV-1-infected cells, although at very low efficiencies.[93]

In another attempt to target retroviral particles to specific cells, Roux *et al.* showed that human cells could be infected with eco-MLV if two different antibodies were added to the virus particle solution. The antibodies were connected at their carboxy termini by streptavidine. One antibody was directed against a cell surface protein, the other against the retroviral envelope protein.[94,95] Although this approach was not practical (infectivity was very inefficient and was performed at 4°C), these data showed that cells which do not have an appropriate receptor for a particular virus can be infected with that virus if binding to the cell surface is facilitated. These data also indicated that antibody-mediated cell targeting with retroviral vectors was possible.

To overcome the technical problems of creating an antibody bridge, it was logical to incorporate the antibody directly into the virus particle. However, complete antibodies are very bulky and are not suitable for this approach. The problem has been solved independently by S. Russell in G. Winter's laboratory and our group using single-chain antibody technology.[96–99] Using hapten model systems, Russell *et al.* and our group showed that retroviral vectors that contained scAs fused to the envelope are competent for infection.[97,98] We were also able to show that retroviral vector particles derived from spleen necrosis virus (SNV, an avian retrovirus) that display various scAs against human cell surface proteins are competent for infection in human cells which express the antigen recognized by the antibody.[98,97,99]

Most recently, the group of K. Cichutek succeeded in using scA-displaying SNV to introduce genes into human T cells with the same high efficiency obtained with vectors containing wild-type envelope.[100] In all such experiments, the wild-type envelope of SNV had to be copresent in the virus particle to enable efficient infection of human cells. However, since SNV vector particles with wild-type SNV envelope do not infect human cells at all, this requirement is not a drawback for using such vector particles for human gene therapy.[101,102]

Although successful gene transfer using scA-displaying MLV vector particles has been reported from one laboratory,[103] further experimentation in the laboratories of several other investigators revealed that MLV-derived vector particles that display various scAs are not competent for infection on human cells.[94,95,104–107] The difference between MLV and SNV cell-targeting vectors is based on the different features of the wild-type envelope and the mode of virus entry (reviewed in ref. 108). Moreover, wild-type ampho-MLV infects human hematopoietic cells extremely poorly, most certainly due to the absence of an ampho-MLV receptor on such cells.[90] Thus, MLV-derived vectors are not the best candidates for human gene therapy of AIDS, and alternative vectors need to be developed.

9. OTHER NEW POTENTIAL VECTOR SYSTEMS

Most recently, other very interesting attempts have been made to combat HIV-1-infected cells. Recombinant vesicular stomatitis virus (VSV) has been engineered which lacks its own glycoprotein gene. Instead, genes coding for the HIV-1 receptor CD4 and a chemokine coreceptor, CXCR4, have been inserted. The corresponding virus was able efficiently to infect HIV-1-infected cells, which display the HIV-1 glycoprotein on the cell surface. Since VSV is a virus which normally kills infected cells, the engineered virus only infects and kills HIV-1-infected cells. It has been reported that this virus indeed reduced HIV-1 replication in tissue culture cells up to 10,000-fold.[109] This novel approach to combat one virus with another will certainly gain a great deal of attention. However, it remains to be shown how effective this approach will be *in vivo*. Will the "antivirus" succeed in eliminating a major portion of HIV-1-infected cells before it is cleared by the immune system? On the other hand, since HIV-1 preferentially kills activated immune cells, will it kill the immune cells which are attempting to clear the body of its own "enemy"? How will the body tolerate a virus that does not "look like" one because it carries human cell surface proteins on the viral surface? The answer to these and other questions are eagerly awaited.[110]

10. OTHER POTENTIAL PROBLEMS

Even if we find a gene transfer system that can transduce enough cells within the body to inhibit virus replication significantly, many other questions remain to be answered. For example, it is not clear whether a cell which has been endowed with an HIV-1-resistance gene will be able to fulfill its normal biological function *in vivo*. It has to be considered that in order to become resistant against HIV-1, the cells usually have to overproduce the corresponding HIV-1-resistance gene. Does this overproduction result in a loss of other functions, e.g., because the cell has to devote a significant part of its energy supply to the production of the HIV-1-resistance gene? Will the body be able to eliminate all HIV-1-infected cells or will the infected person become a lifelong carrier of the virus, which will continue to replicate in the patient's body although at levels that cause no clinical symptoms due to the presence of HIV-1-resistance genes? Will the patient be capable of infecting new individuals?

In summary, considering all facts and problems of current gene transfer technologies, and considering our lack of knowledge regarding many functions of the immune system, how should we move forward with Phase I gene therapy trials to combat HIV-1? In spite of the lack of knowledge of many aspects of this disease, we have to remain optimistic and can only hope that one approach or the other will lead to measurable success toward a "cure," or at least will significantly prolong the life expectancy of infected persons. Clearly, in addition to further exploring novel molecular therapeutics *in vivo*, significant attention must be directed at answering critical basic science questions pertaining to intracellular immunization.

ACKNOWLEDGMENTS. The authors wish to thank M. Schnell and K. Cichutek for helpful discussions. This work was supported in part by grants from the New Jersey Commission on Cancer Research and the March of Dimes to R.D. and VSPHS grants AI36552 and Pediatric AIDS Foundation grant 50605 to R.J.P.

REFERENCES

1. Johnston, M. I., Allaudeen, H. S., and Sarver, N., 1989, HIV proteinase as a target for drug action, *Trends Pharmacol. Sci.* **10:**305–307.
2. Erickson, J. W., and Burt, S. K., 1996, Structural mechanisms of HIV drug resistance, *Annu. Rev. Pharmacol. Toxicol.* **36:**545–571.
3. de Clercq, E., 1996, Non-nucleoside reverse transcriptase inhibitors (NNRTIS) for the treatment of human immunodeficiency virus type 1 (HIV-1 infections: Strategies to overcome drug resistance development, *Med. Res. Rev.* **16:**125–157.
4. Sandstrom, P. A., and Folks, T. M., 1996, New strategies for treating AIDS, *Bioessays* **18:** 343–346.
5. Fulcher, D., and Clezy, K., 1996, Antiretroviral therapy, *Med. J. Aust.* **164:**607.
6. Richman, D., 1992, HIV drug resistance, *AIDS Res. Hum. Retrovir.* **8:**1065–1071.
7. Baltimore, D., 1995, The enigma of HIV infection, *Cell* **82:**175–176.
8. Embretson, J. E., Zupanic, M., Ribas, J. L., Burke, A., Racz, P., Tenner-Racz, K., and Haase, A. T., 1993, Massive covert infection of helper T lymphocytes and macrophages by HIV during incubation period of AIDS, *Nature* **362:**359–362.
9. Pantaleo, G., Graziosi, C., Demarest, J. F., Butini, L., Montroni, M., Fox, C. H., Orenstein, J. M., Kotler, D. P., and Fauci, A. S., 1993, HIV infection is active and progressive in lymphoid tissue during the clinically latent stage of disease, *Nature* **362:**355–358.
10. Baltimore, D., 1988, Intracellular immunization, *Nature* **235:**395–396.
11. Dropulic, B., and Jeang, K.-T., 1994, Gene therapy for human immunodeficiency virus infection: Genetic antiviral strategies and targets for intervention, *Hum. Gene Ther.* **5:** 927–939.
12. Yu, M., Poeschla, E., and Wong-Staal, F., 1994, Progress towards gene therapy for HIV infection, *Gene Ther.* **1:**13–26.
13. Kaplan, J. C., and Hirsch, M. S., 1994, Therapy other than reverse transcriptase inhibitors for HIV infection, *Clin. Lab. Med.* **14:**367–391.
14. Gilboa, E., and Smith, C., 1994, Gene therapy for infectious diseases: The AIDS model, *Trends Genet.* **10:**139–144.
15. Anderson, W. F., 1994, Gene therapy for AIDS, *Hum. Gene Ther.* **5:**149–150.
16. Sarver, N., and Rossi, J., 1993, Gene therapy: A bold direction for HIV-1 treatment, *AIDS Res. Hum. Retrovir.* **9:**483–487.
17. Bahner, I., Zhou, C., Yu, X. J., Hao, Q. L., Guatelli, J. C., and Kohn, D. B., 1993, Comparison of *trans*-dominant inhibitory mutant human immunodeficiency virus type 1 genes expressed by retroviral vectors in human T lymphocytes, *J. Virol.* **67:**3199–3207.
18. Blaese, R. M., 1995, Steps toward gene therapy: 2. Cancer and AIDS, *Hosp. Pract. (Off. Ed.)* **30:**37–45.
19. Bridges, S. H., and Sarver, N., 1995, Gene therapy and immune restoration for HIV disease, *Lancet* **345:**427–432.
20. Bunnell, B. A., and Morgan, R. A., 1996, Gene therapy for AIDS, *Mol. Cells* **6:**1–12.

21. Cohen, J., 1996, Gene therapy—New role for HIV—A vehicle for moving genes into cells, *Science* **272**:195.

22. Ho, A. D., Li, X. Q., Lane, T. A., Yu, M., Law, P., and Wong-Staal, F., 1995, Stem cells as vehicles for gene therapy—Novel strategy for HIV infection, *Stem Cells* **13**:100–105.

23. Pomerantz, R. J., and Trono, D., 1995, Genetic therapies for HIV infections: Promise for the future, *AIDS* **9**:985–993.

24. Stein, C. A., and Cheng, Y. C., 1993, Antisense oligonucleotides as therapeutic agents: Is the bullet really magical? *Science* **261**:1004–1012.

25. Agrawal, S., 1992, Antisense oligonucleotides as antiviral agents, *Trends Biotechnol.* **10**:152–158.

26. Castanotto, D., Rossi, J. J., and Sarver, N., 1994, Antisense catalytic RNAs as therapeutic agents, *Adv. Pharmacol.* **25**:289–317.

27. Rossi, J. J., and Sarver, N., 1992, Catalytic antisense RNA (ribozymes): Their potential and use as anti-HIV-1 therapeutic agents, *Adv. Exp. Med. Biol.* **312**:95–109.

28. Rothenberg, M., Johnson, G., Laughlin, C., Green, I., Cradock, J., Sarver, N., and Cohen, J. S., 1989, Oligodeoxynucleotides as anti-sense inhibitors of gene expression: Therapeutic implications, *J. Natl. Cancer Inst.* **81**:1539–1544.

29. Cohen, J. S., 1991, Antisense oligonucleotides as antiviral agents, *Antiviral Res.* **16**:121–133.

30. Bahner, I., Kearns, K., Hao, Q. L., Smogorzewska, E. M., and Kohn, D. B., 1996, Transduction of human CD34+ hematopoietic progenitor cells by a retroviral vector expressing an RRE decoy inhibits human immunodeficiency virus type 1 replication in myelomonocytic cells produced in long-term culture, *J. Virol.* **70**:4352–4360.

31. Biasolo, M. A., Radaelli, A., Del Pup, L., Franchin, E., De Giuli-Morghen, C., and Palu, G., 1996, A new antisense tRNA construct for the genetic treatment of human immunodeficiency virus type 1 infection, *J. Virol.* **70**:2154–2161.

32. Larsson, S., Hotchkiss, G., Su, J., Kebede, T., Andang, M., Nyholm, T., Johansson, B., Sonnerborg, A., Vahlne, A., Britton, S., and Ahrlund-Richter, L., 1996, A novel ribozyme target site located in the HIV-1 *nef* open reading frame, *Virology* **219**:161–169.

33. Heusch, M., Kraus, G., Johnson, P., and Wong-Staal, F., 1996, Intracellular immunization against sivmac utilizing a hairpin ribozyme, *Virology* **216**:241–244.

34. Poeschla, E. M., and Wong-Staal, F., 1995, Gene therapy and HIV disease, *AIDS Clin. Rev.* 1–45.

35. Chuah, M. K., Vandendriessche, T., Chang, H. K., Ensoli, B., and Morgan, R. A., 1994, Inhibition of human immunodeficiency virus type-1 by retroviral vectors expressing antisense-TAR, *Hum. Gene Ther.* **5**:1467–1475.

36. Yu, M., Poeschla, E., Yamada, O., Degrandis, P., Leavitt, M. C., Heusch, M., Yees, J. K., Wong-Staal, F., and Hampel, A., 1995, *In vitro* and *in vivo* characterization of a second functional hairpin ribozyme against HIV-1, *Virology* **206**:381–386.

37. Leavitt, M. C., Yu, M., Yamada, O., Kraus, G., Looney, D., Poeschla, E., and Wong-Staal, F., 1994, Transfer of an anti-HIV-1 ribozyme gene into primary human lymphocytes, *Hum. Gene Ther.* **5**:1115–1120.

38. Poeschla, E., and Wong-Staal, F., 1994, Antiviral and anticancer ribozymes, *Curr. Opin. Oncol.* **6**:601–606.

39. Sullivan, S. M., 1994, Development of ribozymes for gene therapy, *J. Invest. Dermatol.* **103**:85S–89S.

40. Yu, M., Ojwang, J., Yamada, O., Hampel, A., Rapapport, J., Looney, D., and Wong-Staal, F., 1993, A hairpin ribozyme inhibits expression of diverse strains of human immunodeficiency virus type 1, *Proc. Natl. Acad. Sci. USA* **90**:6340–6344.

41. Weerasinghe, M., Liem, S. E., Asad, S., Read, S. E., and Joshi, S., 1991, Resistance to human immunodeficiency virus type 1 (HIV-1) infection in human CD4+ lymphocyte-derived

cell lines conferred by using retroviral vectors expressing an HIV-1 RNA-specific ribozyme, *J. Virol.* **65**:5531–5534.

42. Leavitt, M. C., Yu, M., Wong-Staal, F., and Looney, D. J., 1996, *Ex vivo* transduction and expansion of CD4(+) lymphocytes from HIV+ donors—Prelude to a ribozyme gene therapy trial, *Gene Ther.* **3**:599–606.

43. Wong-Staal, F., 1995, Ribozyme gene therapy for HIV infection—Intracellular immunization of lymphocytes and CD34+ cells with an anti-HIV-1 ribozyme gene, *Adv. Drug Deliv. Rev.* **17**:363–368.

44. Lisziewicz, J., Sun, D., Lisziewicz, A., and Gallo, R. C., 1995, Antitat gene therapy: A candidate for late-stage AIDS patients, *Gene Ther.* **2**:218–222.

45. Cesbron, J. Y., Agut, H., Gosselin, B., Candotti, D., Raphael, M., Puech, F., Grandadam, M., Debre, P., Capron, A., and Autran, B., 1994, Scid-hu mouse as a model for human lung HIV-1 infection, *C.R. Acad. Sci. III* **317**:669–674.

46. Bahner, I., Kearns, K., Hao, Q. L., Smogorzewska, E. M., and Kohn, D. B., 1996, Transduction of human CD34(+) hematopoietic progenitor cells by a retroviral vector expressing an RRE decoy inhibits human immunodeficiency virus type 1 replication in myelomonocytic cells produced in long-term culture, *J. Virol.* **70**:4352–4360.

47. Lee, S. W., Gallardo, H. F., Gilboa, E., and Smith, C., 1994, Inhibition of human immunodeficiency virus type 1 in human T cells by a potent rev response element decoy consisting of the 13-nucleotide minimal rev-binding domain, *J. Virol.* **68**:8254–8264.

48. Smythe, J. A., Sun, D., Thomson, M., Markham, P. D., Reitz, M. S., Jr., Gallo, R. C., and Lisziewicz, J., 1994, A rev-inducible mutant gag gene stably transferred into T lymphocytes: An approach to gene therapy against human immunodeficiency virus type 1 infection, *Proc. Natl. Acad. Sci. USA* **91**:3657–3661.

49. Matsuda, Z., Yu, X., Yu, Q. C., Lee, T. H., and Essex, M., 1993, A virion-specific inhibitory molecule with therapeutic potential for human immunodeficiency virus type 1, *Proc. Natl. Acad. Sci. USA* **90**:3544–3548.

50. Lisziewicz, J., Sun, D., Smythe, J., Lusso, P., Lori, F., Louie, A., Markham, P., Ross, J., Reitz, M., and Gallo, R. C., 1993, Inhibition of human immunodeficiency virus type 1 replication by regulated expression of a polymeric tat activation response RNA decoy as a strategy for gene therapy in AIDS, *Proc. Natl. Acad. Sci. USA* **90**:8000–8004.

51. Brother, M. B., Chang, H. K., Lisziewicz, J., Su, D., Murty, L. C., and Ensoli, B., 1996, Block of tat-mediated transactivation of tumor necrosis factor beta gene expression by polymeric-TAR decoys, *Virology* **222**:252–256.

52. Chang, H. K., Gendelman, R., Lisziewicz, J., Gallo, R. C., and Ensoli, B., 1994, Block of HIV-1 infection by a combination of antisense tat RNA and TAR decoys: A strategy for control of HIV-1, *Gene Ther.* **1**:208–216.

53. Sullenger, B. A., Gallardo, H. F., Ungers, G. E., and Gilboa, E., 1990, Overexpression of TAR sequences renders cells resistant to HIV replication, *Cell* **63**:601–608.

54. Trono, D., Feinberg, M. B., and Baltimore, D., 1988, HIV-1 gag mutants can dominantly interfere with the replication of wild-type virus, *Cell* **59**:113–120.

55. Buchschacher, G. L., Freed, E. O., and Panganiban, A. T., 1992, Cells induced to express a human immunodeficiency virus type 1 envelope mutant inhibit the spread of the wild-type virus, *Hum. Gene Ther.* **3**:391–397.

56. Freed, E. O., Delwart, E. L., Buchschacher, G. L., and Panganiban, A. T., 1992, A mutation in the human immunodeficiency virus type 1 transmembrane glycosylation gp41 dominantly interferes with fusion and infectivity, *Proc. Natl. Acad. Sci. USA* **89**:70–74.

57. Wu, X., Liu, H., Xiao, H., Conway, J. A., and Kappes, J. C., 1996, Inhibition of human and simian immunodeficiency virus protease function by targeting vpx-protease-mutant fusion protein into viral particles, *J. Virol.* **70**:3378–3384.

58. Ragheb, J. A., Bressler, P., Daucher, M., Chiang, L., Chuah, M. K., Vandendriessche, T., and Morgan, R. A., 1995, Analysis of *trans*-dominant mutants of the HIV type 1 rev protein for their ability to inhibit rev function, HIV type 1 replication, and their use as anti-HIV gene therapeutics, *AIDS Res. Hum. Retrovir.* **11**:1343–1353.

59. Liu, J., Woffendin, C., Yang, Z. Y., and Nabel, G. J., 1994, Regulated expression of a dominant negative form of rev improves resistance to HIV replication in T cells, *Gene Ther.* **1**:32–37.

60. Lori, F., Lisziewicz, J., Smythe, J., Cara, A., Bunnag, T. A., Curiel, D., and Gallo, R. C., 1994, Rapid protection against human immunodeficiency virus type 1 (HIV-1) replication mediated by high efficiency non-retroviral delivery of genes interfering with HIV-1 tat and gag, *Gene Ther.* **1**:27–31.

61. Liem, S. E., Ramezani, A., Li, X., and Joshi, S., 1993, The development and testing of retroviral vectors expressing *trans*-dominant mutants of HIV-1 proteins to confer anti-HIV-1 resistance, *Hum. Gene Ther.* **4**:625–634.

62. Caputo, A., Grossi, M. P., Bozzini, R., Rossi, C., Betti, M., Marconi, P. C., Barbantibrodano, G., and Balboni, P. G., 1996, Inhibition of HIV-1 replication and reactivation from latency by tat transdominant negative mutants in the cysteine rich region, *Gene Ther.* **3**:235–245.

63. Dinges, M. M., Cook, D. R., King, J., Curiel, T. J., Zhang, X. Q., and Harrison, G. S., 1995, HIV-regulated diphtheria toxin α chain gene confers long-term protection against HIV type 1 infection in the human promonocytic cell line U937, *Hum. Gene Ther.* **6**:1437–1445.

64. Curiel, T. J., Cook, D. R., Wang, Y., Hahn, B. H., Ghosh, S. K., and Harrison, G. S., 1993, Long-term inhibition of clinical and laboratory human immunodeficiency virus strains in human T-cell lines containing an HIV-regulated diphtheria toxin α chain gene, *Hum. Gene Ther.* **4**:741–747.

65. Caruso, M., and Klatzmann, D., 1992, Selective killing of CD4+ cells harboring a human immunodeficiency virus-inducible suicide gene prevents viral spread in an infected cell population, *Proc. Natl. Acad. Sci. USA* **89**:182–186.

66. Caruso, M., 1996, Gene therapy against cancer and HIV infection using the gene encoding herpes simplex virus thymidine kinase, *Mol. Med. Today* **2**:212–217.

67. Dinges, M. M., Cook, D. R., King, J., Curiel, T. J., Zhang, X. Q., and Harrison, G. S., 1995, HIV-regulated diphtheria toxin α chain gene confers long-term protection against HIV type 1 infection in the human promonocytic cell line U937, *Hum. Gene Ther.* **6**:1437–1445.

68. Brady, H. J., Miles, C. G., Pennington, D. J., and Dzierzak, E. A., 1994, Specific ablation of human immunodeficiency virus Tat-expressing cells by conditionally toxic retroviruses, *Proc. Natl. Acad. Sci. USA* **91**:365–369.

69. Vandendriessche, T., Chuah, M. K., and Chiang, L., Chang, H. K., Ensoli, B., and Morgan, R. A., 1995, Inhibition of clinical human immunodeficiency virus (HIV) type 1 isolates in primary CD4+ T lymphocytes by retroviral vectors expressing anti-HIV genes, *J. Virol.* **69**:4045–4052.

70. Morgan, R. A., Baler-Bitterlich, G., Ragheb, J. A., Wong-Staal, F., Gallo, R. C., and Anderson, W. F., 1994, Further evaluation of soluble CD4 as an anti-HIV type 1 gene therapy: Demonstration of protection of primary human peripheral blood lymphocytes from infection by HIV type 1, *AIDS Res. Hum. Retrovir.* **10**:1507–1515.

71. Morgan, R. A., Looney, D. J., Muenchau, D. D., Wong-Staal, F., Gallo, R. C., and Anderson, W. F., 1990, Retroviral vectors expressing soluble CD4: A potential gene therapy for AIDS, *AIDS Res. Hum. Retrovir.* **6**:183–191.

72. Bird, R. E., Hardman, K. D., Jacobson, J. W., Johnson, S., Kaufmann, B. M., Lee, S.-M., Lee, T., Pope, S. H., Riordan, G. S., and Whitlow, M., 1988, Single-chain antigen-binding proteins, *Science* **242**:423–426.

73. Whitlow, M., and Filpula, D., 1991, Single-chain Fv proteins and their fusion proteins, *Meth. Enzymol.* **2**:1–9.
74. Duan, L., and Pomerantz, R. J., 1996, Intracellular antibodies for HIV-1 gene therapy, *Sci. Med.* **3**:24–36.
75. Chen, S. Y., Bagley, J., and Marasco, W. A., 1994, Intracellular antibodies as a new class of therapeutic molecules for gene therapy, *Hum. Gene Ther.* **5**:595–601.
76. Chen, S. Y., Khouri, Y., Bagley, J., and Marasco, W. A., 1994, Combined intra- and extracellular immunization against human immunodeficiency virus type 1 infection with a human anti-gp120 antibody, *Proc. Natl. Acad. Sci. USA* **91**:5932–5936.
77. Buonocore, L., and Rose, J. K., 1993, Blockade of human immunodeficiency virus type 1 production in CD4+ T cells by an intracellular CD4 expressed under control of the viral long terminal repeat, *Proc. Natl. Acad. Sci. USA* **90**:2695–2699.
78. Duan, L., Zhang, H., Oakes, J. W., Bagasra, O., and Pomerantz, R. J., 1994, Molecular and virological effects of intracellular anti-rev single-chain variable fragments on the expression of various human immunodeficiency virus-1 strains, *Hum. Gene Ther.* **5**:1315–1324.
79. Levy-Mintz, P., Duan, L., Zhang, H., Hu, B., Dornadula, G., Zhu, M., Kulkosky, J., Bizub-Bender, D., Skalka, A.-M., and Pomerantz, R. J., 1996, Intracellular expression of single-chain variable fragments to inhibit early stages of the viral life cycle by targeting human immunodeficiency virus type 1 integrase, *J. Virol.* **70**:8821–8832.
80. Pomerantz, R. J., Bagasra, O., and Baltimore, D., 1992, Cellular latency of human immunodeficiency virus type 1, *Curr. Opin. Immunol.* **4**:475–480.
81. Muzyczka, N., 1992, Use of adeno-associated virus as a general transduction vector for mammalian cells, *Curr. Top. Microbiol. Immunol.* **158**:97–129.
82. Hallek, M., and Wendtner, C. M., 1996, Recombinant adeno-associated virus (RAAV) vectors for somatic gene therapy—Recent advances and potential clinical applications, *Cytokines Mol. Ther.* **2**:69–79.
83. Miller, A. D., 1990, Retrovirus packaging cells, *Hum. Gene Ther.* **1**:5–14.
84. Morgan, R. A., and Andserson, W. F., 1993, Human gene therapy, *Annu. Rev. Biochem.* **62**:191–217.
85. Dornburg, R., 1995, Reticuloendotheliosis viruses and derived vectors, *Gene Ther.* **2**:301–310.
86. Temin, H. M., 1986, Retrovirus vectors for gene transfer: Efficient integration into and expression of exogenous DNA in vertebrate cell genomes, in: *Gene Transfer* (R. Kucherlapati, ed.), Plenum Press, New York, pp. 144–187.
87. Gilboa, E., 1990, Retroviral gene transfer: Applications to human gene therapy, *Prog. Clin. Biol. Res.* **352**:301–311.
88. Eglitis, M. A., and Anderson, W. F., 1988, Retroviral vectors for introduction of genes into mammalian cells, *Biotechniques* **6**:608–614.
89. Gunzburg, W. H., and Salmons, B., 1996, Development of retroviral vectors as safe, targeted gene delivery systems, *J. Mol. Med. JMM* **74**:171–182.
90. Orlic, D., Girard, L. J., Jordan, C. T., Anderson, S. M., Cline, A. P., and Bodine, D. M., 1996, The level of mRNA encoding the amphotropic retrovirus receptor in mouse and human hematopoietic stem cells is low and correlates with the efficiency of retrovirus transduction, *Proc. Natl. Acad. Sci. USA* **93**:11097–11102.
91. Hunter, E., and Swanstrom, R., 1990, Retrovirus envelope glycoproteins, *Curr. Top. Microbiol. Immunol.* **157**:187–253.
92. Young, J. A. T., Bates, P., Willert, K., and Varmus, H. E., 1990, Efficient incorporation of human CD4 protein into avian leukosis virus particles, *Science* **250**:1421–1423.
93. Matano, T., Odawara, T., Iwamoto, A., and Yoshikura, H., 1995, Targeted infection of a retrovirus bearing a CD4-*env* chimera into human cells expressing human immunodeficiency virus type 1, *J. Gen. Virol.* **76**:3165–3169.

94. Etienne-Julan, M., Roux, P., Carillo, S., Jeanteur, P., and Piechaczyk, M., 1992, The efficiency of cell targeting by recombinant retroviruses depends on the nature of the receptor and the composition of the artificial cell-virus linker, *J. Gen. Virol.* **73**:3251–3255.

95. Roux, P., Jeanteur, P., and Piechaczyk, M., 1989, A versatile and potentially general approach to the targeting of specific cell types by retroviruses: Application to the infection of human by means of major histocompatibility complex class I and class I-antigens by mouse ecotropic murine leukemia virus-derived viruses, *Proc. Natl. Acad. Sci. USA* **86**:9079–9083.

96. Chu, T.-H., and Dornburg, R., 1995, Retroviral vector particles displaying the antigen-binding site of an antibody enable cell-type-specific gene transfer, *J. Virol.* **69**:2659–2663.

97. Chu, T.-H., Martinez, I., Sheay, W. C., and Dornburg, R., 1994, Cell targeting with retroviral vector particles containing antibody-envelope fusion proteins, *Gene Ther.* **1**:292–299.

98. Russell, S. J., Hawkins, R. E., and Winter, G., 1993, Retroviral vectors displaying functional antibody fragments, *Nucleic Acid. Res.* **21**:1081–1085.

99. Chu, T.-H., and Dornburg, R., 1997, Towards highly-efficient cell-type-specific gene transfer with retroviral vectors that display a single chain antibody, *J. Virol.* **71**:720–725.

100. Engelstaedter, M., Bobkova, M., Baier, M., Stitz, J., Holtkamp, N., Tearina Chu, T.-H., Kurth, R., Dornburg, R., Buchholz, C. J., and Cichutek, K., Targeting human T-cells by retroviral vectors displaying antibody domains selected from a phage display library, *Hum. Gene Ther.*, in press.

101. Dougherty, J. P., Wisniewski, R., Yang, S., Rhode, B. W., and Temin, H. M., 1989, New retrovirus helper cells with almost no nucleotide sequence homology to retrovirus vectors, *J. Virol.* **63**:3209–3212.

102. Purchase, H. G., and Witter, R. L., 1975, The reticuloendotheliosis viruses, *Curr. Top. Microbiol. Immunol.* **71**:103–124.

103. Somia, N. V., Zoppe, M., and Verma, I. M., 1995, Generation of targeted retroviral vectors by using single-chain variable fragment: An approach to *in vivo* gene delivery, *Proc. Natl. Acad. Sci. USA* **92**:7570–7574.

104. Marin, M., Noel, D., Valsesia-Wittman, S., Brockly, F., Etienne-Julan, M., Russell, S., Cosset, F.-L., and Piechaczyk, M., 1996, Targeted infection of human cells via major histocompatibility complex class I molecules by Moloney leukemia virus-derived viruses displaying single chain antibody fragment-envelope fusion proteins, *J. Virol.* **70**:2957–2962.

105. Valsesia-Wittman, S., Morling, F., Nilson, B., Takeuchi, Y., Russell, S., and Cosset, F.-L., 1996, Improvement of retroviral retargeting by using acid spacers between an additional binding domain and the N terminus of Moloney leukemia virus SU, *J. Virol.* **70**:2059–2064.

106. Nilson, B. H. K., Morling, F. J., Cosset, F.-L., and Russell, S. J., 1996, Targeting of retroviral vectors through protease–substrate interactions, *Gene Ther.* **3**:280–286.

107. Schnierle, B. S., Moritz, D., Jeschke, M., and Groner, B., 1996, Expression of chimeric envelope proteins in helper cell lines and integration into Moloney murine leukemia virus particles, *Gene Ther.* **3**:334–342.

108. Dornburg, R., 1997, From the natural evolution to the genetic manipulation of the host range of retroviruses, *Biol. Chem.* **378**:457–468.

109. Schnell, M. J., Johnson, J. E., Buonocore, L., and Rose, J. K., 1997, Construction of a novel virus that targets HIV-1-infected cells and controls HIV-1 infection, *Cell* **90**:849–857.

110. Nolan, G. P., 1997, Harnessing viral devices as pharmaceuticals: Fighting HIV-1's fire with fire, *Cell* **90**:821–824.

15

Pediatric HIV

Antiretroviral Therapy

ROBERT P. NELSON, JR., PATRICIA J. EMMANUEL, and MAITE DE LA MORENA

1. INTRODUCTION

Over the past 15 years, human immunodeficiency virus (HIV) infection has become an important cause of morbidity and mortality worldwide. By the year 2000, there will be 5–10 million pediatric HIV-infected patients. Over 95% of these HIV infections are acquired perinatally. One of the most important advances in the pediatric epidemic is the discovery that mother-to-infant transmission can be interrupted with the use of antiretroviral therapy administered during pregnancy, labor, and infant postdelivery. Results of AIDS Clinical Trail Group (ACTG) 076 trial demonstrated a 67% reduction in perinatal transmission, from approximately 25% to 8%.[1] More recently, shorter course regimens have been successful, which is important news for developing countries that cannot afford complicated and expensive regimens.

Based on virologic studies of the newborn, it is estimated that approximately 30% of infants are infected *in utero,* and 70% are infected at the time of birth. Factors that are associated with increased transmission risk include a low maternal CD4 count, advanced maternal disease with high viral burdens, substance abuse, lack of prenatal care, and lack of perinatal antiretroviral therapy.[2] Obstetric factors include chorioamnioitis, prolonged

ROBERT P. NELSON, JR. • Indiana University School of Medicine, Indianapolis, Indiana 46202. PATRICIA J. EMMANUEL • Department of Pediatrics, College of Medicine, University of South Florida, Tampa, Florida 33612. MAITE DE LA MORENA • Washington University School of Medicine, St. Louis, Missouri 63110.

Human Retroviral Infections, edited by Kenneth E. Ugen *et al.*
Kluwer Academic / Plenum Publishers, New York, 2000.

rupture of the membranes, fetal scalp electrode monitoring, and increased exposure to maternal blood. Vaginal deliveries appear to carry a greater risk of transmission than cesarean sections. An additional risk is breast feeding.

Increased understanding of the pathogenesis of HIV, antiretroviral availability, and the capacity to measure viral burden has resulted in new treatment recommendations. The primary factor in HIV pathogenesis is the high level of productive infection, characterized by rapid virion turnover, which occurs in infected individuals. It is estimated that anywhere from 500,000 to 10 billion HIV particles are produced and destroyed each day.[3] High levels of viral replication occur in lymphoid tissues and these levels are reflected in levels of detectable plasma HIV RNA. The ability to quantify HIV RNA has allowed controlled analyses which demonstrate that HIV RNA in plasma is a strong predictor of outcome for adults over a 10-year period.[4,5] HIV infection reduces CD4 half-life to approximately one third of that of healthy HIV-1-seronegative subjects, and this is not compensated for by an increased production rate of CD4+ T cells.[6] Recent studies indicate that children express extremely high levels of viral RNA, approximately 1 log higher on average compared to adults.[7] Antiretroviral therapeutic recommendations were updated after controlled clinical trials demonstrated that reductions in plasma HIV RNA levels are associated with reduced risk of disease progression.

We have witnessed a recent expansion of the pharmaceutical armamentarium of approved antiretroviral drugs, including nucleoside analogue reverse transcriptase inhibitors (NRTIs), nonnucleoside reverse transcriptase inhibitors (NNRTIs), and protease inhibitors (PIs). Clinical studies in adults reveal that *combination* antiretroviral therapy is superior to monotherapy in decreasing the progression to acquired immunodeficiency syndrome (AIDS) and mortality. Combination therapy achieves a greater suppression of plasma viral RNA and improvement in immunological parameters including CD4+ T cell production rates.[6] Three-drug regimens delay the emergence of viral resistance.[8] However, at this time the reconstitution of the CD4 lymphocyte population is incomplete with respect to absolute number, proportions of naive versus memory cells, and the breadth of the T cell receptor repertoire. This fact has led to the recommendation for earlier antiretroviral intervention in adults. Updated recommendations are aimed at suppressing viral replication to levels which are undetected by the standard viral RNA assays and, more recently, an ultrasensitive assay. In this regard, the Office of AIDS Research of the National Institutes of Health (NIH) sponsored an NIH Panel to define the principles of therapy of HIV infection, and these principles are summarized as follows:

1. Ongoing HIV replication leads to immune system damage and progression to AIDS. HIV infection is always harmful, and true

long-term survival free of clinically significant immune dysfunction is unusual.

2. Plasma HIV RNA levels indicate the magnitude of HIV replication and its associated rate of CD4+ T cell destruction, while CD4+ cell counts indicate the extent of HIV-induced immune damage already suffered. Regular, periodic measurement of plasma HIV RNA levels and CD4+ T cell counts is necessary to determine the risk of disease progression in an HIV-infected individual and to determine when to initiate or modify antiretroviral treatment regimens.

3. As rates of disease progression differ among individuals, treatment decisions should be individualized by level of risk indicated by plasma HIV RNA levels and CD4+ T cell counts.

4. The use of potent combination antiretroviral therapy to suppress HIV replication to below the levels of detection of sensitive plasma HIV RNA assays limits the potential for selection of antiretroviral-resistant HIV variants, the major factor limiting the ability of antiretroviral drugs to inhibit virus replication and delay disease progression. Therefore, maximum achievable suppression of HIV replication should be the goal of therapy.

5. The most effective means to accomplish durable suppression of HIV replication is the simultaneous initiation of combinations of effective anti-HIV drugs with which the patient has not been previously treated and that are not cross-resistant with antiretroviral agents with which the patient has been treated previously.

6. Each of the antiretroviral drugs used in combination therapy regimens should always be used according to optimum schedules and dosages.

7. The available effective antiretroviral drugs are limited in number and mechanism of action, and cross-resistance between specific drugs has been documented. Therefore, any change in antiretroviral therapy increases future therapeutic constraints.

8. Women should receive optimal antiretroviral therapy regardless of pregnancy status.

9. The same principles of antiretroviral therapy apply to both HIV-infected children and adults, although the treatment of HIV-infected children involves unique pharmacologic, virologic, and immunologic considerations.

10. Persons with acute primary HIV infection should be treated with combination antiretroviral therapy to suppress viral replication to levels below the limit of detection of sensitive plasma HIV RNA assays.

11. HIV-infected persons, even those with viral loads below detectable

limits, should be considered infectious and should be counseled to avoid sexual and drug use behaviors which are associated with transmission or acquisition of HIV and other infectious pathogens.[9]

A Working Group on Antiretroviral Therapy and Medical Management of HIV Infected Children from the National Pediatric and Family HIV Resource Center (NPHRC), Health Resources and Services, and the National Institutes of Health was convened in 1993 and concluded that antiretroviral therapy was indicated for any child with a definitive diagnosis of HIV infection, who had evidence of significant immunodeficiency, or who had HIV-associated symptoms. At that time, specialists recommended zidovudine monotherapy as the standard for initial therapy. Routine antiretroviral therapy for infected children who were asymptomatic or had only minimal symptoms and normal immune status was not recommended.[10] The rapidity and magnitude of HIV replication during *all* stages of infection has led to new therapeutic strategies for infants, children, and adolescents which focus on early initiation of antiretroviral regimens in an effort to preserve immunologic function and delay development of resistance.[3] The pathogenesis of HIV infection is similar in children and adults. Management and therapeutic considerations in children are unique, however, and include the following: (1) the fact that the acquisition of infection in children is mainly through perinatal exposures; (2) differences in immunologic markers in young children; (3) changes in pharmacokinetic parameters; (4) differences in clinical and virologic manifestations of perinatal infection; and (5) adherence issues, which are special problems for children.

The identification of HIV-infected women is critical to provide optimal therapy for both them and their children. Considerable progress has been made in the development of treatment strategies which improve the quality and duration of life of infants and children with symptomatic HIV or AIDS. Multiple pharmaceutical products which impact HIV are making the transition from bench to bedside and hopefully will open the door for further treatment advances. This is especially true since understanding of the life cycle and pathogenesis of HIV has evolved rapidly in both children and adults. A large network of pediatric investigators are committed to developing more effective approaches for treatment and prevention of HIV. At present, all antiretroviral drugs approved for the treatment of HIV may be used in children, when indicated, irrespective of labeling. There is no current agent which completely eliminates or suppresses viral replication, and therefore antiretroviral therapy is used in an attempt to maintain infants and children in an asymptomatic condition. This emphasizes the necessity to provide preventative therapies for secondary infectious complications. Effective management of HIV-infected infants, children, and adolescents requires a multidisciplinary team approach. Determination of HIV RNA copy numbers and CD4 lymphocyte counts is essential to monitor antiretroviral therapy in children and adolescents, and the health care team must provide education and psychosocial support to ensure that medications are available and adherence is encouraged.

HIV infection in children carries with it a different rate of disease progression and a shorter period of clinical latency when compared to adults. There is a negative impact on growth, weight gain, and neurocognitive function, which allows HIV-disease-specific measures to be assessed during the evaluation of antiretroviral clinical trials. The immune system is normally maturing in infants infected with HIV; although HIV can result in significant immunologic deficiency, the potential for immune recovery may be greater in children than adults. Understanding the HIV life cycle has allowed for the development of agents which selectively inhibit viral replication with a relatively small impact on host cellular function. Ideal agents require high bioavailability with long serum and tissue half-lives and penetration into the central nervous system. Other desirable properties include slow emergence of drug resistance and low toxicity profiles. As combinations of medical regimens are developed, administration should result in additive or synergistic antiretroviral activities without additional toxicity and with a low conferrence of resistance.

Important concepts in the formulation of pediatric treatment recommendations need to be considered in developing effective pediatric antiretroviral modalities. These principles are enunciated by the Working Group and are as follows:

1. Identification of HIV-infected women before or during pregnancy is critical to providing optimal therapy for both infected women and their children and preventing perinatal transmission; therefore, prenatal HIV counseling and HIV testing with consent should be standard of care for all pregnant women in the United States.

2. Enrollment of pregnant HIV-infected women, their HIV-exposed newborns, and infected infants, children, and adolescents into clinical trials affords the best means of determining safe and effective therapies.

3. It is important that pharmaceutical companies and the federal government work together to ensure that drug formulations suitable for administration to infants and children are available at the time that new agents are being investigated in adults.

4. Although some information regarding the efficacy of antiretroviral drugs for children can be extrapolated from clinical trials in adults, concurrent clinical trials are needed in children to determine the

impact of the drug on specific manifestations of HIV infection in children, including growth, development, and neurologic disease. However, the absence of clinical trials addressing pediatric-specific manifestations of HIV infection does not preclude the use of any approved antiretroviral drug in children.

5. All antiretroviral drugs approved for treatment of HIV infection may be used in children when indicated, irrespective of labeling notations.

6. Management of infants, children, and adolescents with HIV/AIDS is rapidly evolving and increasingly complex and therefore, wherever possible, management of HIV-infected children and adolescents should be directed by a specialist in the treatment of pediatric and adolescent HIV infection. If this is not possible, then it is important to consult with such experts regularly.

7. Effective management of the complex and diverse needs of HIV-infected infants, children, adolescents, and their families requires a multidisciplinary team approach that includes physicians, nurses, social workers, psychologists, nutritionists, outreach workers, pharmacists, and others.

8. Determination of HIV RNA copy number and CD4+ lymphocyte levels is essential for monitoring and modifying antiretroviral treatment in infected children and adolescents as well as adults, and therefore assays to measure these variables should be available.

9. The choice of antiretroviral regimens for children and adolescents should take into consideration factors influencing adherence to therapy, including (a) availability and palatability of pediatric formulations, (b) impact of the medication schedule on quality of life, including number of medications, frequency of administration, ability to coadminister with other prescribed medications, and need to take with or without food, (c) ability of the child's caregiver or the adolescent to administer complex drug regimens and availability of resources that might be effective in facilitating adherence, and (d) the potential for drug interactions.

10. The choice of antiretroviral regimens should include consideration of factors related to possible limitation of future treatment options, including the potential for the development of antiretroviral resistance.

11. Monitoring growth and development is important for the care of HIV-infected children. Growth failure and neurodevelopmental deterioration are specific manifestations of HIV infection in children. Nutritional support is a therapeutic intervention that affects immune function, quality of life and bioactivity of antiretroviral drugs.[11]

2. ANTIRETROVIRAL AGENTS

Antiretroviral therapy provides significant clinical benefits to HIV-infected children with immunological abnormalities or clinical symptoms of HIV infection. Agents that decrease HIV replication or interfere with its life cycle include nucleoside analogue reverse transcriptase inhibitors (NRTs), nonnucleoside analogue reverse transcriptase inhibitors (NNRTs), protease inhibitors (PTs), and a nucleotide reverse transcriptase inhibitor.

2.1. Nucleoside Analogue Reverse Transcriptase Inhibitors

NRTIs inhibit reverse transcriptase and block the spread of HIV to infected cells.

2.1.1. Zidovudine (Azidothymidine, 2',3'-Azidothymidine, AZT)

Zidovudine is a NRTI which has been studied extensively.[12] It is highly bioavailable and can be given in two daily doses. It crosses the placenta and can be detected in amniotic fluid, fetal tissues, and cord blood. When given to infants and children, it is associated with increases in weight, growth rate, and neurocognitive function.[13–17] It reduces viral copy number, decreases HIV p24 antigen levels, and transiently improves the CD4 counts as a monotherapy. When given to children with advanced disease, it improves growth and survival.[18] The primary toxicity is myelosuppression, which results in a macrocytic anemia (not responsive to B_{12} or folate) and neutropenia. Patients may experience nausea, headache, increase in liver function, and myositis. Cardiomyopathy, a complication of HIV itself, can also occur secondary to zidovudine toxicity. Mutations in the reverse transcriptase gene which result in amino acid changes at positions 41, 67, 70, 215, and 219 confer in a progressive fashion low-level to high-level resistance.[19]

2.1.2. Dideoxyinosine (Didanosine, ddI)

Dideoxyinosine has a short plasma half-life and somewhat erratic bioavailability. It requires antacids to be delivered to neutralize stomach acid, as it is acid-labile. The drug was simultaneously approved in October 1991 for children and adults. Administration improves CD4 counts and T cell helper function, and decreases p24 antigen levels.[20,21] Dideoxyinosine does not penetrate well into the cerebrospinal fluid; therefore, most children do not show improvement in neurocognitive scores when given didanosine monotherapy. It is generally well tolerated, but adverse effects include peripheral neuropathy, retinal depigmentation, pancreatitis, and hyperamylasemia without pancreatitis. It does not suppress bone marrow function. It is particularly useful in patients who experience hematologic toxicity, or as one drug in a combination regimen. A mutation at codon 74 is associated with development of resistance.

2.1.3. Zalcitabine (2',3'-Dideoxycytidine, ddC)

Zalcitabine is a modestly potent nucleoside analogue which has a short plasma half-life and poor penetration into the central nervous system. Adverse effects include painful and sometimes irreversible peripheral neuropathy. Pharmacokinetic information is available for children. It may play a role in a salvage regimen.

2.1.4. Lamivudine (2'-Deoxy-3'-thiacytidine, 3TC)

Lamivudine has been studied by Lewis *et al.*, who defined the pharmacokinetics and demonstrated good tolerability in children.[22] Although resistance develops rapidly (mutation at codon 184) when it is used as a monotherapy, it is particularly well suited as a combination medication when used with nucleosides including zidovudine, didanosine, and stavudine. Lamivudine is well tolerated in children, and has been shown to improve appetite, weight gain, and a general sense of well-being. Occasional pancreatitis occurs.

2.1.5. Stavudine (2',3'-Didehydro-2',3'-Dieoxythymidine, d4T)

Stavudine is bioavailable and penetrates into the central nervous system; its dose-limiting toxicity is peripheral neuropathy. The pharmacokinetics and toxicities have been evaluated in children.[23] It is notable that zidovudine inhibits formation of stavudine phosphate, suggesting that the two agents should not be used together. Preliminary concerns regarding the additive neurotoxicity when didanosine and stavudine are used together have not been substantiated by controlled clinical trials.

The AIDS Clinical Trial Group (ACTG) 240 team compared the safety and tolerance of stavudine versus zidovudine in symptomatic children 3 months to 6 years of age. A total of 212 evaluable children were randomized to receive stavudine or zidovudine in an open-label, comparative study. Stavudine and zidovudine were largely comparable in terms of safety and tolerance. Neutropenia occurred less commonly among those treated with stavudine. There was evidence that weight gain and absolute CD4+ lymphocyte counts were better maintained in the children receiving stavudine.[24]

2.1.6. Abacavir Sulfate

Abacavir sulfate is a recently approved NRTI. It is available in both a tablet and oral solution formulation. A randomized, double-blind study compared abacavir sulfate, 8 mg/kg twice daily, and lamivudine, 4 mg/kg twice daily, and zidovudine 180 mg/m^2 twice daily, versus lamivudine, 4 mg/kg twice daily, and zidovudine 180 mg/m^2 twice daily. Two hundred and five adult patients were enrolled. Preliminary results indicate that at 16 weeks the median CD4 increase from baseline was 69 cells/mm^3 in the group receiving abacavir sulfate and 9 cells/mm^3 in the control group. Abacavir carries with it a unique toxicity, a potentially fatal hypersensitivity reaction, beginning with skin rash, fatigue, fever, gastrointestinal symptoms such as nausea, vomiting, diarrhea, and abdominal pain. This requires discontinuation as soon as the hypersensitivity reaction is first suspected. It is important that abacavir not be restarted following a hypersensitivity reaction because more severe symptoms may occur within hours and may include life-threatening hypotension and death. Hypersensitivity reactions have been reported in approximately 5% of adult and pediatric patients receiving abacavir.

2.1.7. Combination Nucleoside Therapies

The ACTG multicenter study group 152 designed and executed a double-blind study for symptomatic HIV-infected children 3 months to 18 years of age. The children were stratified according to age (less than 30 months or greater than/or equal to 30 months) and randomly assigned to receive zidovudine, didanosine, or zidovudine plus didanosine. The primary endpoint was length of time to death or to progression of HIV disease. Of 831 children evaluated, 92% had never received antiretroviral therapy; 90% had acquired HIV perinatally. An interim analysis demonstrated a significantly higher risk of HIV disease progression or death in children receiving zidovudine alone than in those receiving combination therapy. The study arm with zidovudine alone was stopped and unblinded; the other two treatment arms were continued. At the end of the study, didanosine alone had an efficacy similar to that of zidovudine plus didanosine, and a significantly lower risk of anemia or neutropenia was seen in patients receiving didanosine alone. For symptomatic HIV-infected children, the conclusion from the study is that treatment with either didanosine alone or zidovudine plus didanosine was more effective than treatment with zidovudine alone.[25]

Pediatric AIDS Clinical Trials Group (PACTG) Protocol 300 assessed the clinical efficacy and safety of combination zidovudine/lamivudine compared with either didanosine alone or combination zidovudine/didanosine. A total of 470 children were evaluated, mean age 2.7 years, median CD4 count of 689 cells/mm^3, median HIV RNA loads of 5.1 log$_{10}$ copies/ml. Median follow-up was 9.4 months. Children receiving zidovudine/lamivudine had a lower risk of HIV disease progression or death than those receiving didanosine alone. Weight and height, growth rates, CD4+ cell counts, and RNA concentration showed results favoring zidovudine/lamivudine. Of the patients concurrently randomized to all three treatment arms, both zidovudine/lamivudine and zidovudine/didanosine recipients had a lower risk of HIV disease progression than those that received didanosine alone.[26]

In summary, it is clear that combination regimens are necessary to optimize clinical efficacy and decrease the occurrence of drug resistance. Recent pediatric trials in symptomatic antiretroviral-naive children have demonstrated that combination therapy with either zidovudine and lamivudine, or zidovudine and didanosine, are clinically, immunologically, and virologically superior to monotherapy as initial therapy.

2.2. Nonnucleoside Reverse Transcriptase Inhibitors

NNRTIs differ chemically from NRTIs, but have in common a high degree of antiretroviral activity. NNRTIs bind directly to HIV reverse transcriptase and prevent replication of HIV in infected cells.

2.2.1. Neviripine (Diplyridodiazepinone)

Neviripine is quite potent and has significant impact on p24, CD4 counts, and quantitative viral RNA loads. Its use as a monotherapy is contraindicated due to the rapid development of resistance. Resistance to neviripine is characterized by a substitution of cystine for tyrosine at position 181. Neviripine was studied simultaneously in children and adults and is safe, well tolerated, and now approved.[27] The major side effect is the development of a rash which can be associated with fever and hepatocellular injury, which, in some circumstances, requires discontinuance of the medication.

2.2.2. Efavirenz (Sustiva™, DMP-266)

Efavirenz is a NNRTI available as capsule (50, 100, 200 mg) for oral administration. Its activity against HIV is mediated predominantly by noncompetitive inhibition of HIV-1 reverse transcriptase. Efavirenz has a half-life of 52–76 h after a single dose; therefore it can be given once daily. Thus, it is an attractive medication for adolescent use. One or more reverse transcriptase (RT) mutations at amino acid positions 100, 101, 103, 108, 190 and 225, especially the 103 lysine-to-asparagine mutations, results in a loss to susceptibility to efavirenz. Rapid emergence of HIV-1 strains that are cross-resistant to NNRTIs has been observed. The only available data regarding safety, dosing, and virologic and immunological efficacy of efavirenz in children are from an open label study (PACTG 382) of efavirenz combined with nelfinavir and NRTIs in 57 pediatric patients, some as young as 3 years of age. In a preliminary intent-to-treat analysis, after 20 weeks of therapy, 65% of children had plasma HIV RNA levels <400 copies/ml, and 52% had HIV RNA levels <50 copies/ml.[28,29] However, there are currently no pharmacokinetic data available on the appropriate dosage of efavirenz in children under age 3 years, and although a liquid preparation is currently under study, only a capsular formulation is currently available. A skin rash can develop with efavirenz similar to that seen with neviripine. Prophylaxis with appropriate antihistamine prior to initiation of therapy should be considered. Changing trimethoprim-sulfamethoxazole to dapsone when possible for *Pneumocystis carinii* pneumonia (PCP) prophylaxis may be considered. Other toxicities include hepatocellular enzyme elevation and increases in serum cholesterol.

2.2.3. Delavirdine

One other NNRTI, delavirdine, is approved for adults, but there is little experience in children.

2.3. Protease Inhibitors

The U.S. Food and Drug Administration (FDA) recently approved the protease inhibitors indinavir, ritonavir, saquinavir, and nelfinavir.[30–32] These compounds act as competitive inhibitors of the HIV protease, which cleaves the Gag and Pol polyproteins into individual functional proteins. Recent studies in adults demonstrate decreases in plasma viral RNA, improvements in CD4 counts, and lower AIDS-related mortality with protease inhibitor monotherapy.[30] Protease inhibitors in combination with NRTIs have a significant impact on the level of plasma HIV RNA and in clinical outcome in adults.[8]

2.3.1. Ritonavir

A recent Phase I/II study assessed the safety, tolerability, pharmacokinetic profile, and antiviral and clinical effects of the oral solution ritonavir in HIV-infected children. HIV-infected children between the age of 6 months and 18 years were eligible; a total of 48 children were included in the analysis. Dose-related nausea, diarrhea, and abdominal pain were the most common toxicities and resulted in the discontinuation of ritonavir in 7 children (14.5%). Ritonavir was well absorbed at all dose levels, and plasma concentration reached a peak at 2–4 h after dose. CD4 cell counts increased by a median 79 cells/mm^3 following 4 weeks of monotherapy and were maintained throughout the study. Plasma HIV RNA decreased by 1–2 log$_{10}$ copies/ml within 4–8 weeks of ritonavir monotherapy, and this level was sustained in patients enrolled at the highest dose level of 400 mg/m^2 for the 24-week period. Palatability is a problem with ritonavir liquid formulation. A list of administration suggestions has been compiled by the AIDS Clinical Trials Group (see Appendix).

2.3.2. Indinavir

Indinavir was studied in a similar fashion as above. Fifty-four HIV-infected children were first treated with indinavir monotherapy and then with combination antiretroviral therapy. Indinavir dry-filled capsules were relatively well tolerated, but the liquid formulation was found not to be consistently bioavailable between patients or within a patient. Hematuria and renal calculi complicated therapy in some individuals. Currently, a combination trial of stavudine and lamivudine is on-going utilizing indinavir capsules in older children.[33]

2.3.3. Nelfinavir Mesylate (Viracept®)

Nelfinavir is an inhibitor of HIV protease which is available in 250-mg tablets and 50 mg/gm strength in a powder. In combination with reverse transcriptase inhibitors, nelfinavir demonstrated additive (didanosine or stavudine) to synergistic (zidovudine, lamivudine, or zalcitibine) antiviral activity *in vitro* without enhanced cytotoxicity. Adverse effects in adults are primarily gastrointestinal, with abdominal cramping or diarrhea occurring in 8–12% of recipients. Nelfinavir was studied in one open-label, uncontrolled trial in 38 pediatric patients ranging in age from 2 to 13 years with a goal of achieving plasma concentrations which approximate those observed in adults. The most frequent reported adverse event among patients receiving nelfinavir was diarrhea.

Nineteen HIV-infected children were administered combination antiretroviral regimens, including nelfinavir. The mean age of this group of experienced, protease-naive children was 8.37 years. Through 48 weeks, the participants treated with nelfinavir-containing regimens had positive CD4+

responses despite viral RNA concentrations which were close to baseline. The apparent "disconnection" between viral failure and immunologic success may last at least 48 weeks.[34] The mechanisms of this disparity are unknown.

2.4. Nucleotide Reverse Transcriptase Inhibitor

2.4.1. Adefovir Dipivoxil (PMEA, Preveon™)

Adefovir is a nucleotide analogue with activity against a broad spectrum of retroviruses and herpes simplex virus 1 and 2, cytomegalovirus, and HIV-1 and HIV-2. It is a potent inhibitor of retroviral reverse transcriptase. The active metabolite, which is adefovir diphosphate, inhibits HIV reverse transcriptase and terminates the growing DNA chain, specifically replacing deoxyadenosine triphosphate.[35] A double-blind, randomized, placebo-controlled study in adults is evaluating the effect of treatment on viral load and CD4 count in adults when adefovir dipivoxil is added to an existing regimen of highly treatment-experienced patients. Enrollment was closed in May 1997 with 442 patients. A prospective virology cohort of 142 of the 442 patients enrolled in the study was analyzed to determine the effect of baseline genotype on the reduction of HIV RNA at 24 weeks. The baseline characteristics and prior treatment experience of the virology cohort was consistent with those of the overall trial. Upon entering the trial more than 90% of the patients were zidovudine-experienced, almost 80% were lamivudine-experienced, and approximately 65% were stavudine-experienced. What is interesting about these results is that at 24 weeks, the lamivudine-associated M184V mutation appears to confer greater susceptibility to adefovir dipivoxil than wild-type strains. A trial is currently underway which is assessing adefovir dipivoxil in children, ages 3 months to 16 years, using two dose levels in combination with nelfinavir and an NRTI. There appears to be adefovir dipivoxil-related nephrotoxicity characterized by type II renal tubular acidosis with phosphate wasting, requiring regular monitoring of serum phosphate and renal function. The use of adefovir dipivoxil as a treatment option for protease inhibitor failures is being evaluated.

3. PEDIATRIC ANTIRETROVIRAL TREATMENT

3.1. Identification of Perinatal HIV Exposure

Universal counseling and mandatory offering of HIV testing is now a standard of care for all pregnant women in the United States and is endorsed by the U.S. Public Health Service, the American Academy of Pediatrics, and the American College of Obstetricians and Gynecologists. Early identification of HIV in infected women allows for management of HIV-exposed and HIV-infected children, as well as optimal treatment options for the infected woman. This allows for the provision of antiretroviral therapy during pregnancy and labor and to the newborn to reduce HIV transmission from mother to child.[1,9] Counseling concerning the risk of HIV transmission through breast milk is given, and HIV-exposed infants may receive PCP prophylaxis pending early diagnostic evaluation between 4–6 weeks to 4 months.[36] In the event that the mother has not been tested during pregnancy, the offering of HIV testing, appropriate counseling, obtaining informed consent, and offering of testing to the neonate is the standard.

3.2. HIV Diagnosis in Infants

Since nearly all children born to seropositive mothers are HIV antibody-positive, diagnosis relies on the direct detection of HIV genetic information by DNA polymerase chain reaction (PCR), RNA PCR, or by HIV culture or detection of HIV protein p24. Diagnostic testing, usually with a DNA PCR analysis, should be performed between birth and 48 h, 1–2 months of age, and 3–6 months of age. Data suggest that HIV DNA PCR is positive in approximately 38% of the infected children by 48 h, and in 93% of infected infants by the second week. A birth to 48 h sample in some cases may not be obtainable, in which case a schema of phlebotomies beginning at 14 days of age with a repeat at 2–4 months of age provides accurate diagnosis in most cases. The HIV DNA test is now widely available commercially and is accurate. Two positive confirmatory tests establish the diagnosis of HIV, whereas two negative tests rule it out.[37] HIV RNA in plasma may also be useful; however, it has not been studied as completely as DNA PCR. HIV culture is more expensive and is probably less sensitive.[38,39] Cord blood should not be sampled to avoid false positives.

3.3. Immunological Monitoring in Children

CD4 lymphocyte counts are at their highest level during infancy, but gradually decline, reaching adult levels at approximately 6 years of age. A pediatric immunologic and clinical HIV staging system has been developed (see Table I).

It is important to recognize that CD4 percentage identifies the specific level of immunologic suppression, whereas the CD4 absolute count changes with age; therefore, CD4 percentage may be a more valuable monitoring tool than absolute count in children. CD4 monitoring should be done at 3-

TABLE I
1994 Revised Pediatric HIV Classification System: Immunologic Categories Based on Age-Specific CD4+ Lymphocyte Count and Percentage

Immune category	Age <12 months		Age 1–5 years		Age 6–12 years	
	Number/ μl	(%)	Number/ μl	(%)	Number/ μl	(%)
Category 1: No suppression	≥1500	(≥25%)	≥1000	(≥25%)	≥500	(≥25%)
Category 2: Moderate suppression	750–1499	(15–24%)	500–999	(15–24%)	200–499	(15–24%)
Category 3: Severe suppression	<750	(<15%)	<500	(<15%)	<200	(<15%)

[a]Modified from ref. 48.

month intervals, with increased frequency at the initiation of antiretroviral therapy or change in antiretroviral therapy. Intercurrent illness and immunization affects CD4 count and should be taken into consideration, with repeat samples obtained when there is concern regarding these influences.

3.4. Monitoring HIV RNA in Children

Quantitative HIV RNA assays can be obtained to assess the numbers of copies of RNA and, indirectly, the virion level or viral burden. A balance is reached in HIV-infected individuals between 6 and 12 months following acute infections, such that there is a steady state of on-going viral production and immune elimination/replacement.[40,41] Plasma HIV-1 RNA levels increase rapidly after birth, peak at 1–2 months of age (median values at 1 and 2 months, 318,000 and 256,000 copies/ml, respectively), and then slowly decline to a median of 34,000 copies/ml at 24 months.[7] Adult levels are reached by age 6 years. Extreme levels of HIV RNA (300,000 copies/ml) correlate with disease progression and death within the first 12 months of life. Levels during the first year of life are associated with a 54% reduction in progression for each $\log_{10}$ reduction in HIV RNA level. Baseline as well as changes in HIV RNA copy number and CD4 lymphocyte percentage over time independently predict mortality in infected children. The use of the two markers together more accurately defines prognosis.[25,42–44] It is important to note that the absolute HIV RNA copy number obtained from a single specimen tested by two different assays can differ as much as twofold (0.3 $\log_{10}$); therefore, a single assay method should be chosen and consistently used for individual patient monitoring. Assays currently available include the Amplicor Monitor (Roche Diagnostics Systems, Branchburg, NJ), the branch-chain DNA assay (Quantiplex, Chiron Corporation, Emeryville, CA), and plasma RNA measure by nucleic acid sequence-based amplification (Organon Technike, Durham, NC). Initiation of antiretroviral therapy and alteration of therapy should ideally be based on two RNA measurements.

The virological course and immunopathogenesis of HIV during adolescence are not clearly defined. Older adolescents should be treated similarly to adults.[9] Antiretroviral medications and prophylactic medications for opportunistic infection prophylaxis are prescribed according to Tanner staging rather than age.[45]

3.5. Initiation of Antiretroviral Therapy

Initiation of therapy early in the course of HIV infection, including during the period of primary infection in the neonate, is theoretically advantageous, although data from clinical trials that address the effectiveness in asymptomatic infants and children with normal immune function are not available. Immunologic control of the viral replication in the perinatally infected infant is inadequate, as demonstrated by the high levels of HIV RNA that are observed in the first 1–2 years of life. A number of firm criteria or indications for the initiation of therapy in infected children include the following:

1. Clinical symptoms associated with HIV infection.
2. Evidence of immunologic suppression.
3. Age <12 months, regardless of clinical immunologic or virologic status.
4. Asymptomatic children, age >1 year, with normal immune status (initiation optional in this group).

Situations in which the risk of clinical disease progression is low, especially where there are concerns regarding the durability of the response, safety, or adherence may in some cases favor postponement of treatment. In these individuals, high or increasing HIV RNA copy number, rapidly declining CD4 lymphocyte number or percentage (i.e., >5% drop in CD4+ T cells), or the development of clinical symptoms may prompt treatment.

Indications for initiation of antiretroviral therapy in postpubertal HIV-infected adolescents should follow adult guidelines.[9]

3.6. Choice of Initial Antiretroviral Therapy

Antiretroviral monotherapy with zidovudine, didanosine, lamivudine, and stavudine slows disease progression, improves survival, reduces viral RNA, and delays development of neurocognitive dysfunction. Recent pediatric clinical trials of symptomatic children who have not previously received antiretroviral therapy demonstrate that combination therapy with either zidovudine and lamivudine or zidovudine and didanosine is superior to monotherapy with respect to clinical, immunologic, and virologic parameters. Combination antiretroviral therapy that includes a protease inhibitor is superior to dual nucleoside therapy. In this regard, monotherapy with the currently available antiretroviral drugs is no longer recommended to treat HIV. Many infants born to seropositive mothers are receiving zidovudine monotherapy; therefore, those infants who are infected will require reassessment for more aggressive antiretroviral therapy when discovered to be infected. The goal of antiretroviral therapy is to suppress viral replication, i.e., to bring viral RNA levels below 50 copies; however, this is somewhat more difficult in children than adults owing to the initial higher RNA levels in children.

There are two protease inhibitors available for infants and children who cannot swallow pills: nelfinavir (Viracept™), which is available in a powder formulation which can be mixed with water or food, and, ritonavir (Norvir®), which is available in a liquid formulation. Indinavir (Crixovan™) and Saquinavir (Invirase™, hard gel capsule, and Fortivase™, soft gel capsule) are not available in liquid formulations. Saquinavir is the protease inhibitor with the fewest pediatric data available. Indinavir is recommended for children who can swallow pills.

Recommended antiretroviral regimens for initial therapy include one highly active protease inhibitor plus two NRTIs. An alternative regimen would be the combination of neviripine, zidovudine, and didanosine, which has produced substantial and sustained suppression of viral replication in two of six infants who were first treated at 4 months of age.[27] A secondary, alternative regimen might be the use of dual NRTIs, but initial viral suppression with such a regimen may not be sustained. Participation in a controlled clinical trial is recommended when available.

3.7. Changes in Antiretroviral Therapy

Failure of the current regimen is based on virologic, immunologic, or clinical parameters. The regimen may be too toxic, or the patient intolerant. New data which demonstrate another drug regimen to be superior to the current regimen may elicit a change in therapy. Clinicians and families must

TABLE II
Factors Influencing Adherence

Patient education prior to prescribing
Cues and reminders
Patient-focused treatment
Social and community support
Caregiver assessment
Liquid formulations for the young
Disclosure issues
School participation
Nursing, social, and behavioral assessments
Frequent medical follow-up
Family-centered systems
Food and housing availability
Medical establishment trust issues
Peer group identity

be aware of the possible contribution of the lack of adherence to the failure of a regimen. It is of great importance to provide intensive family education, training in the administration of the prescribed medications and discussion of the importance of adherence to the regimen. Frequent patient visits and intensive follow-up during the initial months after a new antiretroviral regimen is started are needed, with psychosocial support mechanisms implemented. Sometimes the use of home health nursing to provide at least partial directly observed therapy is helpful. Reasons for considering changing antiretroviral therapy are stated in Table II.

3.8. Duration of Therapy

It is known that residual HIV RNA and DNA in the lymph node will plateau after 1 year of highly active antiretroviral therapy in adults. At this time, these data suggest that there is not a good rationale for discontinuing therapy once it is begun. Non-protease inhibitor triple therapy combinations are available, and in this regard, antiretroviral efficacy and CD4 response with abacavir/lamivudine/zidovudine is equivalent to indinavir/lamivudine/zidovudine after 24 weeks of therapy in antiretroviral-naive adults. Long-term effectiveness of antiretroviral regimens is strongly dependent on strict daily compliance. Alternative strategies to provide two-drug maintenance therapy after successfully reducing plasma HIV RNA levels to undetectable have been unsuccessful in sustaining a reduced viral load.[46,47]

3.9. Salvage Antiretroviral Therapy

Clinicians should have a comprehensive knowledge of the toxicity profile of each agent before replacing a failed regimen. Clinicians are advised to change an entire regimen, especially when there has been viral rebound associated with resistance. A change in one drug or addition of a drug to a failing regimen is suboptimal. The new regimen should include at least three drugs if possible, and the potential for cross-resistance between antiretroviral drugs must be considered in choosing new medications. It is particularly important when a protease inhibitor is being utilized since there is cross-resistance to multiple protease inhibitors. Multiple drug interactions exist between antiretroviral therapies and other medications used for the treatment of infection and opportunistic prophylaxis; this must be considered when a change in regimen occurs. Considerations for terminal care and the intensity of an antiretroviral regimen need to be taken in the context of a multidisciplinary approach.

APPENDIX

Characteristics of Currently Available Antiretroviral Drugs

Drug	Dosage[a,b]	Major toxicities	Drug interactions	Special instructions
Zidovudine (ZDV, AZT) [Retrovir®] Preparations Syrup: 10 mg/ml syrup Capsules: 100 mg Tablet: 300 mg Concentrations for injection/for IV infusion: 10 mg/ml	Pediatric usual dose: Oral: 160 mg/m² q 8 h IV (intermittent infusion): 120 mg/m² q 6 h IV (continuous infusion): 20 mg/m²/h Pediatric dosage range: 90 mg/m² to 180 mg/m² q 6–12 h Neonatal dose: Oral: 2 mg/kg q 6 h; IV: 1.5 mg/kg q 6 h	Most frequent: hematologic toxicity, including granulocytopenia and anemia, headache Unusual: Myopathy, myositis, liver toxicity	May see increased toxicity with concomitant administration of the following drugs (thus, more intensive toxicity monitoring may be warranted): gancyclovir, interferon-α, trimethoprim-sulfamethoxazole, acyclovir Following may increase ZDV concentration (and therefore potential toxicity): probenecid, atovaquone, methadone, valproic acid Decreased renal clearance may be seen with coadministration of cimetidine (can be significant in patients with renal impairment) Fluconazole interferes with metabolism and clearance of ZDV (increases ZDV area under the curve)	Can be administered with food (although the manufacturer recommends administration 30 min before or 1 h after a meal) Decrease dosage in patients with severe renal impairment Significant granulocytopenia or anemia may necessitate interruption of therapy until marrow recovery is observed; use of erythropoietin, filgastrim, or reduced ZDV dosage may be necessary in some patients

Premature infants: standard neonatal dose may be in excess for premature infants

Under study in PACTG 331:
1.5 mg/kg q 12 h from birth to 2 wk of age; then increase to 2 mg/kg q 8 h after 2 wk of age

Adolescent/adult dose:
200 mg tid or 300 mg bid

ZDV metabolism may be increased with coadministration of rifampin and rifabutin

Clarithromycin decreases concentration of ZDV, probably by interfering with absorption (administer 4 h apart)

Ribavirin decreases intracellular phosphorylation of ADV (conversion to active metabolite)

Phenytoin concentrations may increase or decrease

Should not be administered in combination with d4T (poor antiretroviral effect)

Reduced dosage may be necessary in patients with significant hepatic dysfunction

Infuse loading doses and IV doses over 1 h; dilute with D_5W to concentration ≤4 mg/ml

IV solution: refrigerated, diluted solution stable for 24 h

Some Working Group members use a dose of 180 mg/m² q 12 h when using in drug combinations with other antiretroviral compounds, but data on this dosing in children are limited

Drug	Dosage[a,b]	Major toxicities	Drug interactions	Special instructions
Didanosine (dideoxyinosine; ddI) [Videx®] Preparations: Pediatric powder for oral solution (must be mixed with antacid): 10 mg/ml Chewable tablets with buffers: 25, 50, 100, 150 mg Buffered powder for oral solution: 100, 167, and 250 mg	Pediatric usual dose: in combination with other antiretrovirals: 90 mg/m^2 q 12 h Pediatric dosage range: 90–150 mg/m^2 q 12 h (may need higher dose in patients with CNS disease) Neonatal dose (<90 days old): Based on pharmacokinetic data from PACTG 239: 50 mg/m^2 q 12 h Adolescent/adult dose: >60 kg: 200 mg bid; <60 kg: 125 mg bid	Most frequent: Diarrhea, abdominal pain, nausea, vomiting Unusual more severe: Peripheral neuropathy (dose-related), electrolyte abnormalities, hyperuricemia Uncommon: Pancreatitis (dose-related, appears less common in children than adults); increased liver enzymes, retinal depigmentation (asymptomatic, reported with pediatric administration)	Possible decrease in absorption of ketoconazole, itraconazole, tetracycline; administer at least 2 h before or after ddI Ciprofloxacin absorption significantly decreased (chelation of drug by antacid); administer 2 h before or 6 h after ddI Concomitant administration of ddI and delavirdine may decrease the absorption of these drugs; separate dosing by at least 2 h Ganciclovir may increase the area under the curve and peak levels of ddI and predispose to toxicity Administration with protease inhibitors: Indinavir should be administered at least 1 h apart from ddI on an empty stomach; ritonavir should be administered at least 2 h apart from ddI	ddI formulation contains buffereing agents or antacids Food decreases absorption; administer ddI on an empty stomach (1 h before or 2 h after meal); further evaluation in children regarding administration with meals is under study For oral solution: shake well, keep refrigerated; admixture stable for 30 days When administering chewable tablets, at least 2 tablets should be given to ensure adequate buffering capacity (for example, if the child's dose is 50 mg, administer two 25-mg tablets and not one 50-mg tablet)

| Zalcitabine (ddC) [HIVID®] Preparations Syrup: 0.1 mg/ml (investigational) Tablets: 0.375 and 0.75 mg | Pediatric usual dose: 0.01 mg/kg q 8 h Pediatric dosage range: 0.005–0.01 mg/kg q 8 h Neonatal dose: unknown Adolescent/adult dose: 0.75 mg tid | Most frequent: Headache, malaise Unusual (more severe): Peripheral neuropathy, pancreatitis, hepatic toxicity, skin rashes, oral ulcers, esophageal ulcers, hematologic toxicity | Cimetidine, amphotericin, foscarnet, and aminoglycosides may decrease renal clearance of zalcitabine Antacids decrease absorption Concomitant use with ddI is not recommended due to increased risk of peripheral neuropathy IV pentamidine increases the risk of pancreatitis (do not use concurrently) | Administer on empty stomach (1 h before or 2 h after a meal) Decrease dosage in patients with impaired renal function |
| Lamivudine (3TC) [Epivir®] Preparations: Solution: 10 mg/ml Tablets: 150 mg | Pediatric dose: 4 mg/kg q 12 h Neonatal dose (<30 days old): Under study in clinical trials: 2 mg/kg q 12 h Adolescent/adult dose: 150 mg bid | Most frequent: Headache, fatigue, nausea, diarrhea, skin rash, abdominal pain Unusual (more severe): Pancreatitis (primarily seen in children with advanced HIV infection receiving multiple other medications), decreased neutrophil count, increased liver enzymes | TMP/SMX increases lamivudine blood levels (possibly competes for renal tubular secretion); unknown significance When used with ZDV may prevent emergence of resistance, and for ZDV-resistant virus, revision to phenotypic ZDV sensitivity may be seen | Can be administered with food For oral solution: store at room temperature Decrease dosage in patients with impaired renal function |

Drug	Dosage[a,b]	Major toxicities	Drug interactions	Special instructions
Stavudine (d4T) [Zerit®] Preparations: Solution: 1 mg/ml Capsules: 15, 20, 30, and 40 mg	Pediatric dose: 1 mg/kg q 12 h (up to weight of 30 kg) Neonatal dose: Under study in PACTG 332 Adolescent dose: >60 kg: 40 mg bid; 30–60 kg: 30 mg bid	Most frequent: Heachache, GI disturbances, skin rash Uncommon (more severe): Peripheral neuropathy, pancreatitis Other: Increased liver enyzmes	Drugs that decrease renal function could decrease clearance Should not be administered in combination with zidovudine (poor antiretroviral effect)	Can be administered with food Need to decrease dose in patients with renal impairment For oral solution: shake well and keep refrigerated; solution stable for 30 days
Ritonavir [Norvir®] Preparation: Oral solution: 80 mg/ml Capsules: 100 mg	Pediatric dose: 400 mg/m² q 12 h To minimize nausea and/oro vomiting, initiate therapy starting at 250 mg/m² q 12 h and increase stepwise to full dose over 5 days as tolerated Pediatric dosage range: 350–400 mg/m³ q 12 h Neonatal dose: Under study in PACTG 354 (single dose pharmacokinetics)	Most frequent: Nausea, vomiting, diarrhea, headache, abdominal pain, anorexia Less common: Circumoral paresthesias, increase in liver enzymes Rare: Spontaneous bleeding episodes in hemophiliacs, pancreatitis, increased levels of triglycerides and cholesterol, hyperglycemia and diabetes	Ritonavir is extensively metabolized in the liver by cytochrome P450 enzyme 3A (CYP3A); Potential for multiple-drug interactions (see Note). Before administering carefully review patient medication profile for potential drug interactions Not recommended for concurrent use: analgesics (meperidine, peroxicam, propoxyphene); antihistamines (astemizole, terfenadine); certain cardiac drugs (amioderone, bepridil, encainide, flecainide, propafenone, quinidine); ergot alkaloid derivatives; cisapride; sedative-hypnotics (clorazepate, diazepam, estazolam, flurazepam, midazolam, triazolam, zoipidem); certain psychotropic drugs (bupropion, clozapine); rifamin; rifabutin	Administering with food increases absorption If administered with ddI, should be administered 2 h apart Oral capsules must be kept in refrigerator For oral solution: keep refrigerated and store in original container; can be kept at room temp if used within 30 days To minimize nausea, therapy should be initiated at a low dose and increased to full dose over 5 days as tolerated

Adolescent/adult
dose:
400 mg q 12 h × 1 wk,
then 600 mg q 12 h

Oral contraceptives—estradiol levels are reduced by ritonavir, and alternative or additional methods of birth control should be used if coadministering with hormonal methods of birth control

Ritonavir decreases levels of sulfamethoxazole

Ritonavir increases metabolism of theophylline (levels should be monitored and dose may need to be increased)

Ritonavir increases levels of clarithromycin (dose adjustment may be necessary in patients with impaired renal function); desipramine, warfarin (monitoring of anticoagulant effect necessary)

Ritonavir may increase or decrease digoxin levels (monitoring of levels is recommended)

Drugs that increase CYP3A activity can lead to increased clearance and therefore lower levels of ritonavir, such as carbamazepine, dexamethasone, phenobarbital, phenytoin (anticonvulsant drugs should be monitored, as ritonavir can affect the metabolism of these drugs as well)

Administration with other protease inhibitors: decreases the metabolism of indinavir and saquinavir and results in greatly increased concentrations of these drugs; increases nelfinavir concentration 1.5 fold

Techniques to increase tolerance in children: mix oral solution with milk, chocolate milk, vanilla or chocolate pudding, or ice cream

Dull taste buds before administering by having patient chew ice, giving popsicles or partially frozen orange or grape juice concentrate

Coat mouth by giving peanut butter before dose

Administer strong-tasting foods (maple syrup, cheese, strong-flavored chewing gum) immediately after dose

Drug	Dosage[a,b]	Major toxicities	Drug interactions	Special instructions
Indinavir [Crixivan®] Preparations: Capsules: 200 and 400 mg	Pediatric dose: 350 mg/m^2 q 8 h to 500 mg/m^2 q 8 h Neonatal dose: Unknown; due to side effect of hyperbilirubinemia, should not be given to neonates until further information is available Adolescent/adult dose: 8800 mg q 8 h	Most frequent: Nausea, abdominal pain, headache, asymptomatic hyperbilirubinemia (10%) Unusual (more severe): Nephrolithiasis (4%) Rare: Spontaneous bleeding episodes in hemophiliacs, hyperglycemia and diabetes	Cytochrome P450 3A4 responsible for meta; potential for multiple-drug interactions (Note) Before administering, the patient's medication profile should be carefully reviewed for potential drug interactions Not recommended for concurrent use: Indinavir decreases the drug metabolism, resulting in increased drug levels and toxicity: antihistamines (astemizole, terfenedine), cisapride, ergot alkaloid derivatives, sedative-hypnotics (triazolam, midazolam) Indinavir levels significantly reduced with concurrent use of rifampin Rifabutin concentrations are increased and a dose redution of rifabutin to half the usual daily dose is recommended Ketoconazole, itraconazole increase indinavir concentrations (consider adoles/adult indinavir dose reduction to 600 mg q 8 h) Clarithromycin coadministration increases serum concentration of both drugs Nevirapine coadministration may decrease indinavir serum concentration Administration of other protease inhibitors decreases the metabolism of indinavir and results in greatly increased indinavir concentrations	Give on empty stomach 1 h before or 2 h after meal (or take with a light meal) Adequate hydration required to minimize risk of nephrolithiasis If coadministering with ddI, give at least 2 h apart on an empty stomach Decrease dose in patients with cirrhosis Capsules are sensitive to moisture and should be stored in original container with desiccant

Nelfinavir [Viracept®] Preparations: Powder for oral suspension: 50 mg per one level scoop (200 mg per one level teasponon) Tablets: 250 mg tablet	Pediatric dose: 20–30 mg/kg tid Neonatal dose: Under study in PACTG 353: 10 mg/kg tid (note: no preliminary data available, investigational) Adolescent/adult dose: 750 mg tid	Most frequent: Diarrhea Less common: Asthenia, abdominal pain, rash Rare: Hyperglycemia and diabetes	Nelfinavir is in part metabolized by cytochrome P450 3A4 (CYP3A4); potential for multiple-drug interactions (see Note) Before administering, the patient medication profile should be carefully reviewed for potential drug interactions Not recommended for concurrent use: nelfinavir decreases drug metabolism resulting in increased drug levels and potential toxicity: antihistamine (astemizole, terfenedine), cisapride, ergot alkaloid derivatives, sedative-hypnotics (triazolam, midazolam) Nelfinavir levels are greatly reduced with concurrent use: rifampin Rifabutin causes less decline in nelfinavir concentration: if coadministering with nelfinavir, rifabutin should be reduced to one-half usual dose Oral contraceptives—estradiol levels are reduced by nelfinavir and alternative or additional methods of birth control should be used if coadministered with hormonal methods of birth control	Give with meal or light snack For oral solution: powder may be mixed with water, milk, pudding, ice cream, or formula (for up to 6 h) Do not mix with any acidic food or juice because of resultant poor taste Do not add water to bottles of oral powder; a special scoop is provided Tablets dissolve in water for a dispersion and can mix with milk or chocolate milk; tablets can also be crushed and given with pudding

Drug	Dosage[a,b]	Major toxicities	Drug interactions	Special instructions
Nelfinavir [Viracept®] (*cont.*)			Coadministering with delavirdine increases nelfinavir concentration by 2-fold and decreases delavirdine concentration by 50%; no data on coadministering with nevirapine; some experts use higher doses of nelfinavir if used in combination with nevirapine Administering with other protease inhibitors; nelfinavir increases levels of saquinavir and indinavir; coadministering with ritonavir increases nelfinavir levels 1.5-fold without change in ritonavir concentrations	
Saquinavir [Invirase®] Preparations: Capsules: 200 mg Soft gel capsules (investigational): 200 mg	Pediatric dose: Unknown Neonatal dose: Unknown Adolescent/adult dose: 600 mg tid	Most frequent: Diarrhea, abdominal discomfort, headache, nausea Rare: Spontaneous bleeding episodes in hemophiliacs, hyperglycemia and diabetes	Saquinavir is metabolized by the cytochrome P450 3A (CYP3A) system in the liver and there are numerous potential drug interactions Before administration carefully review patient medication profile for potential drug interactions Not recommended for concurrent use: saquinavir decreases drug metabolism, resulting in increased drug levels and potential toxicity: antihistamine (astemizole, terfenedine), cisapride, ergot alkaloid derivatives	Administer within 2 h of a full meal to increase absorption Concurrent administration of grapefuit juice increases saquinavir concentration Sun exposure can cause photosensitivity reactions and sunscreen or protective clothing is recommended

Saquinavir levels are significantly reduced with concurrent use: rifampin and rifabutin decrease saquinavir levels by 80% and 40%, respectively; nevirapine decreases saquinavir levels by 25%

Saquinavir levels are decreased by carbamazepine dexamethasone, phenobarbital, and phenytoin

Saquinavir levels are increased by delavirdine, ketoconazole

Saquinavir may increase levels of calcium channel blockers, dapsone, quinidine, triazolam

Administration with other protease inhibitors: coadministering of ritonavir decreases the metabolism of saquinavir and results in greatly increased saquinavir concentrations

---

[a]Adolescent dosing by Tanner stage: Adolescents in early puberty (Tanner I–II) should be dosed using pediatric schedules, while those in late puberty (Tanner stage V) should be dosed using adult schedules. Youth who are in the midst of their growth spurt (Tanner III females and Tanner IV males) should be closely monitored for medication efficacy and toxicity when choosing adult or pediatric dosing guidelines.

[b]Data in children are limited, and doses may change as more information is obtained about the pharmacokinetics of these drugs in children.

Note. Drugs metabolized by the hepatic cytochrome P450 3A (CYP3A) enzyme system have the potential for significant interactions with multiple drugs, some of which may be life-threatening. These interactions are outlined in detail in the Guidelines for Use of Antiretroviral Agents on HIV-Infected Adults and Adolescents (PanelRec 97)[9] and in prescribing information available from the drug companies. These interactions will not be reiterated here and the health care provider should review those documents for detailed information. Before therapy with these drugs is initiated, the patient's medication profile should be carefully reviewed for potential drug interactions.

REFERENCES

1. Connor, E. M., Sperling, R. S., Gelber, R., *et al.*, 1994, Reduction of maternal–infant transmission of human immunodeficiency virus type 1 with zidovudine treatment, *N. Engl. J. Med.* **331**:1173–1180.
2. Borkowsky, W., Krasinski, K., Cao, Y., *et al.*, 1994, Correlation of perinatal transmission of human immunodeficiency virus type 1 with maternal viremia and lymphocyte phenotypes, *J. Pediatr.* **125**:345.
3. Perelson, A. S., Neumann, A. U., Markowitz, M., *et al.*, 1996, HIV-1 dynamics *in vivo*: Virion clearance rate, infected cell life-span, and viral generation time, *Science* **271**:1582–1586.
4. Mellors, J. W., Kingsley, L. A., Rinaldo, C. R., *et al.*, 1995, Quantitation of HIV-1 RNA in plasma predicts outcome after seroconversion, *Ann. Intern. Med.* **122**:573–579.
5. Mellors, J. W., Munoz, A., Giorgi, J. V., *et al.*, 1997, Plasma viral load and CD4+ lymphocytes as prognostic markers of HIV-1 infection, *Ann. Intern. Med.* **126**:946–954.
6. Hellerstein, M., Hanley, M. B., Cesar, D., Siler, S., *et al.*, 1999, Directly measured kinetics of circulating T lymphocytes in normal and HIV-1-infected humans, *Nature Med.* **5**:83–89.
7. Shearer, W. T., Quinn, T. C., LaRussa, P., *et al.*, 1997, Viral load and disease progression in infants infected with human immunodeficiency virus type 1, *N. Engl. J. Med.* **336**:1337–1342.
8. Hammer, S. M., Squires, K. E., Hughes, M. D., Grimes, J. M., *et al.*, 1997, A controlled trial of two nucleoside analogues plus indinavir in persons with human immunodeficiency virus infection and CD4 cell counts of 200 per cubic millimeter or less. AIDS Clinical Trials Group 320 Study Team, *N. Engl. J. Med.* **337**:725–733.
9. The National Institutes of Health Panel to Define Principles of Therapy of HIV Infection, 1997, *Report of the NIH Panel to Define Principles of Therapy of HIV Infection.* U.S. Public Health Service, Washington, D.C.
10. Working Group on Antiretroviral Therapy: National Pediatric HIV Resource Center, 1993, Antiretroviral therapy and medical management of the human immunodeficiency virus-infected child, *Pediatr. Infect. Dis.* **12**:513–522.
11. CDC, 1998, Guidelines for the use of antiretroviral agents in pediatric HIV infection, *Morbid. Mortal. Weekly Rep.* **47**:RR-4.
12. Yarchoan, R., Mitsuya, H., Myers, C. E., and Broder, S., 1989, Clinical pharmacology of 3′-azido 2′,3′ dideoxythymiden (zidovudine and related dideoxynucleosides), *N. Engl. J. Med.* **321**:726–738.
13. Pizzo, P. A., Eddy, J., Falloon, J., *et al.*, 1988, Effect of continuous intravenous infusion of zidovudine (AZT) in children with symptomatic HIV infection, *N. Engl. J. Med.* **31**:889–896.
14. Pizzo, P., 1989, Emerging concept in the treatment of HIV infection in children, *JAMA* **26**:1989–1992.
15. Brouwers, P., Mocs, H., Loolfus, P., *et al.*, 1990, Effect of continuous-infusion zidovudine therapy on neurophysiologic functioning in children with symptomatic human immuno-deficiency virus infection, *J. Pediatr.* **117**:980–985.
16. McKinney, R. E., Maha, M. A., Connor, E. M., *et al.*, 1991, A multicenter trial of oral zido-vudine in children with advanced human immunodeficiency virus disease, *J. Engl. J. Med.* **324**:1018–1025.
17. Blanche, S., Duliege, A.-M., Navarette, M. S., *et al.*, 1991, Low-dose zidovudine in children with an human immunodeficiency virus type I infection acquired in the perinatal period, *Pediatrics* **88**:364–370.
18. Day, R. E., Mahu, M. A., Connor, E. M., *et al.*, 1991, A multicenter trial of oral zidovudine in children with advanced human immunodeficiency virus disease, *N. Engl. J. Med.* **324**:1018–1025.

19. Larder, B., and Kemp, S. C., 1989, Multiple mutations in HIV-1 reverse transcriptase confer high-level resistance to zidovudine (AZT), *Science* **246**:1155–1158.

20. Butler, K. M., Husson, R. N., Balis, F. M., *et al.*, 1991, Dideoxyinosine in children with symptomatic human immunodeficiency virus infection, *N. Engl. J. Med.* **324**:137–144.

21. Clerici, M., Roilides, E., Butler, K. M., DePalma, L., Shearer, G. M., and Pizzo, P. A., 1992, Changes in T-helper cell function in human immunodeficiency virus-infected children during didanosine therapy as a measure of antiretroviral activity, *Blood* **80**:2196–2202.

22. Lewis, L. L., Venzon, D., Church, J., *et al.*, 1995, Lamivudine in children with human immunodeficiency virus infection, *Pediatrics* **96**:247–252.

23. Kline, M. W., Dunkle, L. M., Church, J. A., *et al.*, 1995, A phase I/II evaluation of stavudine (d4T) in children with human immunodeficiency virus infection, *Pediatrics* **96**:247–252.

24. Kline, M. W., Van Dyke, R. B., Lindsey, J. C., Gwynne, M., *et al.*, 1998, A randomized comparative trial of stavudine (d4T) versus zidovudine (ZDV, AZT) in children with human immunodeficiency virus infection. AIDS Clinical Trials Group 240 Team, *Pediatrics* **101**:214–220.

25. Palumbo, P. E., Raskino, C., Fiscus, S., *et al.*, 1997, Correlation of HIV plasma RNA levels with clinical outcome in a large pediatric trial (ACTH 152), in *Proceedings of the 4th Conference on Retroviruses and Opportunistic Infections, Washington, D.C.*, p. 208 [Abstract #LB14].

26. McKinney, R. E., Jr., Johnson, G. M., Stanley, K., Yong, F. H., *et al.*, 1998, A randomized study of combined zidovudine lamivudine versus didanosine monotherapy in children with symptomatic therapy-naive HIV-1 infection. Pediatric AIDS Clinical Trails Group 300 Team, *J. Pediatr.* **133**:500–508.

27. Luzuriaga, K., Bryson, Y., Krogstad, P., *et al.*, 1997, Combination treatment with zidovudine, didanosine and nevirapine in infants with human immunodeficiency virus type 1 infection, *N. Engl. J. Med.* **336**:1343–1349.

28. Starr, S. E., Fletcher, C., Spector, S. A., *et al.*, 1998, *Efavirenz in combination with nelfinavir and nucleoside reverse transcriptase inhibitors (NRTIs) is safe and virologically effective in HIV-infected children,* presented at the 38th Interscience Conference on Antimicrobial Agents and Chemotherapy, San Diego, California, September 24–27, 1998 [Late breaker abstract LB-6].

29. Hsia, K., Yong, F., Starr, S., *et al.*, 1999, HIV-1 plasma RNA load in children declines slowly to undetectable levels during effective antiretroviral therapy: implications of the Roche standard (std) and Ultrasensitive (ultra) assays, in *Proceedings of the 6th Conference on Retroviruses and Opportunistic Infections*, Chicago, Illinois [Abstract 270].

30. Danner, S. A., Carr, A., Leonard, J. M., *et al.*, 1995, A short-term study of the safety, pharmacokinetics, and efficacy of ritonavir, an inhibitor of HIV-1 protease, *N. Engl. J. Med.* **333**:1528–1533.

31. Collier, A. C., Coombs, R. W., Schoenfeld, D. A., *et al.*, 1996, Treatment of human immunodeficiency virus infection with saquinavir, zidovudine, and zalcitabine, *N. Engl. J. Med.* **334**:1011–1017.

32. Stein, D. S., Fish, D. G., Bilello, J. A., Preston, S. L., *et al.*, 1996, A 24-week open-label phase I/II evaluation of the HIV protease inhibitor MK-639 (indinavir), *AIDS* **10**:485–492.

33. Nelson, R. P., Jr., Sleasman, J., Cervia, J., Scott, G., *et al.*, 1999, Indinavir (IDV) in combination with stavudine (d4T) and lamivudine (3TC) in HIV-infected children, in *Proceedings of the 6th Conference on Retroviruses and Opportunistic Infections*, Chicago, Illinois [Abstract #425].

34. Martel, L., Valentine, M., Ferguson, L., Muelenaer, P., Fisher, R., Katz, S. L., and McKinney, R. E., 1999, Virologic and CD4 response to treatment with nelfinavir in therapy experienced, protease inhibitor naive HIV-infected children: 48 week follow-up, in: *Proceedings of*

the 6th Conference on Retroviruses and Opportunistic Infections. Chicago, Illinois, p. 148 [Abstract 429].

35. Balzarini, J., Naesens, L., Slachmuylders, J., Niphuis, H., Rosenberg, I., *et al.*, 1991, 9-(2-Phosphonylmethoxyethyl) adenine (PMEA) effectively inhibits retrovirus replication *in vitro* and simian immunodeficiency virus infection in rhesus monkeys, *AIDS* **5**:21–28.

36. American Academy of Pediatrics Committee on Pediatric AIDS, 1997, Evaluation and medical treatment of the HIV-exposed infant, *Pediatrics* **99**:909–916.

37. Nelson, R. P., Jr., Price, L. J., Halsey, A. B., Graven, S. N., Resnick, L., Day, N. K., Lockey, R. F., and Good, R. A., 1996, Diagnosis of pediatric human immunodeficiency virus infection utilizing a commercially-available polymerase chain reaction gene amplification, *Arch. Pediatr. Adolesc. Med.* **150**:40–45.

38. McIntosh, K., Pitt, J., Brambilla, D., *et al.*, 1994, Blood culture in the first 6 months of life for diagnosis of vertically transmitted human immunodeficiency virus infection, *J. Infect. Dis.* **170**:996–1000.

39. Nesheim, S., Lee, F., Kalish, M. L., *et al.*, 1997, Diagnosis of perinatal human immunodeficiency virus infection by polymerase chain reaction and p24 antigen detection after immune complex dissociation in an urban community hospital, *J. Infect. Dis.* **175**:1333–1336.

40. Katzenstein, T. L., Pedersen, C., Nielson, C., Lundgren, J. D., Jakobsen, P. H., and Gerstoft, J., 1996, Longitudinal serum HIV RNA quantification: Correlation to viral phenotype at seroconversion and clinical outcome, *AIDS* **10**:167–173.

41. Henrard, D. R., Phillips, J. F., Muenz, L. R., *et. al.*, 1995, Natural history of cell-free viremia, *JAMA* **274**:554–558.

42. Mofenson, L. M., Korelitz, J., Meyer, W. A., *et al.*, 1997, The relationship between serum human immunodeficiency virus type 1 (HIV-1) RNA level, CD4 lymphocyte percent, and long-term mortality risk in HIV-1-infected children, *J. Infect. Dis.* **175**:1029–1038.

43. McIntosh, K., Shevitz, A., Zaknun, D., *et al.*, 1996, Age- and time-related changes in extracellular viral load in children vertically infected by human immunodeficiency virus, *Pediatr. Infect. Dis. J.* **15**:1087–1091.

44. Abrams, E. J., Weedon, J. C., Lambert, G., Steketee, R., and the NYC Perinatal HIV Transmission Collaborative Study, 1996, HIV viral load early in life as a predictor of disease progression in HIV-infected infants, in: *Proceedings of the XI International Conference on AIDS*, Vancouver, Canada, p. 25 [Abstract We.B.311].

45. El-Sadar, W., Oleske, J. M., Agins, B. D., *et al.*, 1994, Evaluation and Management of Early HIV Infection, Clinical Practice Guideline No. 7, AHCPR Publication No. 94-0572, Agency for Health Care Policy and Research Public Health Service, U.S. Department of Health and Human Services, Rockville, Maryland.

46. Pialoux, G., Raffi, F., Brun-Vezinet, F., Meiffredy, V., Flandre, P., *et al.*, 1998, A randomized trial of three maintenance regimens given after three months of induction therapy with zidovudine, lamivudine, and indinavir in previously untreated HIV-1 infected patients. Trilege (Agence Nationale de Recherches sur le SIDA 072) Study Team, *N. Engl. J. Med.* **339**:1269–1276.

47. Havlir, D. V., Marschner, I. C., Hirsch, M. S., Collier, A. C., Tebas, P., *et al.*, 1998, Maintenance antiretroviral therapies in HIV infected patients with undetectable plasma HIV RNA after triple-drug therapy. AIDS Clinical Trials Group Study 343 Team, *N. Engl. J. Med.* **339**:1261–1268.

48. Centers for Disease Control and Prevention, 1994, Revised classification system for human immunodeficiency virus infection in children less than 13 years of age, *Morbid. Mortal. Weekly Rep.* **43**(RR-12):1–10.

Index

An "*f*" or "*t*" suffix indicates that the term may be found in a figure or table on the page indicated.